CONTEMPORARY MATHEMATICS

BRUCE E. MESERVE

Professor of Mathematics
University of Vermont

MAX A. SOBEL

Professor of Mathematics
Montclair State College

PRENTICE-HALL, INC.
ENGLEWOOD CLIFFS, N. J.

CONTEMPORARY MATHEMATICS

CONTEMPORARY MATHEMATICS

Meserve / Sobel

ISBN: 0–13–170100–2

Library of Congress Catalog Card Number 72–37633

10 9 8 7 6 5 4 3 2 1

Printed in the United States of America

PRENTICE-HALL INTERNATIONAL, INC., London
PRENTICE-HALL OF AUSTRALIA, PTY. LTD., Sydney
PRENTICE-HALL OF CANADA, LTD., Toronto
PRENTICE-HALL OF INDIA PRIVATE LIMITED, New Delhi
PRENTICE-HALL OF JAPAN, INC., Tokyo

Contents

INTRODUCTION

Fun with Mathematics

part one
NUMBER SYSTEMS

1 Sets

7   The Set of Real Numbers   229

part two
ALGEBRA, GEOMETRY, STATISTICS, PROBABILITY

8   Sentences in One Variable   261

9   Sentences in Two Variables   288

10   Nonmetric Geometry   330

Preface

Contemporary Mathematics has been written to serve the needs of both liberal arts students and prospective teachers of elementary mathematics in grades K through 8. It is designed primarily as a one year text at the undergraduate level, although it may be used effectively for in-service courses as well.

The authors have prepared this material on the basis of their many years of experience at both the pre-service and in-service levels. They are grateful to the many students whom they have taught, and who have both directly and indirectly contributed to the conception and formulation of this text. Indeed, significant portions of this book have been taught in one form or another by both authors over the past 10 years.

In keeping with the development of modern programs of mathematics during the decade of the 1960's, this text stresses a liberal arts approach suitable for the twentieth century citizen. It also emphasizes the body of mathematics that a prospective teacher needs to know in order to be able to competently teach contemporary courses as found in recent elementary textbooks. However, this text does not neglect those important aspects of mathematics that are as important today as they were prior to the recent "revolution" in school mathematics. Thus adequate attention is given to the development of the number system, and to the various algorithms representing fundamental operations of arithmetic.

Contemporary Mathematics has been prepared in two parts. Part I deals with number systems, and can effectively be used as the content of a one-semester course or a one-quarter course. It features basic concepts of sets and logic, a discussion of systems of numeration, and a development of the properties of the set of real numbers and its various subsets. Part II, also suitable for a one-semester (or one-quarter) course, provides background material from the areas of algebra, geometry, statistics, and probability. Parts I and II are suitable for three-quarter courses with the precise division of topics depending upon the emphasis desired. For example, sets, logic, mathematical systems, numeration systems, and integers (Chapters 1 through 5) might be considered the first quarter; rational numbers, real numbers, and elementary algebra (Chapters 6 through 9) considered the second quarter; and geometry, statistics, and probability (Chapters 10 through 14) considered the third quarter.

A significant and unique feature of this book is the material entitled "Pedagogical Considerations" that appear in each chapter, and at the end of almost every section. These inserts are specifically addressed to the prospective teacher. Here one may find discussions related to the teaching of elementary mathematics, suggestions for additional explorations and laboratory experiments, introductions to various professional organizations such as the National Council of Teachers of Mathematics, and other such items that help provide insight into the full role of a successful teacher of elementary mathematics. In most cases these items relate to the material being covered in the preceding section and thus offer a close and obvious relationship between subject matter and pedagogy.

Although most students will have had some high school mathematics prior to enrolling in a course for which this text will be used, no such assumption is made herein. Thus this book has been prepared with the expectation that the student will only have a minimal amount of past experience in this subject matter. Furthermore, it has been anticipated that a significant number of students will have some fear of mathematics. Thus every effort has been made to provide an interesting approach that will motivate the learning of the subject. To this end, the introduction is entitled "Fun with Mathematics" and is offered as one means of starting a course with topics that can be studied independently of the remainder of the text and just for fun. It has been the experience of the authors that such an approach can do much at the outset of a semester to relieve anxieties and to begin in a light and somewhat informal manner.

It is the fond hope of the authors that both the instructors and students who will use this book will enjoy it as much as the authors have enjoyed its preparation. To their many students—past, present, and future—the authors sincerely dedicate this book and hope that it may serve to stimulate interest in the study of mathematics.

BRUCE E. MESERVE
MAX A. SOBEL

Introduction

Fun
with
Mathematics

Mathematics has numerous practical applications ranging from everyday household usage to the charting of astronauts through outer space. Many people study the subject because of such applications as they relate to their particular fields of interest. Mathematics may also be considered as part of our great cultural heritage, for it has a history that dates back many thousands of years. As such, it should be studied by the average citizen who wishes to be literate in this twentieth century.

Many people study mathematics just for fun! These individuals would rather solve a mathematical puzzle than read a book, watch television, or go to a movie. Admittedly, not everyone does have, or can have, this type of disposition. On the other hand, most of us use mathematical concepts in a variety of ways and have never been given an opportunity to explore some of the more interesting aspects of mathematics.

Elementary school teachers have a special need for understanding mathematical concepts. In a wide variety of classroom situations mathematical concepts are involved in about one out of every five hours of their teaching. Many of these hours of teaching mathematics can be fun for both the teacher and the students. This introduction contains a smorgasbord of items that have been fun for many students and teachers. The topics are not sequential and an understanding of them is not necessary for later chapters in this text, although some of the topics presented here will be pre-

1

sented in greater detail in later chapters. Thus this introduction is expendable in the formal structure of this text. However, it is significant for elementary school teachers as an indication of the spirit of a modern approach to the teaching of elementary school mathematics.

1
explorations with number patterns

Mathematicians love to search for patterns and generalizations in all branches of their subjects—in arithmetic, in algebra, and in geometry. A search for such patterns may not only be interesting but may also help one develop insight into mathematics as a whole.

Many patterns that often escape notice may be found in the structure of arithmetic. For example, consider the multiples of 9:

$$1 \times 9 = 9$$
$$2 \times 9 = 18$$
$$3 \times 9 = 27$$
$$4 \times 9 = 36$$
$$5 \times 9 = 45$$
$$6 \times 9 = 54$$
$$7 \times 9 = 63$$
$$8 \times 9 = 72$$
$$9 \times 9 = 81$$

What patterns do you notice in the column of multiples on the right? You may note that the sum of the digits in each case is always 9. You should also see that the units digit decreases (9, 8, 7, . . .), whereas the tens digit increases (1, 2, 3, . . .). What lies behind this pattern?

Consider the product

$$5 \times 9 \times 45.$$

To find 6×9 we need to add 9 to 45. Instead of adding 9, we may add 10 and subtract 1.

$$
\begin{array}{cc}
45 & 55 \\
+10 & -1 \\
\hline
55 & 54 = 6 \times 9
\end{array}
$$

That is, by adding 1 to the tens digit, 4, of 45, we are really adding 10 to 45. We then subtract 1 from the units digit, 5, of 45 to obtain 54 as our product.

Most people speak of "adding digits," "subtracting from the units digit," and so forth as we have done. Many elementary school teachers are more precise in their terminology and recognize that digits are numerals, that is, symbols for numbers, rather than numbers. Numerals can be written. Only numbers can be added or subtracted. However, unless the more precise terminology is needed to avoid major confusion, we use the commonly accepted phraseology.

The number 9, incidentally, has other fascinating properties. Of special interest is a procedure for multiplying by 9 on one's fingers. For example,

2

to multiply 9 by 3, place both hands together as in the figure, and bend the third finger from the left. The result is read as 27.

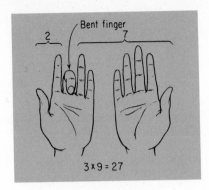

3 × 9 = 27

The next figure shows the procedure for finding the product 7 × 9. Note that the seventh finger from the left is bent, and the result is read in terms of the tens digit, to the left, and the units digit to the right of the bent finger. (Note that a thumb is considered to be a finger.)

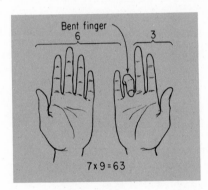

7 × 9 = 63

What number fact is shown in the next figure?

Here is one more pattern related to the number 9. You may, if you wish, verify that each of the following is correct:

$$1 \times 9 + 2 = 11$$
$$12 \times 9 + 3 = 111$$
$$123 \times 9 + 4 = 1,111$$
$$1,234 \times 9 + 5 = 11,111$$
$$12,345 \times 9 + 6 = 111,111$$

Try to find a correspondence of the number of 1's in the number symbol on the right with one of the numbers used on the left. Now see if you can supply the answers, without computation, to the following:

$$123,456 \times 9 + 7 = ?$$
$$1,234,567 \times 9 + 8 = ?$$

Let us see *why* this pattern works. To do so we shall examine just one of the statements. A similar explanation can be offered for each of the other statements. Consider the statement:

$$12,345 \times 9 + 6 = 111,111$$

We can express 12,345 as a sum of five numbers as follows:

$$
\begin{array}{r}
11,111 \\
1,111 \\
111 \\
11 \\
1 \\
\hline
12,345
\end{array}
$$

Next we multiply each of the five numbers by 9:

$$11,111 \times 9 = 99,999$$
$$1,111 \times 9 = 9,999$$
$$111 \times 9 = 999$$
$$11 \times 9 = 99$$
$$1 \times 9 = 9$$

Finally, we add 6 by adding six ones as in the following array, and find the total sum:

$$
\begin{array}{r}
99,999 + 1 = 100,000 \\
9,999 + 1 = 10,000 \\
999 + 1 = 1,000 \\
99 + 1 = 100 \\
9 + 1 = 10 \\
1 = 1 \\
\hline
111,111
\end{array}
$$

Here is another interesting pattern. After studying the pattern, see if you can add the next four lines to the table.

$$1 \times 1 = 1$$
$$11 \times 11 = 121$$
$$111 \times 111 = 12{,}321$$
$$1{,}111 \times 1{,}111 = 1{,}234{,}321$$
$$11{,}111 \times 11{,}111 = 123{,}454{,}321$$

Do you think that the pattern displayed will continue indefinitely? Compute the product $1{,}111{,}111{,}111 \times 1{,}111{,}111{,}111$ to help you answer this question.

Interesting discoveries can often be made by studying arithmetic patterns. A famous German mathematician by the name of Karl Gauss (1777–1855) is said to have been a precocious child who would often drive his teachers to despair. The story is told that on one occasion his teacher asked him to add a long column of figures, hoping to keep him suitably occupied for some time. Instead, young Gauss recognized a pattern, and gave the answer immediately. He is said to have found the sum of the first 100 counting numbers as indicated in this array:

$$1 + 2 + 3 + \cdots + 98 + 99 + 100$$

He reasoned that there would be 50 pairs of numbers, each with a sum of 101 (consider $100 + 1$, $99 + 2$, $98 + 3$, $\ldots$, $50 + 51$). Thus the sum is 50×101, that is, 5050.

Another interesting pattern emerges from the story of the man who was offered a month's employment at the rate of 1¢ for the first day, 2¢ for the second day, then 4¢, 8¢, 16¢, and so forth. He wished to determine what his total wages would be for four weeks of five days each; that is, for 20 working days.

Of course, one could list the 20 daily salaries and add, but this would be a tedious job indeed. Let us, rather, consider the big problem by exploring smaller tasks first. This is frequently a helpful technique in problem solving. Thus we consider total salaries in cents, for five, for six, for seven, and for eight days.

For 5 days	For 6 days	For 7 days	For 8 days
1	1	1	1
2	2	2	2
4	4	4	4
8	8	8	8
16	16	16	16
	32	32	32
31		64	64
	63		128
		127	
			255

Do you see a pattern emerging? Compare the total for the first five days with the salary for the sixth day; compare the total for six days with the salary for the seventh day. Notice that the total for five days is one cent less than the salary for the sixth day; the total for six days is one cent less than the salary for the seventh day; and so forth. Thus his total salary for ten days will be one cent less than the salary for the eleventh day. That is,

$$1 + 2 + 4 + 8 + 16 + 32 + 64 + 128 + 256 + 512$$
$$= 2(512) - 1, \text{ or } 1023.$$

His salary for the eleventh day would be 1024 cents; in 10 days he will earn a *total* of 1023 cents, or $10.23.

To use this approach to answer the original question will require a good deal more work, but will still be easier than the addition of 20 amounts. By a doubling process we first need to find the salary for the twentieth day. The total salary for all 20 days can then be found by doubling this amount (to find the salary for the twenty-first day), and subtracting one.

To generalize, we note that the doubling process starting with one gives powers of two:

$$2^0 = 1, \quad 2^1 = 2, \quad 2^2 = 4, \quad 2^3 = 8, \quad 2^4 = 16, \quad 2^5 = 32, \quad \cdots$$

Then, by our discovery of a pattern, we may say

$$2^0 + 2^1 + 2^2 + 2^3 + \cdots + 2^n = 2^{n+1} - 1.$$

When $n = 5$, we have

$$2^0 + 2^1 + 2^2 + 2^3 + 2^4 + 2^5 = 1 + 2 + 4 + 8 + 16 + 32$$
$$= 2^6 - 1 = 63.$$

To find the total salary for 20 working days, we need to find the sum:

$$\underset{\text{1st day}}{2^0} + \underset{\text{2nd day}}{2^1} + \underset{\text{3rd day}}{2^2} + \cdots + \underset{\text{20 th day}}{2^{19}}$$

From the preceding discussion we know that this sum is equal to $2^{20} - 1$. We can compute 2^{20} by a doubling process, or can use a shortcut such as the following:

$$2^{10} = 2 \times 2^9 = 2 \times 512 = 1024$$

Alternatively, we may say

$$2^{10} = 2^5 \times 2^5 = 32 \times 32 = 1024.$$

In a similar manner we compute

$$2^{20} = 2^{10} \times 2^{10} = 1024 \times 1024 = 1,048,576$$

and thus $2^{20} - 1 = 1,048,575$.

In 20 days, our worker would earn a total of $10,485.75.

In the following exercises the reader will have an opportunity to make other discoveries of his own through careful explorations of patterns.

EXERCISES

1. Verify that the process for finger multiplication shown in this section will work for each of the multiples of nine from 1×9 through 9×9.

2. Follow the procedure outlined in this section and show that

$$1234 \times 9 + 5 = 11,111.$$

3. Study the following pattern and use it to express the squares of 6, 7, 8, and 9 in the same manner.

$$1^2 = 1$$
$$2^2 = 1 + 2 + 1$$
$$3^2 = 1 + 2 + 3 + 2 + 1$$
$$4^2 = 1 + 2 + 3 + 4 + 3 + 2 + 1$$
$$5^2 = 1 + 2 + 3 + 4 + 5 + 4 + 3 + 2 + 1$$

4. Study the entries that follow and use the pattern that is exhibited to complete the last four rows.

$$1 + 3 = 4 \text{ or } 2^2$$
$$1 + 3 + 5 = 9 \text{ or } 3^2$$
$$1 + 3 + 5 + 7 = 16 \text{ or } 4^2$$
$$1 + 3 + 5 + 7 + 9 = ?$$
$$1 + 3 + 5 + 7 + 9 + 11 = ?$$
$$1 + 3 + 5 + 7 + 9 + 11 + 13 = ?$$
$$1 + 3 + 5 + \cdots + (2n - 1) = ?$$

5. An addition problem can be checked by a process called "casting out nines." To do this, you first find the sum of the digits of each of the addends (that is, numbers that are added), divide by 9, and record the remainder. Digits may be added again and again until a one-digit remainder is obtained. The sum of these remainders is then divided by 9 to find a final remainder. This should be equal to the remainder found by considering the sum of the addends (that is, the answer), adding its digits, dividing the sum of these digits by 9, and finding the remainder. Here is an example:

Addends	Sum of digits	Remainders
4,378	22	4
2,160	9	0
3,872	20	2
1,085	14	5
11,495		11

When the sum of the remainders is divided by 9, the final remainder is 2. This corresponds to the remainder obtained by dividing the sum of the digits in the answer $(1 + 1 + 4 + 9 + 5 = 20)$ by 9.

Try this procedure for several other examples and verify that it works in each case.

6. Try to discover a procedure for checking multiplication by casting out nines. Verify that this procedure works for several cases.

7. There is a procedure for multiplying a two-digit number by 9 on one's fingers provided that the tens digit is smaller than the ones digit. The accompanying diagram shows how to multiply 28 by 9.

Reading from the left, put a space after the second finger and bend the eighth finger. Read the product in groups of fingers as 252.

Use this procedure to find: **(a)** 9×47; **(b)** 9×39; **(c)** 9×18; **(d)** 9×27. Check each of the answers you have obtained.

8. John offered to work for 1¢ the first day, 2¢ the second day, 4¢ the third day, and so forth, doubling the amount each day. Bill offered to work for $30 a day. Which boy would receive the most money for a job that lasted **(a)** ten days? **(b)** fifteen days? **(c)** sixteen days?

9. Using a method similar to that of Gauss, find:
   (a) The sum of the first 80 counting numbers.
   (b) The sum of the first 200 counting numbers.
   (c) The sum of all the odd numbers from 1 through 49.
   (d) The sum of all the odd numbers from 1 through 199.
   (e) The sum of all the even numbers from 2 through 400.

10. Use the results obtained in Exercise 9 and try to find a formula for the sum of:
   (a) The first n counting numbers, that is,
   $$1 + 2 + 3 + \cdots + (n - 1) + n.$$
   (b) The first n odd numbers, that is,
   $$1 + 3 + 5 + \cdots + (2n - 3) + (2n - 1).$$

11. Find the sum:
   $$\frac{1}{2} + \frac{1}{2^2} + \frac{1}{2^3} + \frac{1}{2^4} + \cdots + \frac{1}{2^n}$$

   To help you discover this sum, complete these partial sums and search for a pattern.

(a) $\dfrac{1}{2} + \dfrac{1}{2^2} = \dfrac{1}{2} + \dfrac{1}{4} = $?

(b) $\dfrac{1}{2} + \dfrac{1}{2^2} + \dfrac{1}{2^3} = \dfrac{1}{2} + \dfrac{1}{4} + \dfrac{1}{8} = $?

(c) $\dfrac{1}{2} + \dfrac{1}{2^2} + \dfrac{1}{2^3} + \dfrac{1}{2^4} = \dfrac{1}{2} + \dfrac{1}{4} + \dfrac{1}{8} + \dfrac{1}{16} = $?

*12. Try to discover the rule that is being used in each case to obtain the answer given. For example, the given information

$$2, 5 \to 6, \qquad 3, 10 \to 12, \qquad 7, 8 \to 14, \qquad 5, 3 \to 7$$

should lead you to the rule $x, y \to x + y - 1$.

(a) $3, 4 \to 8;$ $\quad 3, 3 \to 7;$ $\quad 1, 5 \to 7;$ $\quad 2, 8 \to 11.$
(b) $2, 4 \to 8;$ $\quad 3, 5 \to 15;$ $\quad 1, 7 \to 7;$ $\quad 3, 9 \to 27.$
(c) $1, 5 \to 1;$ $\quad 5, 2 \to 2;$ $\quad 3, 9 \to 3;$ $\quad 6, 5 \to 5.$
(d) $4, 8 \to 4;$ $\quad 5, 1 \to 5;$ $\quad 6, 6 \to 6;$ $\quad 8, 2 \to 8.$
(e) $3, 4 \to 5;$ $\quad 4, 4 \to 4;$ $\quad 5, 7 \to 0;$ $\quad 9, 2 \to 1.$

PEANUTS | THE ANSWER IS TWELVE! | IT ISN'T? HOW ABOUT SIX? | FOUR? NINE? TWO? TEN? | DO YOU HAVE THE FEELING THAT I'M GUESSING?

© 1971 United Feature Syndicate, Inc. Reproduced by permission.

2

explorations with geometric patterns

In the study of geometry we frequently form conclusions on the basis of a small number of examples, together with an exhibited pattern. Consider, for example, the problem of determining the number of triangles that can be formed from a given convex polygon by drawing diagonals from a given vertex P. First we draw several figures and consider the results in tabular form as follows.

Number of sides of polygon	4	5	6
Number of diagonals from each vertex	1	2	3
Number of triangles formed	2	3	4

---

* An asterisk preceding an exercise indicates that the exercise is more difficult or challenging than the others.

From the pattern of entries in the table it appears that the number of triangles formed is two less than the number of sides of the polygon. Thus we expect that we can form 10 triangles for a dodecagon, a polygon with 12 sides, by drawing diagonals from a given vertex. In general, then, for a polygon with n sides, called an n-gon, we can form $n - 2$ triangles.

This is reasoning by *induction*. We formed a generalization on the basis of several specific examples and an obvious pattern. This procedure does not, however, constitute a proof. In order to *prove* that $n - 2$ triangles can be formed in this way for a polygon with n sides, we must observe that two of the n sides intersect at the common point of the diagonals and that each of the other $n - 2$ sides is used to form a different triangle.

It is important to recognize that not all patterns lead to valid generalizations. Patterns offer an opportunity to make reasonable guesses, but these need to be proved before they can be accepted with certainty. Consider, for example, the maximum number of regions into which a circular region can be divided by line segments joining given points on a circle in all possible ways.

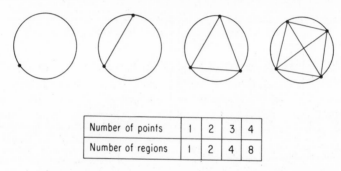

Number of points	1	2	3	4
Number of regions	1	2	4	8

Would you agree that a reasonable guess for the number of regions derived from five points is 16? Draw a figure to confirm your conjecture. What is your guess for the maximum number of regions that can be derived from six points? Again, draw a figure to confirm your conjecture. You may be in for a surprise!

EXERCISES

1. Take a piece of notebook paper and fold it in half. Then fold it in half again and cut off a corner that does not involve an edge of the original piece of paper.

Step 1   Step 2

Your paper, when unfolded, should look like the preceding sketch.
That is, with two folds we produced one hole. Repeat the same
process but this time make three folds before cutting off an edge. Try
to predict the number of holes that will be produced. How many holes
will be produced with four folds? With n folds?

2. We wish to color each of the pyramids in the accompanying figure so
that no two of the faces (sides and base) that have a common edge are
of the same color.

(a) What is the minimum number of colors required for each pyramid?
*(b) What is the relationship between the minimum number of colors
required and the number of faces of a pyramid?

3. Consider the following set of figures. In each figure we count the number
V of vertices, the number A of arcs, and the number R of regions into
which the figure divides the plane. A square, for example, has 4 vertices,
4 arcs, and divides the plane into 2 regions (inside and outside of the
square). See if you can discover a relationship between V, R, and A
that holds for each case. Confirm your generalization by testing it on
several other figures.

	V	R	A
□	4	2	4
◸	4	3	5
⊠	5	5	8
☆	10	7	15

*4. What is the maximum number of pieces that can be obtained by slicing
an orange with two cuts? With three cuts? With four cuts?

5. Consider the problem of arranging sets of squares that are joined along

their edges. In general, these figures are referred to as **polyominoes**. A single square, a **monomino**, can be arranged in only one way. Two squares, a **domino**, can also be arranged in only one way, since the position of the 2 × 1 rectangle does not affect the arrangement. That is, any two such figures are congruent. There are two distinct arrangements for a **tromino**, that is, three squares.

Monomino   Domino   Trominoes

Next consider a **tetromino**, a set of four squares. Here we find that there are five possible arrangements such that no two are congruent. As in the case of the domino and other arrangements, we exclude any rearrangement that merely consists of a rotation which places the same squares in a congruent position. Here are three of the five possible tetrominoes:

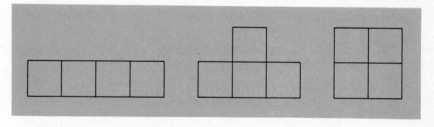

Draw the two remaining tetrominoes.

6. Consider a polyomino formed with five squares, called a **pentomino**. There are twelve such arrangements possible. Try to draw all twelve.

7. How many triangles can you count in the figure? Name them by their vertices.

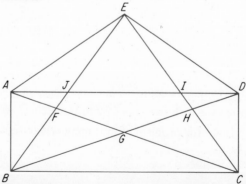

8. Twelve matchsticks are arranged to form the figure shown. By removing only two of the matches, form a figure that consists of two squares.

9. Copy the given figure and form a figure consisting of two triangles by removing three of the line segments.

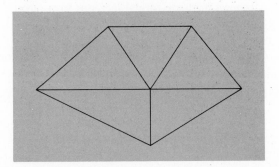

10. Twenty matchsticks are arranged to form the figure shown. By rearranging only three matchsticks, form a new figure that consists of five congruent squares.

11. Arrange four 1-inch sticks and four $\frac{1}{2}$-inch sticks to form three congruent squares.

12. A square lake has a tree planted at each of its four corners. Show that the lake can be enlarged to twice its original size without replanting the trees, having the trees surrounded by water, or changing the shape of the lake.

3

mathematical recreations

1	12	7	14
8	13	2	11
10	3	16	5
15	6	9	4

The popularity of mathematics as a means of recreation and pleasure is evidenced by the frequency with which it is found in popular magazines and newspapers. In this section we shall explore several of these recreational aspects of mathematics.

How many of us have ever failed to be impressed by a magic square? The magic square shown here is arranged so that the sum of the numbers in any row, column, or diagonal is always 34.

Although there are formal methods to complete such an array, we will not go into them here. Suffice it to say that such arrangements have fascinated man for many centuries. Indeed, the first known example of a magic square is said to have been found on the back of a tortoise by the Emperor Yu in about 2200 B.C.! This was called the "lo-shu" and appeared as an array of numerals indicated by knots in strings as in the figure on the left below. Black knots were used for even numbers and white ones for odd numbers. In modern times this appears as a magic square of third order. The sum along any row, column, or diagonal is 15 as in the figure below.

4	9	2
3	5	7
8	1	6

+	3	4	1
7			
2			
5			

Closely akin to magic squares are square arrays of numbers useful in "mathemagic." Let us "build" a trick together. We begin by forming a square array and placing any six numerals in the surrounding spaces as in the figure. The numbers 3, 4, 1, 7, 2, and 5 are chosen arbitrarily. Next find the sum of each pair of numbers as in a regular addition table.

Now we are ready to perform the trick. Have someone circle any one of the nine numerals in the box, say 10, and then cross out all the other numerals in the same row and column as 10.

Next circle one of the remaining numerals, say 3, and repeat the process. Circle the only remaining numeral, 9. The sum of the circled numbers is $10 + 2 + 9 = 22$.

+	3	4	1
7	10	11	8
2	5	6	3
5	8	9	6

+	3	4	1
7	(10)	11	8
2	5	6	3
5	8	9	6

+	3	4	1
7	(10)	11	8
2	5	6	(3)
5	8	(9)	6

The interesting item here is that the sum of the three circled numbers will always be equal to 22, regardless of where you start! Furthermore, note that 22 is the sum of the six numbers outside the square. Try to explain why this trick works, and build a table with 16 entries.

Another type of mathematical trick that is quite popular is the "think of a number" type. Follow these instructions:

Think of a number.
Add 3 to this number.
Multiply your answer by 2.
Subtract 4 from your answer.
Divide by 2.
Subtract the number with which you started.

If you follow these instructions carefully, your answer will always be 1, regardless of the number with which you start. We can explain why this trick works by using algebraic symbols or by drawing pictures, as shown below.

Try to make up a similar trick of your own.

Think of a number:	n	☐	(Number of coins in a box)
Add 3:	$n+3$	☐ ○ ○ ○	(Number of original coins plus three)
Multiply by 2:	$2n+6$	☐ ○ ○ ○ / ☐ ○ ○ ○	(Two boxes of coins plus six)
Subtract 4:	$2n+2$	☐ ○ / ☐ ○	(Two boxes of coins plus two)
Divide by 2:	$n+1$	☐ ○	(One box of coins plus one)
Subtract the original number, n:	$(n+1)-n=1$	○	(One coin is left)

Optical illusions are vivid reminders of the fact that we cannot always trust our eyes. Can you trust yours? Test yourself and see.

First guess which of the line segments, *a* or *b*, appears to be the longer in the four parts of the above figure. Then use a ruler and check your estimate.

Now see whether you can guess which line segments are parallel, if any, in the three parts of the next figure.

You should have found, in both cases, that looks can be deceiving. In each of the first four drawings, the segments are equal in length; in each of the last three they are parallel!

Here are two other optical illusions that were recently created and that appear most startling at first glance.

Mathematical fallacies have always intrigued both professional and amateur mathematicians alike. Here is an arithmetic fallacy to puzzle you. You might even consider trying this on your local banker. First you need to deposit $50 in the bank and then make withdrawals in the following manner:

Withdraw $20, leaving a balance of $30.
Withdraw $15, leaving a balance of $15.
Withdraw $ 9, leaving a balance of $ 6.
Withdraw $ 6, leaving a balance of $ 0.

Adding, we have: $50 $51

The total withdrawal is $50, whereas the total of the balances is $51. Can you therefore go to the bank to demand an extra dollar?

Here is a "proof" that $1 = 2$. Even though you may have forgotten the algebra you need to follow this, don't let it stop you; see if you can discover the fallacy.

Let $a = b$. Then

$$a^2 = b^2 = b \cdot b.$$

Since $a = b$, we may write $b \cdot b$ as $a \cdot b$. Thus

$$a^2 = a \cdot b.$$

Subtract b^2:

$$a^2 - b^2 = a \cdot b - b^2$$

Factor:

$$(a + b)(a - b) = b(a - b)$$

Divide by $a - b$:

$$\frac{(a + b)(a - b)}{(a - b)} = \frac{b(a - b)}{(a - b)}$$

Thus

$$a + b = b.$$

Since $a = b$, we may write this as

$$b + b = b \quad \text{or} \quad 2b = b.$$

Divide by b:

$$\frac{2b}{b} = \frac{b}{b}$$

Therefore

$$2 = 1.$$

Many people enjoy solving interesting or amusing puzzles. A variety of these, many of which are old-timers, appear in the exercises that follow.

EXERCISES

1. All of the following puzzles have logical answers, but they are not strictly mathematical. See how many you can answer.
   (a) How many two-cent stamps are there in a dozen?
   (b) How many telephone poles are needed in order to reach the moon?
   (c) How far can you walk into a forest?
   (d) Two United States coins total 55¢ in value, yet one of them is not a nickel. Can you explain this?
   (e) How much dirt is there in a hole which is 3 feet wide, 4 feet long, and 2 feet deep?
   (f) There was a blind beggar who had a brother, but this brother had no brothers. What was the relationship between the two?

2. A farmer has to get a fox, a goose, and a bag of corn across a river in a boat which is only large enough for him and one of these three items. Now if he leaves the fox alone with the goose, the fox will eat the goose. If he leaves the goose alone with the corn, the goose will eat the corn. How does he get all items across the river?

3. Three Indians and three missionaries need to cross a river in a boat big enough only for two. The Indians are fine if they are left alone or if they are with the same number or with a larger number of missionaries. They are dangerous if they are left alone in a situation where they outnumber the missionaries. How do they all get across the river without harm?

4. A bottle and cork cost $1.50 together. The bottle costs one dollar more than the cork. How much does each cost?

5. A cat is at the bottom of a 30-foot well. Each day she climbs up 3 feet; each night she slides back 2 feet. How long will it take for the cat to get out of the well?

6. If a cat and a half eats a rat and a half in a day and a half, how many days will it take for 100 cats to eat 100 rats?

7. Ten coins are arranged to form a triangle as shown in the figure. By

rearranging only three of the coins, form a new triangle that points in the opposite direction from the one shown.

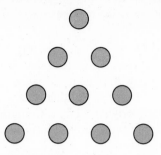

8. Rearrange the eight segments shown so as to form three congruent squares; that is, three squares that are exactly the same size. Each of the four smaller segments is one-half of the length of one of the larger segments.

9. Use six matchsticks, all of the same size, to form four equilateral triangles. (An equilateral triangle has all three sides the same length.)

10. Four matchsticks are arranged to form a "cup" with a coin contained within three of the segments, as shown in the figure. By rearranging only two of the matchsticks, form a new figure that is congruent to the original one (that is, exactly the same in shape but possibly different in position) but with the coin no longer contained within any of the segments.

11. A sailor lands on an island inhabited by two types of people. The *A*'s always lie, and the *B*'s always tell the truth. The sailor meets three inhabitants on the beach and asks the first of these: "Are you an *A* or a *B*?" The man answers, but the sailor doesn't understand him and asks the second person what he had said. The man replies: "He said that he was a *B*. He is, and so am I." The third inhabitant then says: "That's not true. The first man is an *A* and I'm a *B*." Can you tell who was lying and who was telling the truth?

12. A man goes to a well with three cans whose capacities are 3 gallons, 5 gallons, and 8 gallons. Explain how he can obtain exactly 4 gallons of water from the well.

13. Here is a mathematical trick you can try on a friend. Ask someone to place a penny in one of his hands, and a dime in the other. Then tell him to multiply the value of the coin in the right hand by 6, multiply the

value of the coin held in the left hand by 3, and add. Ask for the result. If the number given is even, you then announce that the penny is in the right hand; if the result is an odd number, then the penny is in the left hand and the dime is in the right hand. Can you figure out why this trick works?

14. Consider a house with six rooms and furniture arranged as in the accompanying figure. We wish to interchange the desk and the bookcase, but in such a way that there is never more than one piece of furniture in a room at a time. The other three pieces of furniture do not need to return to their original places. Can you do this? Try it using coins or other objects to represent the furniture.

Cabinet			Desk	
Television set	Sofa		Bookcase	

15. Three men enter a hotel and rent a suite of rooms for $30. After they are taken to their rooms the manager discovers he overcharged them; the suite rents for only $25. He thereupon sends a bellhop upstairs with the $5 change. The dishonest bellhop decides to keep $2 and returns only $3 to the men. Now the rooms originally cost $30, but the men had $3 returned to them. This means that they paid only $27 for the room. The bellhop kept $2. $27 + $2 = $29. What happened to the extra dollar?

16. Write the numbers from 1 through 10 using four 4's for each. Here are the first three completed for you:

$$\frac{44}{44} = 1, \qquad \frac{4}{4} + \frac{4}{4} = 2, \qquad \frac{4 + 4 + 4}{4} = 3$$

17. Arrange two pennies and two dimes as shown below. Try to interchange the coins so that the pennies are at the right and the dimes at the left. You may move only one coin at a time, you may jump over only one coin, and pennies may be moved only to the right while dimes may be moved only to the left. No two coins may occupy the same space at the same time. What is the minimum number of moves required to complete the game?

P	P		D	D

18. Repeat Exercise 17 for three pennies and three dimes, using seven blocks. What is the minimum number of moves required to complete the game?

19. What are the next two letters in the sequence:

$$O, T, T, F, F, S, S, \ldots ?$$

20. What are the next two letters in the sequence:

$$E, F, F, N, O, S, S, \ldots ?$$

PEDAGOGICAL CONSIDERATIONS

Throughout this book explorations of the use of topics in teaching elementary school students are made under the heading **Pedagogical Considerations.**

1. Prepare a bulletin board display of optical illusions.

2. Make a collection of mathematical puzzles for use with elementary school students.

3. Make a collection of mathematical games. Here is one that you can explore by yourself:

Place a half-dollar, a quarter, and a nickel in one position, *A*, as in the figure. Then try to move these coins, one at a time, to position *C*. Coins may also be placed in position *B*. At no time may a larger coin be placed on a smaller coin. This can be accomplished in $2^3 - 1$, that is, 7 moves.

Next add a penny to the pile and try to make the change in $2^4 - 1$, that is, 15 moves.

This is an example of a famous problem called the **Tower of Hanoi.** The ancient Brahman priests were to move a pile of 64 such discs of decreasing size, after which the world would end. This would require $2^{64} - 1$ moves. Try to estimate how long this would take at the rate of one move per second.

4. Here is a game that must be played by two persons:

Two players alternate in selecting one of the numbers 1, 2, 3, 4, 5, or 6. After each number is selected, it is added to the sum of those previously selected. For example, if player *A* selects 3 and player *B* selects 5, then the total is 8. If *A* selects 3 again, then the total is 11 and player *B* takes his turn. The object of the game is to be the first one to reach 50. There is a way to always win if you are permitted to go first. See if you can discover this method for winning, and then try to play the game with a classmate and with some elementary school students. Incidentally, games such as these are excellent means of providing for practice on number facts in an interesting and disguised manner.

5. Make a collection of mathematical tricks. Many tricks of magic have their basis in elementary mathematics and may be found in books on mathematical recreations. Here is one example of such a trick:

Have someone place three dice on top of one another while you turn your back. Then instruct him to look at and find the sum of the values shown on the two faces that touch each other for the top and middle dice, the two faces that touch each other for the middle and bottom dice, and the value of the bottom face of the bottom dice. You then turn around and at a glance tell him this sum. The trick is this: You merely subtract the value showing on the top face of the top die from 21. Stack a set of three dice in the manner described and try to figure out why the trick works as it does.

part
one

NUMBER SYSTEMS

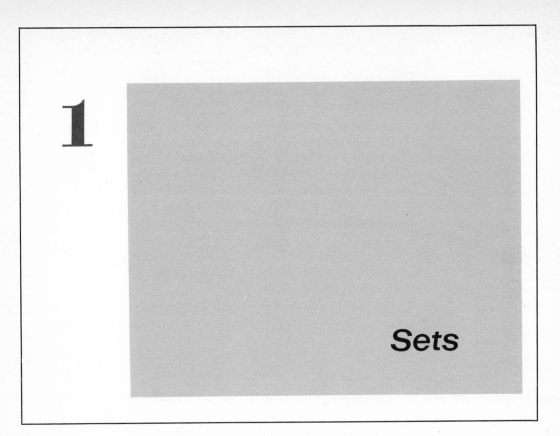

1

Sets

The concept of a set pervades all of mathematics and is used in many of our daily activities. Think of the many times that you refer to a pen and pencil set, a set of dishes, a set of encyclopedias, or in some other way identify a collection of items as a set.

In elementary school classes number concepts are based upon student experiences with sets of elements. Many properties of numbers and relations among numbers are developed from sets of elements. Each algebraic statement is true for a certain set of numbers, called its solution set; each geometric figure consists of a set of points. The use of sets of elements is a unifying theme throughout all branches of mathematics.

Elementary school teachers have a special need for understanding relations among sets of elements since these relations are used extensively in the introduction of number concepts in arithmetic.

1-1

set notation

Before man ever began to count he probably used a matching process to determine the size of a collection. The caveman who had seen five animals roaming in the fields might have shown this by sketching a picture on the wall of his abode as in the figure at the top of the next page.

He was, in fact, using a very sophisticated idea—that of a **one-to-one correspondence.** For each animal he saw in the field he placed a mark on the wall. For each mark his spouse saw, she knew he had seen one animal. They could communicate in this manner without ever knowing about or using the **numeral** 5, which is merely a symbol invented at a later date to represent the size of a collection.

Later, man became more sophisticated and created a set of numbers with which he could count. We call this collection the set C of **counting numbers** and represent them in **set notation** as follows:

$$C = \{1, 2, 3, 4, 5, \ldots\}$$

The three dots indicate that the set continues indefinitely in the pattern indicated, that is, that there are infinitely many elements in the set. The set C is said to be an **infinite set.** A set is usually named by a capital letter, and the **elements** or **members** of the set are indicated within a pair of braces.

There are many words in our language that indicate sets: *school* of fish, *swarm* of bees, *gaggle* of geese, *herd* of cattle, *squadron* of planes, etc. A **set** is a collection of things. Thus we may speak of

> The set of letters of the English alphabet.
> The set of states of the United States of America.
> The set of English names of the days of the week.
> The set of counting numbers 1 through 5.

Each of the preceding sets is said to be **well-defined.** That is, from the description you can tell whether or not any given element belongs to the set. For example, we know that r, s, and t are members of the first set listed above, whereas π and 7 clearly are not members.

Can you tell, from the description given, which of the following numbers belong to the set of counting numbers 1 through 5?

$$3, \quad 2\tfrac{1}{2}, \quad 8, \quad 1, \quad 12$$

You should note that 3 and 1 are members of the set, whereas $2\tfrac{1}{2}$, 8, and 12 are not. The set of counting numbers 1 through 5 is a well-defined set when the set of counting numbers is defined as the set 1, 2, 3, 4, 5, 6, 7, 8, 9, 10, and so forth.

Not all sets are well-defined. See if you can explain why each set in the next list is not a well-defined one.

> The set of good tennis players.
> The set of interesting numbers.
> The set of beautiful movie stars.
> The set of well-written books.

In this text we shall be concerned primarily with well-defined sets. Furthermore, we shall agree to name a set by an arbitrarily chosen capital letter and to list the elements of the set within a pair of braces as in the following examples.

Example 1: Write in set notation: the set W of English names of the days of the week.

Solution:

$$W = \{\text{Sunday, Monday, Tuesday,}\\ \text{Wednesday, Thursday, Friday, Saturday}\}$$

Example 2: List the elements in the set T of counting numbers 1 through 5.

Solution:

$$T = \{1, 2, 3, 4, 5\}$$

The **membership symbol** $\in$ is often used as in

$$2 \in \{1, 2, 3, 4, 5\} \quad \text{and} \quad 7 \notin \{1, 2, 3, 4, 5\}$$

to indicate that elements are or are not members of sets. We read $\in$ as "is a member of" and $\notin$ as "is not a member of" a given set. Where it becomes tedious to list individually all of the members of a set, we frequently use three dots to indicate missing elements.

Example 3: List the elements in the set A of letters of the English alphabet.

Solution:

$$A = \{a, b, c, \ldots, z\}$$

Notice that when three dots are used to show that some elements are missing, it is necessary to state the elements in some order so that a pattern can be observed and used to identify the missing elements. Any ordered set of elements is a **sequence.** In Example 3 the sequence has a last element and is a **finite set.**

We also use three dots at the end of a sequence of elements to indicate that the set continues indefinitely in the pattern indicated; that is, there is an infinite number of elements in the set.

Example 4: List the elements in the set G of counting numbers greater than 100.

Solution:

$$G = \{101, 102, 103, \ldots\}$$

It is interesting, at times, to seek verbal descriptions for given sets of elements which have already been listed. The next example illustrates this point.

Example 5: Write a verbal description for the set

$$Y = \{1, 3, 5, 7, 9\}.$$

Solution: There are several correct responses that might be given. Two of these are: "The set of odd numbers 1 through 9"; "The set of odd numbers between 0 and 11." Note that the word "between" implies that the first and last numbers (that is, 0 and 11) are not included as members of the given set.

The basic property of a set of elements is the identification of its members. A set is well-defined if its members can be identified. The sets $A = \{1, 2, 3\}$ and $B = \{3, 1, 2\}$ are the same since they have the same members. When two sets consist of precisely the same elements, they are said to be **equal sets,** that is, **identical sets.** We write $\{1, 2, 3\} = \{3, 1, 2\}$, since order of listing does not affect membership, $\{1, 1, 2\} = \{1, 2\}$, since listing an element more than once does not affect membership, and in general $A = B$ to show that sets A and B have the same members. We write $A \neq B$ to show that sets A and B do not have the same members.

EXERCISES

Replace the asterisk (∗) by $\in$ or $\notin$ to obtain a true statement.

1. $7 * \{1, 3, 5, 7, 9\}$ 2. $7 * \{2, 4, 6, 8, 0\}$

3. $\triangle * \{., |, \square\}$ 4. $\triangle * \{/, \wedge, \triangle\}$

Tell whether or not each of the following sets is well-defined.

5. The set of U.S. astronauts who have orbited the earth.

6. The set of cities that are state capitals in the United States of America.

7. The set of large states in the United States.

8. The set of states in the United States with good climates.

State whether or not the sets of letters are equal as sets.

9. $\{r, a, t\}$, $\{t, a, r\}$

10. $\{t, r, a, p\}$, $\{a, p, a, r, t\}$

11. $\{w, o, l, f\}$, $\{f, o, l, l, o, w\}$

12. $\{p, r, o, f, e, s, s, o, r\}$, $\{f, r, a, p, p, e, s\}$

List the elements in each of the following sets, using the notation developed in this section.

13. The set of English names of the months in the year.

14. The set of letters in the English alphabet that precede *j*.

15. The set of counting numbers between 1 and 10.

16. The set of odd counting numbers less than 25.

Write a verbal description for each of the following sets.

17. $N = \{1, 2, 3, 4, 5, 6\}$

18. $R = \{1, 2, 3, \ldots, 99\}$

19. $S = \{51, 52, 53, \ldots\}$

20. $M = \{5, 10, 15, 20, \ldots\}$

21. $K = \{10, 20, 30, \ldots, 150\}$

22. $T = \{1, 4, 9, 16, 25, 36\}$

*23. $P = \{0, 2, 6, 12, 20, \ldots, 90\}$

*24. $A = \{8, 5, 4, 9, 1, 7, 6, 3, 2\}$

PEDAGOGICAL CONSIDERATIONS

For elementary school teachers there are two major aspects of any mathematical topic:

What is its mathematical content?
How can it be used in teaching elementary school children?

Our present topic is *sets.* The mathematical content has been introduced. The exercises have provided an opportunity for you to check and reinforce your own understanding. We now explore how the topic can be used in teaching elementary school children. We shall do this for many of the topics that we study. Among the aspects of teaching that will be considered at various times are:

contemporary uses of a topic
suggested teaching procedures (promising practices)
suggested activities for pupils at various grade levels
construction and use of models
construction and use of visual aids
collection of resources for teaching
useful historical information
useful references
development of evaluation instruments
development of review instruments

How are *sets* used in teaching elementary school children? The basic use of sets is in the identification and description of elements that are being discussed.

Example 1: The meanings of many words are learned by identifying the sets of elements to which they refer. For example, from a collection of toys on the floor a child is asked to select (a) the balls; (b) the blocks; (c) the red blocks.

Example 2: Properties of elements of a set are learned by grouping the elements with a specified property (for example, selecting the red blocks) and by matching elements with a particular property. For example, draw lines to match the figures that have the same shape:

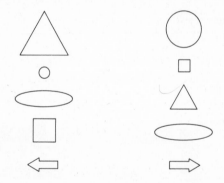

Example 3: Understanding of membership in a set under discussion is explored at all grade levels. For example, note the need for definitions of sets of elements in order to answer these questions.

(a) Is *C* a point of the given circle?
(b) Is 51 a prime number?
(c) Is 6 a perfect number?
(d) Is this curve a simple curve?

The following exercises are provided to help you develop your own teaching skills.

1. Describe one or more teaching situations to help pupils develop their understanding of **(a)** the color green; **(b)** the meaning of the word *chair*; **(c)** membership in a set; **(d)** membership in each of two sets; **(e)** membership in at least one, possibly both, of two sets.

2. Make patterns for cutouts that could be used on a felt board, and describe the use that you have in mind, to help pupils develop their understanding of **(a)** triangles; **(b)** right triangles; **(c)** the shape of a geometric figure; **(d)** geometric figures of the same size and shape.

3. Look through some contemporary elementary school textbooks and describe other ways in which membership in sets is effectively used in teaching elementary school pupils.

4. There are many synonyms for "set" that are commonly used for different types of elements. List at least fifty (50) of these terms to help you convince yourself, and your students, of the universality of the concept of a set.

1-2

subsets The set containing the totality of elements for any particular discussion is called the **universal set** U, and may vary for each discussion. For example, let us agree, for the present, to talk about the set of counting numbers 1 through 9. Then our universal set is

$$U = \{1, 2, 3, 4, 5, 6, 7, 8, 9\}.$$

A set A is a **subset** of U if the set A has the property that each element of A is also an element of U. We write $A \subseteq U$ (read "A is included in U"). There are many subsets of U; a few of them are listed below:

$$A_1 = \{1, 2, 3\}$$
$$A_2 = \{1, 5, 7, 8, 9\}$$
$$A_3 = \{2\}$$
$$A_4 = \{1, 2, 3, 4, 5, 6, 7, 8, 9\}$$

Note in particular that the set A_4 contains each of the elements of U and is classified as a subset of U. Any set is said to be a subset of itself.

A set A is said to be a **proper subset** of U if A is a subset of U and there is at least one element of U that is not an element of A. We write $A \subset U$ (read "A is properly included in U"). Intuitively we speak of a proper subset as part of, but not all of, a given set. Each of the subsets A_1, A_2, A_3, and A_4 are subsets of U; the sets A_1, A_2, and A_3 are proper subsets of U; the set A_4 is not a proper subset of U. We may write this in symbols as follows:

$$A_1 \subset U, \qquad A_2 \subset U, \qquad A_3 \subset U, \qquad A_4 \subseteq U$$

Example 1: List three proper subsets of the set R where

$$R = \{1, 2, 3, 4\}.$$

Solution: Here are three proper subsets:

$$A = \{1, 2\}, \qquad B = \{1, 3, 4\}, \qquad C = \{1\}.$$

Do you see that there are others? Note that the choice of letters to name each of the subsets is completely arbitrary.

The **empty** set or **null** set is the set that contains no elements and is denoted by the symbol $\emptyset$ or by the symbol $\{\ \}$. Note that the first symbol given for the empty set is not included within a pair of braces. Some examples of empty sets are:

> The set of counting numbers between 2 and 3.
> The set of states of the United States with borders on
> both the Atlantic Ocean and the Pacific Ocean.

A set A has been defined to be a subset of a set B if each element of A is also an element of B. If we wish to prove that a set A is not a subset of a set B, we must find an element of A that is not an element of B. For example, if $G = \{1, 3, 5\}$ and $H = \{1, 2, 3, 4, 5\}$, then $G \subseteq H$ but $H \not\subseteq G$, since $2 \in H$ and $2 \notin G$; also $4 \in H$ and $4 \notin G$. Notice that since the empty set has no elements, there are no elements of the empty set that can fail to be elements of a set B. Thus the empty set is a subset of every set B; $\emptyset \subseteq B$ for every set B. This property of the empty set is often emphasized by defining a set A to be a *subset* of a set B if there are no elements of A that are not elements of B.

Example 2: List all the possible subsets of the set $\{1, 2\}$.

Solution:

$$\emptyset, \qquad \{1\}, \qquad \{2\}, \qquad \{1, 2\}.$$

Notice in the solution of Example 2 that the subsets of $\{1, 2\}$ are the sets obtained by selecting or not selecting each of the elements of the given set. If we select both elements, we obtain $\{1, 2\}$; if we select exactly one of the elements, we obtain $\{1\}$ or $\{2\}$; if we do not select either element, we obtain the empty set. Since there are two elements and two choices (take it or leave it) for each element, there are 2×2 subsets, as we obtained.

The subsets obtained in Example 2 may be paired as

$$\{1, 2\} \text{ with } \emptyset, \quad \text{and} \quad \{1\} \text{ with } \{2\}.$$

Then the two sets of each pair have no common elements and contain all elements of the given set $\{1, 2\}$. Relative to the given set, if we had the set $\{1\}$, we could find the corresponding set $\{2\}$; if we had the set $\{2\}$, we could find the corresponding set $\{1\}$; if we had the set $\{1, 2\}$ we could find the set $\varnothing$; and if we had the set $\varnothing$, we could find the set $\{1, 2\}$. This process may be described as finding the complement of a set relative to the given set $\{1, 2\}$.

For any given universal set U, each set A has a complement A' (also written $\overline{A}$) that consists of the elements of U that are not elements of A and is called the **complement of A relative to U**. For example, if $U = \{1, 3, 5, 7, 9\}$ and $A = \{1, 3, 7\}$, then $A' = \{5, 9\}$. If $A = U$, then $A' = \varnothing$; if $A = \varnothing$, then $A' = U$. In other words, the complement of the universal set is the empty set; the complement of the empty set is the universal set.

Example 3: List the subsets of $\{1, 2, 3\}$ and pair each set with its complementary set.

Solution:

$$\{1, 2, 3\} \qquad \{2, 3\} \qquad \{1, 3\} \qquad \{1, 2\}$$
$$\varnothing \qquad\quad \{1\} \qquad\; \{2\} \qquad\; \{3\}$$

It seems reasonable to expect that all possible subsets should be obtained by considering the elements three at a time, two at a time, one at a time, and zero at a time. As a check that we have found all possible subsets, notice that the original set had three elements; we have two choices (take it or leave it) for each element; $2 \times 2 \times 2 = 8$, and we found eight subsets.

EXERCISES

1. Describe three empty sets.

2. Let $U = \{c, a, r, t\}$ and list the elements of A' when A is defined as **(a)** $\{c, a, r\}$; **(b)** $\{r, a, t\}$; **(c)** $\{a, t\}$; **(d)** $\varnothing$.

3. Let $U = \{p, r, o, f, e, s, s, o, r\}$ and list the elements of A' when A is defined as **(a)** $\varnothing$; **(b)** $\{r, o, p, e\}$; **(c)** $\{r, o, s, e\}$; **(d)** $\{p, r, o, s, e\}$.

4. Let U be the set of counting numbers $\{1, 2, 3, \ldots\}$. Let A be the set of even counting numbers and describe the set A'.

5. If $U = \{1, 2, 3, 4, 5, 6, 7, 8, 9\}$, $A = \{1, 3, 5, 7, 9\}$, and $B = \{3, 6, 9\}$, find **(a)** A'; **(b)** B'.

6. List all possible subsets of each set.

(a) $\{a\}$ **(b)** $\{a, b\}$ **(c)** $\{a, b, c\}$

(d) $\{a, b, c, d\}$ **(e)** $\{a, b, c, d, e\}$ **(f)** $\varnothing$

7. Use the results obtained in Exercise 6, copy, and complete this array.

Number of elements	0	1	2	3	4	5	6
Number of subsets	1						

8. Make an array as in Exercise 7 for the number of proper subsets of the sets in Exercise 6.

9. Use your answer for Exercise 7 and conjecture a formula for the number N of subsets that can be formed from a set consisting of n elements.

10. Use your answer for Exercise 8 and conjecture a formula for the number N of proper subsets that can be formed from a set consisting of n elements.

PEDAGOGICAL CONSIDERATIONS

1. Describe how you could use sets of elements to explain: **(a)** 3 + 2; **(b)** 5 − 2; **(c)** the impossibility of 2 − 5 with counting numbers; **(d)** 6 ÷ 2; **(e)** 4 × 3.

2. Explain how one of the operations +, ×, −, ÷ can be associated with the taking of the complement of a set.

3. Give five examples of the empty set that can be used with elementary school children.

4. Make a set of suitable materials for doing one of the problems suggested in Exercise 1 of these pedagogical considerations.

5. Look at a series of contemporary elementary school textbooks and describe some of the uses of subsets, often starting in the first grade, in that series.

6. Select a daily paper or news magazine and list several of the references to sets and subsets that are made in that paper or magazine. Don't miss the committees and subcommittees.

1-3

equivalent sets   If you have a set of books, you can count them and use a number to tell how many books there are in the set. If you have a set of bookmarks, you can count them and use a number to tell how many bookmarks there are in the set. If your purpose was to be sure that you had exactly one bookmark for each of your books, you could have done this without counting either the books or the bookmarks. Instead you could have placed one bookmark in each book. If this one-to-one matching of books and bookmarks could

be continued until all books and bookmarks were used, you would know that you had the same number of books in the set of books as you had bookmarks in the set of bookmarks. We would then say that there is a one-to-one correspondence between the books and the bookmarks.

Two sets, $X = \{x_1, x_2, \ldots\}$ and $Y = \{y_1, y_2, \ldots\}$, are said to be in **one-to-one correspondence** if we can find a pairing of the x's and y's such that each x corresponds to one and only one y and each y corresponds to one and only one x. Consider the sets $R = \{a, b, c\}$ and $S = \{\$, +, \%\}$. We can match the element $a \in R$ with any one of the elements in S (three choices); we can then match $b \in R$ with any one of the other two elements of S (two choices), and then match $c \in R$ with the remaining element of S (one possibility). Thus a one-to-one correspondence of the sets R and S can be shown in $3 \times 2 \times 1$ (that is, 6) ways:

$$\{a, \quad b, \quad c\} \qquad \{a, \quad b, \quad c\}$$
$$\updownarrow \quad \updownarrow \quad \updownarrow \qquad\qquad \updownarrow \quad \updownarrow \quad \updownarrow$$
$$\{\$, \quad +, \quad \%\} \qquad \{\$, \quad \%, \quad +\}$$

$$\{a, \quad b, \quad c\} \qquad \{a, \quad b, \quad c\}$$
$$\updownarrow \quad \updownarrow \quad \updownarrow \qquad\qquad \updownarrow \quad \updownarrow \quad \updownarrow$$
$$\{+, \quad \%, \quad \$\} \qquad \{+, \quad \$, \quad \%\}$$

$$\{a, \quad b, \quad c\} \qquad \{a, \quad b, \quad c\}$$
$$\updownarrow \quad \updownarrow \quad \updownarrow \qquad\qquad \updownarrow \quad \updownarrow \quad \updownarrow$$
$$\{\%, \quad \$, \quad +\} \qquad \{\%, \quad +, \quad \$\}$$

Two sets A and B that can be placed in a one-to-one correspondence are said to be **equivalent sets** (written $A \leftrightarrow B$). Any two equivalent sets have the same number of elements, that is, the same **cardinality.**

You may use the concept of a one-to-one correspondence to show whether or not any two sets of elements have the same number of members. You also use this concept of a one-to-one correspondence when you count the elements of a set. To count, you use the set of counting numbers

$$\{1, 2, 3, 4, 5, 6, \ldots\}.$$

As in §1-1, the dots indicate that the pattern continues indefinitely. If you were to count the books in the set shown in the previous figure, you would use the set of numbers $\{1, 2, 3, 4, 5\}$. You would form a one-to-one correspondence of the books with the numbers. If you were to count the bookmarks, you would use the same set of numbers.

Two sets that may be placed in one-to-one correspondence with the same set of numbers may be placed in one-to-one correspondence with each other and thus have the same number of elements.

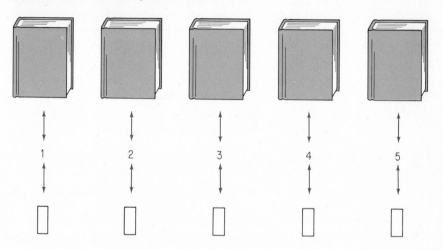

Any set A of elements that may be placed in one-to-one correspondence with the set of elements $\{1, 2, 3, 4, 5\}$ is said to have five elements; we write $n(A) = 5$ to show that the number of elements in the set A is 5. Any set B of elements that may be placed in one-to-one correspondence with the set of elements $\{1, 2, 3, 4, \ldots, k - 1, k\}$ is said to have k elements; $n(B) = k$. Notice this use of the set of counting numbers and one-to-one correspondences to determine how many elements there are in a set; that is, to determine the **cardinal number** of a set. When the counting numbers are taken in order, the last number used in the one-to-one correspondence is the cardinal number of the set. We define the cardinality of the empty set to be zero; that is, $n(\varnothing) = 0$.

We have used one-to-one correspondence to determine the cardinal number of a set of numbers. Specifically, we asserted that the last counting number used would be the cardinal number of the set. However, is there always a "last" counting number used? Consider each of the following sets:

$$A = \{1, 2, 3, 4, 5, \ldots, n, \ldots\}$$

$$B = \{2, 4, 6, 8, 10, \ldots, 2n, \ldots\}$$
$$C = \{1, 3, 5, 7, 9, \ldots, 2n - 1, \ldots\}$$
$$D = \{2, 2^2, 2^3, 2^4, 2^5, \ldots, 2^n, \ldots\}$$
$$E = \{1, \frac{1}{2}, \frac{1}{3}, \frac{1}{4}, \frac{1}{5}, \ldots, \frac{1}{n}, \ldots\}$$
$$F = \{10, 10^2, 10^3, 10^4, \ldots, 10^n, \ldots\}$$

In each case the set fails to have a last element; consequently, the set does not have a counting number as its cardinal number.

Example: Consider the sets A and B:

$$A = \{1, 2, 3, 4, 5, \ldots, n, \ldots\}$$
$$B = \{2, 4, 6, 8, 10, \ldots, 2n, \ldots\}$$

Explain: (a) $B \subseteq A$; (b) $B \neq A$; (c) $B \subset A$; (d) B is equivalent to A.

Solution: (a) Each element of B is also an element of A; that is, B is a subset of A, $B \subseteq A$.

(b) There is at least one element of A (for example, 1) that is not an element of B; thus B and A are different sets, $B \neq A$.

(c) The set B is a subset of A [see part (a)] that does not contain all elements of A [see part (b)]; that is, B is a proper subset of A, $B \subset A$.

(d) The one-to-one correspondence of n to $2n$ between the elements of A and the elements of B establishes the equivalence of A and B as shown in this array.

$$A: \{1, \quad 2, \quad 3, \quad 4, \quad \ldots \quad, \quad n, \quad \ldots\}$$
$$\quad \updownarrow \quad \updownarrow \quad \updownarrow \quad \updownarrow \qquad\qquad \updownarrow$$
$$B: \{2, \quad 4, \quad 6, \quad 8, \quad \ldots \quad, \quad 2n, \quad \ldots\}$$

Notice that we have shown that the set A is equivalent to one of its proper subsets.

A set that is equivalent to one of its proper subsets is an *infinite set*. Any infinite set fails to have a last element and fails to have a counting number as its cardinal number. The empty set and any set that has a counting number as its cardinal number are called *finite sets*.

The cardinal number of an infinite set is called a **transfinite cardinal number.** However, as in the case of the books and bookmarks, we do not need to find the cardinal numbers of two sets in order to show that they have the same number of elements. We need only demonstrate that a one-to-one correspondence exists between the sets. In the example the sets A and B have the same number of elements, $n(A) = n(B)$, because there is a one-to-one correspondence of n to $2n$ between the elements of the sets. In other words, there are just as many even counting numbers as there are counting numbers.

Each of the sets identified as A, B, C, D, E, and F may be placed in one-to-one correspondence with the set A. Each of these sets is an infinite set and each has the same transfinite cardinal number as the set of counting numbers.

The properties of infinite sets confused mathematicians for centuries. Indeed, it is only within the last century that infinite sets have been reasonably well understood. The symbol $\aleph_0$ (read "aleph-null") is used to represent the transfinite cardinal number of the set of counting numbers. It is correct then to say that there are $\aleph_0$ counting numbers, just as you might say that there are 7 days of the week or 10 fingers on your hands. Furthermore, any set that can be matched in a one-to-one correspondence with the set of counting numbers is also of size $\aleph_0$. We shall consider a few properties of transfinite cardinal numbers and some of their apparent paradoxes in the next section and in §7-6.

EXERCISES

Show all possible one-to-one correspondences of the two sets:

1. $\{1, 2\}$ and $\{p, q\}$

2. $\{1, 2, 3\}$ and $\{x, y, z\}$

3. $\{1, 2, 3, 4\}$ and $\{r, e, s, t\}$

Describe a set with cardinal number:

4. 5   *5.* 6

6. 7   *7.* 0

Find the cardinal number of each of the sets in Exercises 8 through 13:

8. $\{\triangle, >, =, \square\}$ *9.* $\{3, 5, 7\}$

10. $\{x\}$ *11.* $\{11, 12, \ldots, 18\}$

12. $\{6\}$ *13.* $\varnothing$

14. Give a set that may be placed in one-to-one correspondence with the set of counting numbers.

15. Give a set that may be placed in one-to-one correspondence with the set of even counting numbers.

16. Consider the set $A = \{a, b, c, d, e\}$ and find a set that is **(a)** equivalent and identical (equal) to A; **(b)** equivalent but not equal to A.

17. Explain whether or not **(a)** any two equal sets are necessarily equivalent; **(b)** any two equivalent sets are necessarily equal.

Show a one-to-one correspondence between:

18. The set of counting numbers 4 through 8 and the set of vowels in the English language.

19. The set of vowels in the English language and the set of even counting numbers less than or equal to 10.

20. The set of counting numbers greater than 100 and the set of all counting numbers.

*21. The set of positive odd integers and the set of positive integral multiples of 5.

PEDAGOGICAL CONSIDERATIONS

First grade children may be asked questions such as:

(i) Use lines to match the set of circles with the set of triangles.

(ii) The squares represent cards. Match each picture card with the proper numeral card.

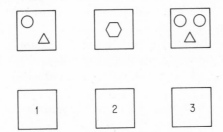

(iii) Draw a circle around the numeral for the set.

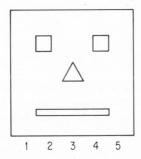

In the second and third grades children are asked to write the numeral for a given set, that is, to tell how many elements are in the set.

1. Use pictures, colored pieces of paper, etc. and make $8\frac{1}{2} \times 11$ sheets as cards that could be used to help primary school children understand what is meant by the cardinal numbers 4, 5, and 6.

2. Specify the grade level that you are considering and then describe how you might use the materials developed in Exercise 1.

3. Make up five questions regarding cardinal numbers and suitable for use with **(a)** second graders; **(b)** third graders; **(c)** fifth graders.

4. Look at a series of contemporary elementary school textbooks and describe how cardinal numbers are developed in that series.

1-4

uses of numbers

The set of numbers that we use in counting

$$\{1, 2, 3, 4, 5, \ldots\}$$

is called the set of *counting numbers* and has been considered in several earlier sections. The counting numbers provide the basis for the development of other sets of numbers and also for consideration of various uses of numbers.

Counting numbers and zero may be used as *cardinal numbers* of finite sets of elements. Counting numbers may also be used to assign an order to the elements of a finite set. For example, you expect page 15 of this book to follow page 14. Numbers used to assign an order to the elements of a set are often called **ordinal numbers.**

Numbers are also used in other ways. For example, neither cardinality nor order is significant for the number of your driver's license, your telephone number, or your social security number. Such numbers are used solely for **identification.**

The use of numbers as cardinal numbers was considered in the previous section. At that time the symbol $\aleph_0$ was introduced to represent the transfinite cardinal number of the set of counting numbers. We now list a few of the unusual arithmetic properties of this number and shall base the discussion upon the following sets:

$$A = \{1, 2, 3, 4, \ldots, n, \ldots\}$$
$$B = \{2, 4, 6, 8, \ldots, 2n, \ldots\}$$
$$C = \{1, 3, 5, 7, \ldots, 2n - 1, \ldots\}$$

1. $\aleph_0 + 1 = \aleph_0$. The set A has the transfinite cardinal number $\aleph_0$; the set A with one element 0 added has cardinal number $\aleph_0 + 1$. The equivalence of these two sets may be shown as follows:

$$\{0, \quad 1, \quad 2, \quad 3, \quad 4, \ldots, \quad n, \ldots\}$$
$$\updownarrow \quad \updownarrow \quad \updownarrow \quad \updownarrow \quad \updownarrow \qquad \updownarrow$$
$$\{1, \quad 2, \quad 3, \quad 4, \quad 5, \ldots, \quad n+1, \ldots\}$$

2. $\aleph_0 - 5 = \aleph_0$. The set A has the transfinite cardinal number $\aleph_0$;

the set A with the numbers 1, 2, 3, 4, 5 removed has the cardinal number $\aleph_0 - 5$. The equivalence of the two sets may be shown as follows:

$$\{6, \quad 7, \quad 8, \quad 9, \quad 10, \ . \ . \ . \ , \qquad n+5, . \ . \ .\}$$
$$\downarrow \quad \downarrow \quad \downarrow \quad \downarrow \quad \downarrow \qquad\qquad \downarrow$$
$$\{1, \quad 2, \quad 3, \quad 4, \quad 5, \ . \ . \ . \ , \qquad n, \ . \ . \ .\}$$

3. $\aleph_0 + \aleph_0 = \aleph_0$. The sets B and C each have the transfinite cardinal number $\aleph_0$; the union of these sets has the cardinal number $\aleph_0 + \aleph_0$. However, sets B and C together form set A which has cardinal number $\aleph_0$.

The example of the previous section demonstrates the *apparent* paradox that there are just as many even counting numbers as there are counting numbers altogether. Nevertheless, this is not a paradox but one of the amazing facts of transfinite arithmetic. Indeed, an infinite set has been defined as one that can be placed in a one-to-one correspondence with a proper subset of itself.

There are still many unanswered questions about transfinite numbers. For example, it can be shown that

$$\aleph_0^{\aleph_0}$$

leads to another transfinite cardinal number that is larger than $\aleph_0$ and is often called $\aleph_1$. The problem of determining whether or not there is a cardinal number between $\aleph_0$ and $\aleph_1$ has occupied mathematicians up to recent times and has still not been completely resolved. Indeed, you may be reading about further discoveries on this topic at any time, inasmuch as a number of mathematicians are currently at work on this famous problem.

EXERCISES

In each of the following tell whether the number is used as a cardinal number, an ordinal number, or merely for identification:

1. This is the first problem on the list.

2. There are 20 volumes in the set of encyclopedias.

3. Mathematics is discussed in the 12th volume.

4. My license plate number is EZT 133.

5. Dorothy is in the eighth row.

6. There are 35 students in the class.

7. I am listening to 104 on the FM dial.

8. It takes 9 men to field a baseball team.

9. This is the third time he is taking the course.

10. The card catalog lists the book as 510.7.

Show a one-to-one correspondence between:

11. The set of counting numbers and the set of odd counting numbers.
12. The set of counting numbers and the set of counting numbers greater than 25.
13. The set of even counting numbers and the set of odd counting numbers.
14. The set of even counting numbers and the set of even counting numbers greater than 100.
15. The set of odd counting numbers greater than 100 and the set of even counting numbers greater than 100.
16. The set of odd counting numbers greater than 50 and the set of odd numbers greater than 150.

Classify each set as finite or infinite:

17. $A = \{1, 2, 3, \ldots\}$
18. $B = \{2, 4, 6, \ldots, 1000\}$
19. $C = \{2, 4, 6, \ldots, 1{,}000{,}000\}$
20. $D = \{1, 3, 5, \ldots, 2n - 1, \ldots\}$
21. The number of inhabitants in the world.
22. The number of words in this book.
23. The number of counting numbers greater than 1000.
24. The number of counting numbers less than a billion.

PEDAGOGICAL CONSIDERATIONS

1. Select an elementary school grade level and make a worksheet (a set of ten or more problems designed for student explorations or practice) for helping students recognize the wide variety of common uses of numbers.

2. Make a worksheet for establishing firmly the dependence of finite cardinal numbers upon one-to-one correspondences.

3. As a sequel to the worksheet developed in Exercise 2, make a worksheet to show the existence of at least one cardinal number that is not a finite cardinal number.

4. Paradoxes may be used to provide elementary school students with insight into the properties of infinite sets. Here is an example from geometry. Consider the fact that a line segment contains an infinite number of points. From this one may demonstrate the fact that two line segments of unequal length nevertheless contain the same number of points. Thus consider segments *AB* and *CD* where lines *CA* and *DB* meet at a point *P*.

We can show that there is a one-to-one correspondence between the points of segments *AB* and *CD*. Given any point *M* on *CD*, draw line segment *PM*. This segment crosses *AB* at a point *M'* that corresponds to *M*. Also given any point *N* on segment *AB*, draw line segment *PN* and extend it until it meets *CD* at a point *N'* that corresponds to *N*. Thus there is a one-to-one correspondence between the points of segments *AB* and *CD*, and there are just as many points on segment *AB* as there are on segment *CD*.

1-5
intersection and union

Let us consider two sets A and B, defined as follows:

$$A = \{1, 2, 3, 4, 5, 6, 7\}$$
$$B = \{2, 5, 7, 8, 9\}$$

From these two sets let us form another set C, whose members are those elements that appear in each of the two given sets:

$$C = \{2, 5, 7\}$$

The set C consists of the elements that the sets A and B have in common and is called the **intersection** of the sets A and B. Formally we say that the intersection of two sets A and B (written $A \cap B$) is the set of elements that are members of both of the given sets.

Example 1: If $A = \{a, r, e\}$ and $B = \{c, a, t\}$, find $A \cap B$.

Solution: $A \cap B = \{a\}$; that is, a is the only letter that appears in each of the two given sets.

43

Example 2: If $X = \{1, 2, 3, 4, 5\}$ and $Y = \{4, 5, 6, 7\}$, find $X \cap Y$.

Solution:

$$X \cap Y = \{4, 5\}$$

Example 3: If $X = \{1, 2, 3\}$ and $Y = \{4, 5, 6\}$, find $X \cap Y$.

Solution: Sets X and Y have no elements in common and are said to be **disjoint sets;** their intersection is the null set. Thus we may write $X \cap Y = \varnothing$.

From the two given sets A and B of this section let us next form another set D, whose members are those elements that are elements of at least one of the two given sets:

$$D = \{1, 2, 3, 4, 5, 6, 7, 8, 9\}$$

The set D is called the **union** of sets A and B. Formally we say that the union of two sets A and B (written $A \cup B$) is the set of elements that are members of at least one of the given sets.

Example 4: Let A represent the set of names of boys on a particular committee, $A = \{$Bill, Bruce, Max$\}$. Let B represent the set of names of boys on another committee, $B = \{$Bruce, John, Max$\}$. Find $A \cup B$.

Solution: $A \cup B = \{$Bill, Bruce, John, Max$\}$. Here the union of the two sets is the set of names of the boys who are on at least one of the two committees. Note that the names Bruce and Max appear only once in the set $A \cup B$ even though these names are listed for both committees.

Example 5: If $A = \{1, 2, 3, 4, 5\}$ and $B = \{3, 5, 7, 9\}$, find $A \cup B$.

Solution:

$$A \cup B = \{1, 2, 3, 4, 5, 7, 9\}$$

Example 6: If $U = \{1, 2, 3, \ldots, 10\}$, $A = \{1, 2, 3, 4, 5\}$, and $B = \{3, 4, 5, 6, 7\}$, list the elements in $A' \cap B'$.

$A' = \{6, 7, 8, 9, 10\}$, $B' = \{1, 2, 8, 9, 10\}$, and $A' \cap B' = \{8, 9, 10\}$.

EXERCISES

For each of the following sets list the elements in **(a)** $A \cup B$; **(b)** $A \cap B$:

1. $A = \{1, 2, 3\}$, $B = \{1, 3, 5, 7\}$.

2. $A = \{3, 4, 5\}$, $B = \{4, 5, 6, 7\}$.

3. $A = \{1, 3, 5, 7\}$, $B = \{7, 9\}$.

4. $A = \{2, 4, 6, 8\}$, $B = \{4, 6, 7, 8\}$.

5. $A = \{1, 5, 10\}$, $B = \{3, 7, 8\}$.

6. $A = \{1, 3, 5, \ldots\}$, $B = \{2, 4, 6, \ldots\}$.

7. $A = \varnothing$, $B = \{1, 2, 3, \ldots\}$.

8. $A = \{1, 2, 3, \ldots\}$, $B = \{1, 3, 5, \ldots\}$.

For each of the given universal sets list the elements in (a) A'; (b) B';
(c) $A' \cup B'$; (d) $A' \cap B'$:

9. $U = \{1, 2, 3, 4, 5\}$, $A = \{1, 2\}$, $B = \{1, 3, 5\}$.

10. $U = \{1, 2, 3, \ldots, 10\}$, $A = \{1, 3, 5, 7, 9\}$, $B = \{2, 4, 6, 8, 10\}$.

11. $U = \{1, 2, 3, \ldots\}$, $A = \{1, 3, 5, \ldots\}$, $B = \{2, 4, 6, \ldots\}$.

12. $U = \{1, 2, 3, 4, 5, 6, 7\}$, $A = \varnothing$, $B = \{1, 2, 3, 4, 5, 6, 7\}$.

13. $U = \{1, 2, 3\}$, $A = \{1\}$, $B = \{3\}$.

14. $U = \{1, 2, 3, \ldots, 10\}$, $A = \{1, 3, 5, 7, 9\}$, $B = \{1, 2, 3, 4, 5\}$.

For Exercises 15 through 20, let $U = \{1, 2, 3, 4, 5, 6, 7, 8, 9, 10\}$, $A = \{1, 3, 4, 5, 7, 8\}$, *and* $B = \{2, 4, 5, 9, 10\}$, *and list the elements in:*

15. $A' \cap B'$

16. $(A \cup B)'$

17. $A' \cup B$

18. $(A' \cap B)'$

19. $A \cap B$

20. $(A \cup B')'$

21. If $A \subseteq B$, describe (a) $A \cap B$; (b) $A \cup B$; (c) $A \cap B'$; (d) $A' \cup B$.

22. If $A \subseteq B$ and $B \subseteq A$, describe (a) $A \cup B$; (b) $A \cap B$; (c) $A \cap B'$;
(d) $A' \cup B$.

PEDAGOGICAL CONSIDERATIONS

1. Name at least two countries that have names indicating a union of provinces or other districts.

2. Addition is presented to first grade children using disjoint sets. Give examples indicating why it is essential for the sets to be disjoint.

3. Make appropriate visual aids and describe at least one way of showing first grade students that (a) $3 + 2 = 5$; (b) $1 + 4 = 5 = 4 + 1$.

4. As in Exercise 3, but for second or third grade, explain the statement:
If $3 + 2 = 5$, then $2 + 3 = 5$, $5 - 2 = 3$, and $5 - 3 = 2$.

5. Let A be the set of factors of 36 and B the set of factors of 24. (See §5-3.)

  (a) Describe $A \cap B$.

  (b) Describe a use of members of $A \cap B$ in reducing the fraction $\frac{24}{36}$.

6. Let C be the set of multiples of 15 and D the set of multiples of 12.

  (a) Describe $C \cap D$.

  (b) Describe a use of a member of $C \cap D$ in finding $\frac{2}{15} + \frac{7}{12}$.

1-6

sets of points

Relationships among sets are often represented by sets of points. For example, we may represent the universal set U as a set of points on a line, a set A by the shaded subset of U, and the set A' by the remaining points of U.

The sets of points that are most frequently used to represent sets of elements are rectangular regions and circular regions as identified below.

Rectangular region   Circular region

Rectangular regions and circular regions may be formally introduced but are used here simply as the sets of points that first grade children would color in coloring the points of a rectangle (or circle) and its interior points. Since there is no confusion about which points are meant, we shall not concern ourselves with formal definitions.

We shall use rectangular regions to represent universal sets and frequently shade the regions under consideration. As in the case of A', we use a dashed line for the part of the boundary of a region that does not belong to the region.

The shading is often omitted when the meaning is clear without the shading. With or without the shading, the figures are called **Euler diagrams.** We may use Euler diagrams to show the intersection of two sets.

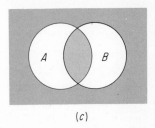

(a)   (b)   (c)

Note above that in Figure (a), $A \cap B$ is the empty set; in Figure (b), $A \cap B = B$. We may also use Euler diagrams to show the union of two sets. Note below that in Figure (b), $A \cup B = A$.

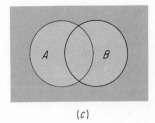

(a)   (b)   (c)

The figures for intersection and union may also be used to illustrate the following properties of any two sets A and B:

$$(A \cap B) \subseteq A, \qquad (A \cap B) \subseteq B$$
$$A \subseteq (A \cup B), \qquad B \subseteq (A \cup B)$$

We consider only well-defined sets (§1-1), and thus each element of the universal set U is a member of exactly one of the sets A and A'. When two sets A and B are considered, an element of U must belong to exactly one of these four sets:

$$A \cap B, \qquad \text{i.e., both } A \text{ and } B$$
$$A \cap B', \qquad \text{i.e., } A \text{ and not } B$$
$$A' \cap B, \qquad \text{i.e., } B \text{ and not } A$$
$$A' \cap B', \qquad \text{i.e., neither } A \text{ nor } B$$

An Euler diagram in which each of these four regions is represented is often called a **Venn diagram.**

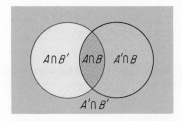

Venn diagrams may be used to show that two sets refer to the same set of points, that is, are equal.

Example 1: Show by means of a Venn diagram that $(A \cup B)' = A' \cap B'$.

Solution: We make separate Venn diagrams for $(A \cup B)'$ and $A' \cap B'$. The diagram for $(A \cup B)'$ is made by shading the region for A horizontally, shading the region for B vertically, identifying the region for $A \cup B$ as consisting of the points in regions that are shaded in any way (horizontally, vertically, or both horizontally and vertically), and identifying the region for $(A \cup B)'$ as consisting of the points in the region without horizontal or vertical shading.

$(A \cup B)'$: Region outside both circles

The diagram for $A' \cap B'$ is made by shading the region for A' horizontally, shading the region for B' vertically, and identifying the region for $A' \cap B'$ as consisting of all points in regions that are shaded both horizontally and vertically.

The solution is completed by observing that the region for $(A \cup B)'$ in the first Venn diagram is identical with the region for $A' \cap B'$ in the second Venn diagram; the two sets have the same elements and therefore are equal.

We may also use Venn diagrams for three sets. In this case there are eight regions that must be included, and the figure is usually drawn as follows:

48

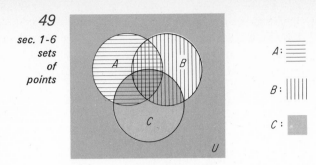

Venn diagrams for three sets are considered in these two examples and in Exercises 19 through 28.

Example 2:   Show that $A \cap (B \cup C) = (A \cap B) \cup (A \cap C)$.

Solution: Set A is shaded with vertical lines; $B \cup C$ is shaded with horizontal lines. The intersection of these sets, $A \cap (B \cup C)$, is the subset of U which has both vertical and horizontal shading.

The set $A \cap B$ is shaded with horizontal lines; $A \cap C$ is shaded with vertical lines. The union of these sets is the subset of U, which is shaded with lines in either or in both directions.

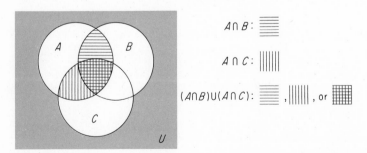

Note that the final results in the two diagrams are the same, thus showing the equivalence of the set $A \cap (B \cup C)$ and the set $(A \cap B) \cup (A \cap C)$.

Example 3: In a group of 35 students, 15 are studying algebra, 22 are studying geometry, 14 are studying trigonometry, 11 are studying both algebra and geometry, 8 are studying geometry and trigonometry, 5 are studying algebra and trigonometry, and 3 are studying all three subjects. How many of these students are not taking any of these subjects? How many are taking only geometry?

Solution: This problem can easily be solved by means of a Venn diagram with three circles to represent the set of students in each of the listed subject-matter areas. It is helpful to start with the information that there are 3 students taking all three subjects. We write the number 3 in the region that is the intersection of all three circles. Then we work backwards; since 5 are taking algebra and trigonometry, there must be 2 in the region representing algebra and trigonometry but *not* geometry. Continuing in this manner, we enter the given data in the figure.

Since the total of the numbers in the various areas is 30, there must be 5 students not in any of the classes listed in the various regions. Also, reading directly from the figure, we find that there are 6 students taking geometry only.

The terms Euler diagram, Euler circle, and Venn diagram are used synonymously in some elementary books. Their use in this book is the historical one and should provide a basis for comprehension of any other special uses encountered.

EXERCISES

Consider the given diagram and find:

1. (a) $n(A \cap B)$ (b) $n(A)$
    (c) $n(B \cap A')$ (d) $n(B \cup A)$

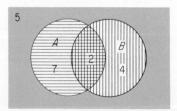

2. (a) $n(P \cup Q)$ (b) $n(P' \cap Q')$
   (c) $n(P' \cup Q)$ (d) $n(P \cup Q')$

Use an Euler diagram to represent:

3. $A \subset B$.
4. A and B are disjoint sets.
5. $A \cup B$ when $A \cap B = \varnothing$.
6. $A \subseteq B$ and $B \subseteq A$.
7. $(A \cup B)'$ when A and B are disjoint sets.
8. $(A \cup B \cup C)'$ when A, B, and C are disjoint sets.

Use a Venn diagram to represent:

9. $A' \cup B$ 10. $A' \cap B$
11. $A \cap B'$ 12. $A \cup B'$

Venn diagrams to show that:

13. $A \cup B = B \cup A$
14. $A \cap B = B \cap A$
15. $(A \cap B) \subseteq A$
16. $A \subseteq (A \cup B)$
17. $(A \cap B)' = A' \cup B'$
18. $A \cup B' = (A' \cap B)'$

Consider the given diagram and find:

19. (a) $n(A \cap B \cap C)$ (b) $n(A \cap B \cap C')$
    (c) $n(A \cap B' \cap C')$ (d) $n(A)$
    (e) $n(A \cup B)$ (f) $n(B \cup C)$

20. (a) $n(R' \cap S \cap T)$ (b) $n(R')$
(c) $n(R' \cup S)$ (d) $n(S' \cup T')$
(e) $n(R' \cup S' \cup T')$ (f) $n(R \cup S' \cup T)$

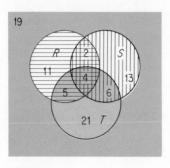

21. (a) $n(A \cup B)$ (b) $n(B \cap C)$
(c) $n(A \cap B')$ (d) $n(A \cup B \cup C)$
(e) $n(A \cup B' \cup C')$ (f) $n(A \cap B' \cap C')$

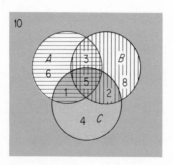

22. (a) $n(X \cup Y)$ (b) $n(X \cup Z)$
(c) $n(X')$ (d) $n(X \cup Y')$
(e) $n(X' \cap Y \cap Z)$ (f) $n(X \cap Y' \cap Z')$

Use Venn diagrams to solve each problem:

23. In a survey of 50 students, the following data were collected: There were 19 taking biology, 20 taking chemistry, 19 taking physics, 7 taking physics and chemistry, 8 taking biology and chemistry, 9 taking biology and physics, 5 taking all three subjects. How many of the group are not taking any of the three subjects? How many are taking only chemistry? How many are taking physics and chemistry, but not biology?

24. A survey was taken of 30 students enrolled in three different clubs, A, B, and C. Show that the following data that were collected are inconsistent: 18 in A, 10 in B, 9 in C, 3 in B and C, 6 in A and B, 9 in A and C, 2 in A, B, and C.

Use a Venn diagram to represent:

25. (a) $A \cap B \cap C$ (b) $A \cap B \cap C'$
    (c) $A \cap B' \cap C$ (d) $A \cap B' \cap C'$

26. (a) $A' \cap B \cap C$ (b) $A' \cap B \cap C'$
    (c) $A' \cap B' \cap C$ (d) $A' \cap B' \cap C'$

Use Venn diagrams to show that:

27. $A \cup (B \cup C) = (A \cup B) \cup C$
28. $A \cap (B \cap C) = (A \cap B) \cap C$

PEDAGOGICAL CONSIDERATIONS

1. Third and fourth graders may be asked to copy the given numerals in the appropriate region of a given Venn diagram. For example, complete this diagram using the numbers given:

Odd numbers

Numbers less than 10

Odd numbers less than 10

8   11   15   17

7   6   5   1

2. Select sets A and B with $n(A) = 5$, $n(B) = 4$, and such that
   (a) $n(A \cup B) = 7$;   (b) $n(A \cup B) = 5$;
   (c) $n(A \cup B) = 9$;   (d) $n(A \cap B) = 2$.

3. In Exercise 2 above find:
   (a) $n(A \cap B)$ if $n(A \cup B) = 7$.
   (b) $n(A \cap B)$ if $n(A \cup B) = 5$.
   (c) $n(A \cap B)$ if $n(A \cup B) = 9$.
   (d) $n(A \cup B)$ if $n(A \cap B) = 2$.
   (e) A formula for $n(A \cup B)$ in terms of $n(A)$, $n(B)$, and $n(A \cap B)$.

4. Probability is now often introduced in the upper elementary school grades. The probability of obtaining a 7 on a single throw of two ordinary dice is $\frac{6}{36}$; the probability of obtaining an 11 is $\frac{2}{36}$.

  (a) What is the probability of obtaining a 7 and an 11 on a single throw of two dice?

  (b) What is the probability of obtaining a 7 or an 11 on a single throw of two dice? *Hint:* Use the formula obtained in Exercise 3(e).

5. Study textbooks for each elementary school grade and list at least three specific examples of the use of sets at each grade level.

CHAPTER TEST

1. State whether or not the sets $\{b, a, t\}$ and $\{l, a, b\}$ are **(a)** equal; **(b)** equivalent.

2. Let $U = \{m, e, m, o, r, a, b, l, e\}$ and list the elements of A' when A is defined as **(a)** $\{m, o, r, e\}$; **(b)** $\{l, a, b, o, r\}$.

3. Show all possible one-to-one correspondences of the two sets $\{t, o, p\}$ and $\{r, a, t\}$.

4. List all possible subsets of $\{h, a, t\}$.

Let $U = \{0, 1, 2, 3, 4, 5, 6, 7, 8, 9\}$, $A = \{2, 4, 6\}$, *and* $B = \{0, 3, 6, 9\}$. *Then find:*

5. $A' \cap B$

6. $(A \cup B)'$

Use a Venn diagram to represent:

7. $A \cap B'$

8. $A' \cap B \cap C$

Consider the given diagram and find:

9. $n(A \cap B \cap C')$

10. $n(A \cup B \cup C')$

54

2

Concepts
of
Logic

You undoubtedly make numerous conscious and subconscious decisions. Many of these decisions are based upon reasoning. Your satisfactions from your decisions often depend upon the validity of your reasoning and thus upon logical concepts. In this sense an understanding of logical concepts can improve your personal opportunities to find the kind of life that you want and to enjoy the kind of life that you find.

Reasoning is based upon the handling of facts, usually in the form of statements. You may think of reasoning as an "algebra" of statements. In this chapter truth values of statements are considered in an effort to recognize relations among statements. Then some of these relations are selected to provide the basis for the discussion of valid arguments, logical reasoning, and proofs. In the process you will learn how to recognize several different forms of the same statement in the English language. Many people feel that the study of logical concepts is worthwhile solely for the purpose of increasing their facility with the English language.

2-1
sets of statements

A sentence that is either *true* or *false* is often called a **statement.** Each of the following is an example of a **simple statement:**

> Today is Friday in Chicago.
> This is a sunny day.

Notice that a command such as "Stand up and be counted" is neither true nor false and is not considered to be a statement as we are using the word here.

A **compound statement** is formed by combining two or more simple statements. An example is the following:

Today is Friday in Chicago and this is a sunny day.

In this illustration the two simple statements are combined by the connective "and." Other connectives could also have been used. Consider the same simple statements using the connective "or":

Today is Friday in Chicago or this is a sunny day.

We shall consider such compound statements and determine the conditions under which they are true or false, assuming that the simple statements are true. In doing this we use letters or variables to represent statements and symbols to represent connectives. For example, we may use p and q to represent these simple statements:

p: Today is Friday in Chicago.
q: This is a sunny day.

The following connectives are commonly used:

$\wedge$: and
$\vee$: or
$\sim$: not

We may use p and q as previously defined and write these statements in symbolic form together with their translations in words:

$p \wedge q$: Today is Friday in Chicago and this is a sunny day.
$p \vee q$: Today is Friday in Chicago or this is a sunny day.
$\sim p$: Today is not Friday in Chicago.

Example 1: Translate $p \wedge (\sim q)$, where p and q are as given in this section.

Solution: Today is Friday in Chicago and this is not a sunny day.

Example 2: Write, in symbolic form: Today is not Friday in Chicago or this is not a sunny day.

Solution:

$$(\sim p) \vee (\sim q)$$

EXERCISES

1. Use p: Jim is tall; q: Bill is short. Think of "short" as "not tall," and write each of these statements in symbolic form:

(a) Jim is short and Bill is tall.

(b) Neither Jim nor Bill is tall.

(c) Jim is not tall and Bill is short.

(d) It is not true that Jim and Bill are both tall.

(e) Either Jim or Bill is tall.

2. Assume that Bill and Jim are both tall. Which of the statements in Exercise 1 are true?

3. Use p: Joan is happy; q: Mary is sad. Think of "sad" as "not happy," and write each of these statements in symbolic form:

(a) Joan and Mary are both happy.

(b) Either Joan is happy or Mary is happy.

(c) Neither Joan nor Mary is happy.

(d) It is not true that Joan and Mary are both sad.

(e) It is not true that neither Joan nor Mary is happy.

4. Assume that Joan and Mary are both happy. Which of the statements in Exercise 3 are true?

5. Use p: I like this book; q: I like mathematics. Give each of these statements in words:

(a) $p \wedge q$ **(b)** $\sim q$

(c) $\sim p$ **(d)** $(\sim p) \wedge (\sim q)$

6. Continue as in Exercise 5 with the statements:

(a) $(\sim p) \wedge q$ **(b)** $p \vee q$

(c) $\sim (p \wedge q)$ **(d)** $\sim [(p) \wedge q]$

7. Assume that you like this book and that you like mathematics. Which of the statements in Exercises 5 and 6 are true for you?

8. Assume that you like this book but that you do not like mathematics. Which of the statements in Exercises 5 and 6 are true for you?

*9. Assume that two given statements p and q are both true and indicate whether or not you would expect each of the following statements to be true:

(a) $p \wedge q$ **(b)** $p \vee q$

(c) $p \vee (\sim q)$ **(d)** $(\sim p) \vee q$

*10. Repeat Exercise 9 under the assumption that p is true and q may be true or false.

PEDAGOGICAL CONSIDERATIONS

Elementary school teachers find many ways to help students understand *and* or *or*. This is necessary, for example, to enable students to follow instructions:

> Put the blocks in the box *and* the paper in the basket.
> Bring a sandwich *or* a piece of fruit for recess.
> List the nouns on page 4 *or* page 5.
> Please sharpen the pencil *and* give it to me.
> Read the story *and* write a report on it by Friday.

Students also need to understand compound sentences as they extend their vocabulary.

> This chain is large *and* heavy.
> The rain hats on the pegs are all black *or* yellow.
> A polygon that is a rhombus *and* also a rectangle is a square.

Teachers need to understand the use of compound statements so that they can determine which students follow instructions and also so that they can avoid giving ambiguous instructions. For example, one teacher told the class to draw a picture and read a book or write a story. Sally read a book and wrote a story. Linda drew a picture and read a book. Jack drew a picture and wrote a story. Joe wrote a story. Bill played the piano. Which students completed the assignment? Actually the instructions are subject to two different interpretations. Under one interpretation everyone is to draw a picture and then read a book or write a story or do both. Under the other interpretation writing a story is an alternative for both drawing a picture and reading a book. If the first interpretation is intended, the instructions should be restated for that interpretation. If the second interpretation is intended, the instructions should be restated, possibly in two parts, for that interpretation.

1. Give five examples of instructions to first graders that require the students to understand *and* and *or*.

2. Repeat Exercise 1 for fifth graders.

3. For fourth graders give two examples of ambiguous instructions and restate each instruction for each of its interpretations.

2-2

truth values of statements

We first consider the **negation** ($\sim p$) of a statement p. The symbol was introduced in the last section. As we would expect from the meaning of the word "negation," if p is true, then $\sim p$ is false; if p is false, then $\sim p$ is true. The truth values of the statement $\sim p$ are given in the accompanying **truth table**, where T stands for "true" and F stands for "false."

We often use Euler diagrams to represent situations in which a statement p is true. See the diagram at the top of page 59.

Next consider the simple statements:

p	$\sim p$
T	F
F	T

> p: It is snowing.
> q: This is the month of December.

The truth of any compound statement, such as $p \wedge q$, is determined by the truth of each of the simple statements. Since each of the statements p and q may be either true or false, there are four distinct possibilities:

p true,	q true
p true,	q false
p false,	q true
p false,	q false

For each of these possibilities we wish to determine the truth value of the statement $p \wedge q$. When we say that $p \wedge q$ is true, we mean that both p and q are true. We may define the truth values of $p \wedge q$ by means of the truth table at the left.

p	q	$p \wedge q$
T	T	T
T	F	F
F	T	F
F	F	F

The statement $p \wedge q$ is called the **conjunction** of p and q. The truth values of $p \wedge q$ are independent of the *meanings* assigned to the variables. For example, if you are told that each of two simple statements is true, it follows immediately that the conjunction of these statements is also true; in any other circumstance, it follows immediately that the conjunction of the statements is false.

Venn diagrams may be used to represent the truth values of two statements p and q. Note that in the following figure we have four possible regions indicated. These regions correspond to the four possibilities for truth and falsity for two general statements (also to the four rows of the truth table) and may be labeled by conjunctions of statements. Notice the similarity of the meanings of $\wedge$ for statements and $\cap$ for sets.

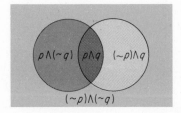

We next consider the compound statement $p \vee q$, called the **disjunction** of p and q. We translate "$\vee$" as "or" even though we shall use the word "or" more precisely than in the ordinary English language. As we shall use it here, "p or q" means that at least one of the statements p, q is true; that is, either p is true, or q is true, or both p and q are true. In other words, the meaning of $\vee$ for statements corresponds to the meaning of $\cup$ for sets.

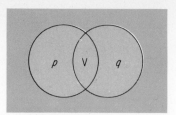

p	q	p∨q
T	T	T
T	F	T
F	T	T
F	F	F

The truth values of $p \lor q$ are defined by the accompanying truth table. From the table we deduce the fact that the compound statement $p \lor q$ is true unless both p and q are false.

We can use truth tables to summarize the truth values of various compound statements. To illustrate this procedure we shall construct a truth table for the statement $p \land (\sim q)$.

First set up a table with the appropriate headings as follows:

p	q	p	∧	(~q)
T	T			
T	F			
F	T			
F	F			

Now complete the column headed "p" by using the truth values that appear under p in the first column. In the column headed "$\sim q$" write the negation of the values given under q in the second column. (Why?) Our table now appears as follows:

p	q	p	∧	(~q)
T	T	T		F
T	F	T		T
F	T	F		F
F	F	F		T

Finally we find the conjunctions of the values given in the third and fifth columns. The completed table appears as follows, with the order in which the columns were considered indicated by (a), (b), and (c) and the final results in bold print:

p	q	p	∧	(~q)
T	T	T	**F**	F
T	F	T	**T**	T
F	T	F	**F**	F
F	F	F	**F**	T

(a)   (c)   (b)

We can summarize this table by saying that the statement $p \wedge (\sim q)$ is true only in the case when p is true and q is false. Thus the statement "Today is Wednesday and it is not snowing" is a true statement on a hot Wednesday in July. (We must assume, of course, that it will not snow on a hot day in July.)

EXERCISES

Construct truth tables for:

1. $(\sim p) \wedge q$ *2.* $(\sim p) \vee q$

3. $(\sim p) \vee (\sim q)$ *4.* $(\sim p) \wedge (\sim q)$

5. $\sim(p \wedge q)$ *6.* $p \vee (\sim q)$

In Exercises 7 through 10 copy and complete each truth table:

7.

p	q	$\sim$	$\big[p$	$\vee$	$(\sim q)\big]$
T	T				
T	F				
F	T				
F	F				

(d)   (a)   (c)   (b)

8.

p	q	$\sim$	$\big[(\sim p)$	$\vee$	$q\big]$
T	T				
T	F				
F	T				
F	F				

(d)   (a)   (c)   (b)

9.

p	q	$\sim$	$\big[(\sim p)$	$\wedge$	$(\sim q)\big]$
T	T				
T	F				
F	T				
F	F				

(d)   (a)   (c)   (b)

10.

p	q	$\sim$	$\big[(\sim p)$	$\vee$	$(\sim q)\big]$
T	T				
T	F				
F	T				
F	F				

(d)   (a)   (c)   (b)

11. Define $p \veebar q$ to mean "p or q but not both" and construct a truth table for this connective.

12. Construct a truth table for $p \mid q$, which we define to be true when p and q are not both true and to be false otherwise.

13. In Exercise 12 express $p \mid q$ in terms of other connectives that we have previously defined.

14. Use p: I like this book; q: I like mathematics. Tell the conditions under which each of the statements in Exercises 1 through 6 is true.

1. Elementary school students need to learn to identify statements as true or false. Suppose that there are dark clouds, the sun is not shining, and it is not raining. Identify each statement as true or false:
   (a) The sun is not shining.
   (b) The sky is clear and it is not raining.
   (c) Either the sun is not shining or it is raining.

2. Describe a situation and make up at least five compound statements such as those in Exercise 1 which could be used for elementary school students to identify as true or false.

3. Examine at least three different sets of junior high school mathematics books for grades 7 and 8. Report on their use, or lack of use, of truth tables.

4. Prepare a set of transparencies for use with an overhead projector to show $p \wedge q$ and $p \vee q$. Use a rectangular region for the universe of discussion. Prepare two overlays, each with a circular region to represent the truth value of a simple statement. Then place these overlays in various positions to show the conjunction and disjunction of statements.

2-3

conditional statements

Many of the statements that we make in everyday conversation are based upon a condition. For example, consider the following:

> If the sun shines, then I will cut the grass.
> If I have no homework, then I will go bowling.
> If I bribe the instructor, then I will pass the course.

Each of these statements is expressed in the **if-then** form:

> If p, then q.

Any if-then statement can be expressed in symbols as $p \rightarrow q$, which is read either as "if p, then q" or as "p implies q." The symbol "$\rightarrow$" is called the **implication symbol**; it is a connective used to form a compound statement called a **conditional statement**. Then the statement p is the **premise** and q is the **conclusion** of the conditional statement $p \rightarrow q$.

Our first task is to consider the various possibilities for p and q in order to define $p \rightarrow q$ for each of these cases. One way to do this would be to present a completed truth table and to accept this as our definition of $p \rightarrow q$. Let us, however, attempt to justify the entries in such a table. Consider again the conditional statement:

> If the sun shines, then I will cut the grass.

Now if the sun shines and I do cut the grass, then the statement is obviously true. On the other hand, the statement is false if the sun shines and I do not cut the grass. Assume now that the sun is not shining; then, whether I cut the grass or not, the original statement is true in that I only declared my intentions under the condition that the sun shines. We can summarize these assertions by means of the truth table at the left.

p	q	$p \rightarrow q$
T	T	T
T	F	F
F	T	T
F	F	T

Consider the statement:

If it rains, then I will give you a ride home.

Have I lied to you:

1. If it rains and I give you a ride home?

2. If it rains and I do not give you a ride home?

3. If it does not rain and I give you a ride home?

4. If it does not rain and I do not give you a ride home?

According to the accepted meanings of the words used, you have a right to feel that I lied to you only if it rains and I do not give you a ride home. In other words, the conditional statement is true unless the premise is true and the conclusion is false.

Many people feel uneasy about some of the truth values of $p \rightarrow q$. Actually truth values must be assigned to the conditional statement for each of the four possible combinations of truth values of p and q. Thus the only debatable question is:

Under what conditions is a statement of the form
$p \rightarrow q$ considered to be a true statement?

In other words, we are concerned with the use of words in the English language. If we are to communicate with each other, we must accept definitions of the meanings of the words that we use. The accepted meaning of $p \rightarrow q$ is given in the truth table. Other meanings for the symbols could have been used, but since this meaning has been accepted we should use some other symbol if we wish to ascribe another meaning to a compound statement in terms of p and q.

Example 1: Give the truth value of each statement:
  (a) If $5 + 7 = 12$, then $6 + 7 = 13$.
  (b) If $5 \times 7 = 35$, then $6 \times 7 = 36$.
  (c) If $5 + 7 = 35$, then $6 + 7 = 13$.
  (d) If $5 + 7 = 35$, then $6 \times 7 = 36$.

Solution: Think of each statement in the form $p \rightarrow q$.
  (a) For p true and q true, the statement $p \rightarrow q$ is *true*.
  (b) For p true and q false, the statement $p \rightarrow q$ is *false*.
  (c) For p false and q true, the statement $p \rightarrow q$ is *true*.
  (d) For p false and q false, the statement $p \rightarrow q$ is *true*.

According to the truth table for $p \rightarrow q$, any statement of the form "if p, then q" is false only when p is true and q is false. On the other hand, if p is false, then the statement $p \rightarrow q$ is accepted as true regardless of the truth value of q. Thus each of the following statements is true by this definition:

> If $2 + 3 = 7$, then George Washington is now the President of the United States.
>
> If $2 + 3 = 7$, then the moon is made of green cheese.
>
> If $2 + 3 = 7$, then July follows June.

If you have difficulty accepting any of these statements as true, then you should review the definition of the truth values of if-then statements. Remember also that there need be no relationship between p and q in an if-then statement, although we tend to use such statements in this way in everyday life.

Example 2: Find all possible replacements for x such that the following statement is true:

> If $3 \times 4 = 34$, then $x - 3 = 5$.

Solution: Think of the statement in the form $p \rightarrow q$ for

$$p: \quad 3 \times 4 = 34$$
$$q: \quad x - 3 = 5$$

Since p is false, the given statement $p \rightarrow q$ is true for all values of x. That is, regardless of the value assigned to x, the statement

> If $3 \times 4 = 34$, then $x - 3 = 5$

is always true.

EXERCISES

In Exercise 1 through 4 consider the statements:

> p: You will study hard.
> q: You will get an A.

Then translate each of the following symbolic statements into an English sentence:

1. $p \rightarrow q$
2. $q \rightarrow p$
3. $(\sim p) \rightarrow (\sim q)$
4. $(\sim q) \rightarrow (\sim p)$

5. Repeat Exercises 1 through 4 for the statements:

p: The triangle is equilateral.

q: The triangle is isosceles.

6. Give the truth value of each statement:
(a) If $2 \times 3 = 5$, then $2 + 3 = 6$.
(b) If $2 \times 3 = 5$, then $2 + 3 = 5$.
(c) If $2 + 3 = 5$, then $2 \times 3 = 5$.

7. Give the truth value of each statement:
(a) If $5 \times 6 = 56$, then $5 + 6 = 11$.
(b) If $5 \times 6 = 42$, then $5 + 6 = 10$.
(c) If $5 \times 6 = 42$, then $5 + 5 = 10$.

8. Assume that $a \times b = c$, $b \times c = d$, and $c \neq d$. Then give the truth value of each statement:
(a) If $a \times b = c$, then $b \times c = d$.
(b) If $a \times b = d$, then $b \times c = c$.
(c) If $a \times b = d$, then $b \times c = d$.
(d) If $a \times b = c$, then $b \times c = c$.

9. Show by means of truth tables that $(\sim p) \vee q$ has the same truth values as $p \rightarrow q$.

10. Show that $\sim[p \wedge (\sim q)]$ has the same truth values as $p \rightarrow q$.

struct truth tables for:

11. $(p \rightarrow q) \wedge (q \rightarrow p)$

12. $q \rightarrow [(\sim p) \vee q]$

Copy and complete each truth table:

13.

p	q	$[(\sim p)$	$\wedge$	$q]$	$\longrightarrow$	$(p \vee q)$
T	T					
T	F					
F	T					
F	F					

(a) (c) (b) (e) (d)

14.

p	q	$[(\sim p)$	$\vee$	$q]$	$\longrightarrow$	$(p \vee q)$
T	T					
T	F					
F	T					
F	F					

(a) (c) (b) (e) (d)

Find all possible replacements for x for which each sentence is a true sentence:

15. If $2 + 3 = 5$, then $x + 1 = 8$.

16. If $2 + 3 = 6$, then $x + 1 = 8$.

*17. If $x + 1 = 8$, then $2 + 3 = 5$.

*18. If $x + 1 = 8$, then $2 + 3 = 6$.

*19. If $3 - x = 1$, then $2 \times 5 = 13$.

*20. If $3 - x = 1$, then $2 \times 5 = 10$.

Conditional statements are used extensively in elementary school classrooms. Usually the student needs to determine first whether the "if part" (premise) applies to him. If the premise does not apply to him (is not true for him), then he may disregard the conclusion without disobeying the instruction. Note that the failure of the premise to apply to a particular student corresponds to the last two rows of the truth table for implications.

> Raise your hand if you are six years old or in the first
> grade.
> Bring your paper to the desk if you have done all the
> problems and checked them all.

1. Give five conditional statements that could be used with first graders.

2. Repeat Exercise 1 for fourth graders.

2-4

equivalent statements Two statements are **equivalent statements** if they are either both true or both false. If p and q are equivalent statements, we write

$$p \leftrightarrow q, \quad \text{read as "p is equivalent to q."}$$

Example 1: Show by means of a truth table that

$$(p \rightarrow q) \leftrightarrow [(\sim q) \rightarrow (\sim p)].$$

Solution:

p	q	$[p \rightarrow q]$	$\longleftrightarrow$	$[(\sim q)$	$\longrightarrow$	$(\sim p)]$
T	T	T	T	F	T	F
T	F	F	T	T	F	F
F	T	T	T	F	T	T
F	F	T	T	T	T	T
		(a)	(e)	(b)	(d)	(c)

The desired equivalence is always true, as indicated in column (e). By comparing columns (a) and (d) you will see that both statements have the same truth values.

The following pairs of equivalent statements are considered in Exercises 1 through 4:

1. $\sim(p \wedge q)$ and $(\sim p) \vee (\sim q)$

2. $\sim(p \vee q)$ and $(\sim p) \wedge (\sim q)$

3. $p \rightarrow q$ and $q \vee (\sim p)$

4. $(\sim p) \rightarrow (\sim q)$ and $q \rightarrow p$

The first two of these pairs of equivalent statements correspond to properties of sets and their complements

$$(A \cap B)' = A' \cup B' \qquad (A \cup B)' = A' \cap B'$$

and are called *De Morgan's laws for statements.* The third pair enables us to replace any statement of implication by a disjunction.

There are many other forms in which a conditional statement may appear. We shall explore several of these here. Given any simple statement of implication, $p \rightarrow q$, three other related statements may be identified:

Statement:	$p \rightarrow q$	If p, then q.
Converse:	$q \rightarrow p$	If q, then p.
Inverse:	$(\sim p) \rightarrow (\sim q)$	If not p, then not q.
Contrapositive:	$(\sim q) \rightarrow (\sim p)$	If not q, then not p.

The statement proved in Example 1 may now be expressed in the form:

Any statement of implication is equivalent to its contrapositive.

Here are five examples of statements of implication with their converses, inverses, and contrapositives:

1. *Statement:* If it is snowing, I leave my car in the garage.
Converse: If I leave my car in the garage, it is snowing.
Inverse: If it is not snowing, I do not leave my car in the garage.
Contrapositive: If I do not leave my car in the garage, it is not snowing.

2. *Statement:* If $\triangle ABC \cong \triangle XYZ$, then $\triangle ABC \sim \triangle XYZ$.
Converse: If $\triangle ABC \sim \triangle XYZ$, then $\triangle ABC \cong \triangle XYZ$.
Inverse: If $\triangle ABC$ is not congruent to $\triangle XYZ$, then $\triangle ABC$ is not similar to $\triangle XYZ$.
Contrapositive: If $\triangle ABC$ is not similar to $\triangle XYZ$, then $\triangle ABC$ is not congruent to $\triangle XYZ$.

3. *Statement:* If x is negative, then $x \neq 0$.
Converse: If $x \neq 0$, then x is negative.
Inverse: If x is not negative, then $x = 0$.
Contrapositive: If $x = 0$, then x is not negative.

4. *Statement:* If $x + 2 = 5$, then $x = 3$.
Converse: If $x = 3$, then $x + 2 = 5$.
Inverse: If $x + 2 \neq 5$, then $x \neq 3$.
Contrapositive: If $x \neq 3$, then $x + 2 \neq 5$.

5. *Statement:* $p \rightarrow (\sim q)$.
   Converse: $(\sim q) \rightarrow p$.
   Inverse: $(\sim p) \rightarrow \sim (\sim q)$, which can be simplified as $(\sim p) \rightarrow q$.
   Contrapositive: $\sim (\sim q) \rightarrow (\sim p)$, or simply $q \rightarrow (\sim p)$.

The following truth tables for these variants of a conditional statement $p \rightarrow q$ summarize this discussion and specify the truth values for each statement:

		Statement	Converse	Inverse	Contrapositive
p	q	$p \rightarrow q$	$q \rightarrow p$	$(\sim p) \rightarrow (\sim q)$	$(\sim q) \rightarrow (\sim p)$
T	T	T	T	T	T
T	F	F	T	T	F
F	T	T	F	F	T
F	F	T	T	T	T

From the table you should see again that the contrapositive of a statement is equivalent to the statement. Also note that the converse and inverse of any simple statement of implication are equivalent.

Example 2: Write the contrapositive of the following statement:

   If I work hard, then I will pass the course.

Solution:

   If I do not pass the course, then I have not worked hard.

This statement is equivalent to the original one. If you fail the course, then you have not worked hard, because if you had worked hard then you would have passed the course.

EXERCISES

Show by means of truth tables that the following statements are equivalent:

   1. $\sim(p \wedge q)$ and $(\sim p) \vee (\sim q)$.
   2. $\sim(p \vee q)$ and $(\sim p) \wedge (\sim q)$.
   3. $p \rightarrow q$ and $q \vee (\sim p)$.
   4. $(\sim p) \rightarrow (\sim q)$ and $q \rightarrow p$.

Write the converse, inverse, and contrapositive of each statement:

5. If we can afford it, then we buy a new car.

6. If we play pingpong, then you win the game.

7. If two sides and the included angle of one triangle are congruent to two sides and the included angle of another triangle, then the triangles are congruent.

8. If $x > 2$, then $x \neq 0$.

9. If $x(x - 1) = 0$, then $x = 1$.

10. $(\sim p) \rightarrow q$.

Exercises 11, 12, 13, and 14, refer to the statements given in Exercises 7, 8, and Tell whether or not you accept as always true:

11. The given statement.

12. The converse of the given statement.

13. The inverse of the given statement.

14. The contrapositive of the given statement.

PEDAGOGICAL CONSIDERATIONS

1. From newspapers and magazines prepare a collection of statements that are written in if-then form.

2. Examine at least three different sets of elementary school textbooks. Report on the attention given to the concept of a statement and its converse.

3. See if you can find a junior high school mathematics textbook that presents a discussion of the concept of a converse of a statement.

2-5

**forms
of
statements**

Any two equivalent statements may be considered as different forms of the same statement. The recognition of different forms of a statement is particularly useful when statements of implication are under consideration. Frequently the words *necessary* and *sufficient* are used in conditional statements. For example, consider the statement:

Working hard is a sufficient condition for passing the course.

Let us use p to mean "work hard" and q to represent "pass the course."

We need to decide whether the given statement means "if p, then q" or "if q, then p." The word sufficient can be interpreted to mean that working hard is adequate or enough—but not necessary—for passing. That is, there may be other ways to pass the course, but working hard will do it. Thus we interpret the statement to mean:

If you work hard, then you will pass the course.

The symbolic statement $p \rightarrow q$ may thus be used for each of these statements:

If p, then q.
p is a sufficient condition for q.

Next consider the statement:

Working hard is a necessary condition for passing the course.

Here you are told that working hard is necessary or essential in order to pass. That is, regardless of what else you do, you had better work hard if you wish to pass. However there is no assurance that working hard alone will do the trick. It is necessary, but not sufficient. (You may also have to get good grades.) Therefore, we interpret the statement to mean:

If you pass the course, then you have worked hard.

The symbolic statement $q \rightarrow p$ may thus be used for each of these statements:

If q, then p.
p is a necessary condition for q.

Still another form to consider is the statement "q, only if p." In terms of the example used in this section we may write this as:

You will pass the course only if you work hard.

Note that this does *not* say that working hard will insure a passing grade. It does mean that if you have passed, then you have worked hard. That is, "q, only if p" is equivalent to the statement "if q, then p."

We can also interpret this in another way. This statement "q, only if p" means "if not p, then not q." The contrapositive of this last statement, however, is "if q, then p." In terms of our illustration this means that if you do not work hard, then you will not pass. Therefore, if you pass, then you have worked hard.

To summarize our discussion to date, each of the following statements represents an equivalent form of writing the statement $p \rightarrow q$:

If p, then q.
q, if p.
p implies q.
q is implied by p.
p is a sufficient condition for q.
q is a necessary condition for p.
p, only if q.

The last form is probably the hardest to use. It should appear reasonable when considered as a restatement of the fact that if $p \rightarrow q$, then it is impossible to have p without having q. Many people find it helpful to remember these facts:

The premise is sufficient.
The conclusion is necessary.
If you can identify one part of a conditional statement as necessary, then the other part is sufficient.
If you can identify one part of a conditional statement as sufficient, then the other is necessary.
The premise is the "if part."
The conclusion is the "only-if part."
If you can identify one part of a conditional statement as the premise (if part), then the other part is the conclusion (only-if part).
If you can identify one part of a conditional statement as the conclusion (only-if part), then the other part is the premise (if part).

The seven ways of asserting that a statement p implies a statement q illustrate the difficulty of understanding the English language. We shall endeavor to reduce the confusion by considering statements of implication in the form

$$p \rightarrow q$$

when stated in symbols and, when stated in words, in the form

If p, then q.

Example 1:   Write each statement in if-then form:
(a) Right angles are congruent.
(b) $x > 5$, only if $x \geq 0$.

Solution: (a) If two angles are right angles, then they are congruent.
(b) If $x > 5$, then $x \geq 0$.

Example 2:   Translate into symbolic form, using

p: I will work hard.
q: I will get an A.

(a) I will get an A only if I work hard.
(b) Working hard will be a sufficient condition for me to get an A.
(c) If I work hard, then I will get an A, and if I get an A, then I will have worked hard.

Solution: (a) $q \rightarrow p$; (b) $p \rightarrow q$; (c) $(p \rightarrow q) \wedge (q \rightarrow p)$.

The statement $(p \rightarrow q) \wedge (q \rightarrow p)$ in Example 2(c) is one way of stating that p and q are equivalent, $p \leftrightarrow q$. Any statement of implication is sometimes called a conditional statement; any statement of the equivalence of two statements is then called a **biconditional statement.** The symbol ↔ is referred to as the **biconditional symbol** or the **equivalence symbol.**

Any biconditional statement

$$p \leftrightarrow q$$

is a statement that p is a sufficient condition for q and also p is a necessary condition for q. We may condense this by saying that p is a **necessary and sufficient condition** for q. The biconditional statement may be stated in any one of these forms:

p if and only if q; that is, p iff q.
p implies and is implied by q.
p is a necessary and sufficient condition for q.

Example 3: Complete a truth table for the statement:

$$(p \rightarrow q) \wedge (q \rightarrow p)$$

Solution:

p	q	$(p \rightarrow q)$	$\wedge$	$(q \rightarrow p)$
T	T	T	T	T
T	F	F	F	T
F	T	T	F	F
F	F	T	T	T

(a) (c) (b)

---

p	q	$p \leftrightarrow q$
T	T	T
T	F	F
F	T	F
F	F	T

In Example 3 we constructed a truth table for $(p \rightarrow q) \wedge (q \rightarrow p)$. However, we have previously agreed that this conjunction of statements is equivalent to $p \leftrightarrow q$. This enables us to construct a truth table for $p \leftrightarrow q$, as shown at the left.

From the truth table we see that $p \leftrightarrow q$ is true when p and q are both true or are both false. Thus each of these statements is true:

$2 \times 2 = 4$ if and only if $7 - 5 = 2$. (Both parts are true.)
$2 \times 2 = 5$ if and only if $7 - 5 = 3$. (Both parts are false.)

Each of the following statements is false because exactly one part of each statement is false:

$2 \times 2 = 4$ if and only if $7 - 5 = 3$.
$2 \times 2 = 5$ if and only if $7 - 5 = 2$.

As we have previously noted, we use the symbol "↔" to show that two statements are equivalent. Consider these statements:

p: I cut the grass this afternoon.

q: The sun is shining.

Then the statement "$p \leftrightarrow q$," that is, p if and only if q, is true in these two cases:

1. I cut the grass and the sun is shining.

2. I do not cut the grass and the sun is not shining.

In all other cases, the statement "$p \leftrightarrow q$" is false. Briefly, two statements "p" and "q" are equivalent if each implies the other; in other words, we may write "$p \leftrightarrow q$" if it is true that $p \rightarrow q$ and also that $q \rightarrow p$.

Example 4: Under what conditions is the following statement true?

I will get an A if and only if I work hard.

Solution:

The statement is true in each of two cases:

(a) You get an A and you work hard.
(b) You do not get an A and you do not work hard.

EXERCISES

Write each statement in if-then form:

1. All ducks are birds.

2. Vertical angles are congruent.

3. Complements of the same angle are congruent.

4. Supplements of congruent angles are congruent.

5. Any two parallel lines are coplanar.

6. All triangles are polygons.

7. All circles are round.

8. All mathematics books are dull.

9. All teachers are boring.

10. All p are q.

11. You will like this book only if you like mathematics.

12. A necessary condition for liking this book is that you like mathematics.

13. To like this book it is sufficient that you like mathematics.

14. A sufficient condition for liking this book is that you like mathematics.

15. Liking this book is a necessary condition for liking mathematics.

Write each statement in symbolic form using:

p: The sun shines. q: I cut the grass.

16. If the sun shines, then I cut the grass.

17. I will cut the grass only if the sun shines.

18. If the sun does not shine, then I do not cut the grass.

19. The sun's shining is a necessary condition for me to cut the grass.

20. I cut the grass if and only if the sun shines.

21. If the sun shines, then I do not cut the grass.

22. For me to cut the grass it is sufficient that the sun shine.

23. A necessary and sufficient condition for me to cut the grass is that the sun shine.

Write each statement in symbolic form using:

p: I miss my breakfast. q: I get up late.

24. I miss my breakfast if and only if I get up late.

25. A necessary condition for me to miss my breakfast is that I get up late.

26. For me to miss my breakfast it is sufficient that I get up late.

27. A necessary and sufficient condition for me to miss my breakfast is that I get up late.

28. For me not to miss my breakfast it is necessary that I do not get up late.

Express each statement in if-then form and classify as true or false:

29. $11 - 3 > 8$ if $9 + 3 < 10$.

30. A necessary condition for 2×2 to be equal to 5 is that $8 - 5 = 3$.

31. For 7×4 to be equal to 25 it is sufficient that $5 + 3 = 8$.

32. $7 \times 6 = 40$ only if $8 \times 5 \neq 40$.

33. $7 \times 6 = 42$ only if $8 \times 5 \neq 40$.

Decide relative to each given assertion whether or not a young man who makes the assertion, receives the money, and fails to marry the young lady can be sued for breach of promise:

34. I will marry your daughter only if you give me \$10,000.

35. A sufficient condition for me to marry your daughter is that you give me \$10,000.

PEDAGOGICAL CONSIDERATIONS

All teachers need to be able to state things in a variety of ways.
Restate each sentence in at least four equivalent ways.

1. Learning your part is necessary for you to be in the play.

2. It is sufficient to come to rehearsals to sing in the chorus.

3. Doing the daily work regularly will enable you to pass this course.

4. We have seen the difficulty that arises with the words "only if." Thus the statement "I will get an A only if I work hard" means "If I get an A, then I have worked hard." However, in everyday language, most people tend to interpret the first statement (incorrectly) as "If I work hard, then I will get an A."

   Also a statement such as "You may leave early if you finish the quiz" logically means "If you finish the quiz, then you may leave early." However many tend to interpret this permissive use of *if* as meaning *only if*. Thus they would then incorrectly think of the given statement as meaning "If you leave early, then you have finished the quiz."

   The use of an if-then statement to give permission frequently has the special meaning of "don't . . . unless . . . ." With this usage, the statement

   You may leave early if you finish the quiz

is interpreted as

   Don't leave early unless you have finished the quiz.

Prepare a set of three "if-then" statements that give permission, and rewrite each in the form "don't . . . unless . . . ."

2-6
the nature of proof

How do you "prove" a statement to a friend? Undoubtedly there are several ways, including these three:

   1. Find the statement in an encyclopedia or other reference that he will accept without further proof.

   2. Prove to him that the statement is a necessary consequence of some statement that he accepts.

   3. Prove to him that the statement cannot be false.

In mathematics there are also several ways of "proving" statements. In essence each proof is based upon:

   1. Statements that are accepted as true (assumed).

   2. Sequences of statements (arguments) such that each statement is either assumed or is a *logical consequence* of the preceding statements, and the statement to be proved is included in the sequence.

   3. Proofs that statements cannot be false.

Thus the "key" to our understanding the nature of proof is our understanding of "logical consequences." We shall find in §2-7 that each logical consequence of a statement depends not only upon the given statement but also upon one or more statements that cannot fail to be true. Any statement such as $p \lor (\sim p)$ that is always true is called a **tautology.**

Example: Determine whether or not the statement

$$[(p \to q) \land p] \to q$$

is a tautology.

Solution:

p	q	$[(p \longrightarrow q)$	$\land$	$p]$	$\longrightarrow$	q
T	T	T	T T	T		T
T	F	F	F T	T		F
F	T	T	F F	T		T
F	F	T	F F	T		F

(a) $\quad\quad$ (c) $\quad$ (b) $\quad$ (e) $\quad$ (d)

The truth value of the given statement is T under all possible situations, as shown in column (e); therefore the statement is a tautology.

EXERCISES

Determine whether or not each statement is a tautology.

1. $p \lor (\sim p)$
2. $\sim[p \land (\sim p)]$
3. $[(p \to q) \land (\sim q)] \to (\sim p)$
4. $[(p \to q) \land q] \to p$
5. $[(p \to q) \land (\sim p)] \to (\sim q)$
6. $(p \land q) \leftrightarrow (q \land p)$
7. $[(p \lor q) \land (\sim p)] \to q$
8. $(p \lor q) \to (q \lor p)$
9. $[(p \to q) \land (q \to r)] \to (p \to r)$

PEDAGOGICAL CONSIDERATIONS

Proofs in elementary school classes are usually at the level of giving reasons or explanations. The explanations by the teacher should be logically correct. Most students are in the process of learning to provide logical explanations.

Start a collection of "striking" examples of youthful reasoning such as these two gathered by the Swiss psychologist Jean Piaget:

> When you're walking in the moonlight, the moon moves along with you.
> Trees make the wind blow by waving their branches.

2-7

valid We shall consider proofs in terms of the patterns (arguments) formed by **arguments** the statements used. The assumed statements are identified as **given statements** and called **premises** of the argument. The statements to be proved are called **conclusions** of the argument.

Example 1: Determine whether or not the following argument is valid:
Given: If Mary is a junior, she is taking algebra.
Given: Mary is a junior.
Conclusion: Mary is taking algebra.

Solution: Use

$$p:\quad \text{Mary is a junior.}$$
$$q:\quad \text{Mary is taking algebra.}$$

and think of the argument as:

Given: $p \to q$.
Given: p.
Conclusion: q.

The argument is valid if and only if the statement

$$[(p \to q) \land p] \to q$$

is a tautology. This statement is a tautology (§2-6, Example). Thus the argument is valid.

The argument used in Example 1 illustrates one of the basic rules of inference, the **law of detachment**, also called **modus ponens**:

> If a statement of the form "if p, then q" is assumed to be true, and if p is known to be true, then q must be true.

Example 2: Determine whether or not the following argument is valid:
Given: If $3x - 1 = 8$, then $3x = 9$.
Given: $3x - 1 = 8$.
Conclusion: $3x = 9$.

77

Solution: For

$$p: \quad 3x - 1 = 8$$
$$q: \quad 3x = 9$$

the argument is valid by *modus ponens*. It has the form of the tautology

$$[(p \rightarrow q) \wedge p] \rightarrow q.$$

Example 3: Determine whether or not the following argument is valid:
Given: If you worked hard, then you passed the course.
Given: You passed the course.
Conclusion: You worked hard.

Solution: For

$$p: \quad \text{You worked hard.}$$
$$q: \quad \text{You passed the course.}$$

the argument has the form

$$[(p \rightarrow q) \wedge q] \rightarrow p.$$

This statement is not a tautology (§2-6, Exercise 4). Thus the argument is not valid.

Notice that anyone who uses the argument in Example 3 to convince his friends that he has worked hard is hoping that his friends will think that the converse of any true conditional statement is also true; that is,

$$\text{If } p \rightarrow q, \text{ then } q \rightarrow p.$$

We know from our comparison of truth values of $p \rightarrow q$ and $q \rightarrow p$ that a statement does not necessarily imply its converse. As another example of this type of reasoning consider the advertisement:

$$\text{If you want to be healthy, eat KORNIES.}$$

The advertiser hopes that the consumer will fallaciously assume the converse statement:

$$\text{If you eat KORNIES, then you will be healthy.}$$

The argument is not valid and is called a **fallacy.**

Example 4: Determine whether or not the following argument is valid:
Given: If you worked hard, then you passed the course.
Given: You did not pass the course.
Conclusion: You did not work hard.
Solution: For

$$p: \quad \text{You worked hard.}$$
$$q: \quad \text{You passed the course.}$$

the argument has the form

$$[(p \rightarrow q) \land (\sim q)] \rightarrow (\sim p).$$

This statement is a tautology (§2-6, Exercise 3); the argument is valid.

Notice that the argument in Example 4 is based upon the equivalence of any given conditional statement $p \rightarrow q$ and its contrapositive statement $(\sim q) \rightarrow (\sim p)$ as we observed in §2-4.

Example 5: Determine whether or not the following argument is valid:
Given: If you worked hard, then you passed the course.
Given: You did not work hard.
Conclusion: You did not pass the course.

Solution: For

        p: You worked hard. q: You passed the course.

the argument has the form

$$[(p \rightarrow q) \land (\sim p)] \rightarrow (\sim q).$$

This statement is not a tautology (§2-6, Exercise 5), and the argument is not valid. It is another example of a fallacy.

Notice that anyone who uses the argument in Example 5 expects his listeners to assume that the inverse $(\sim p) \rightarrow (\sim q)$ of any true statement $p \rightarrow q$ must be true. As in the case of a conditional statement and its converse, we know that a statement does not necessarily imply its inverse. As another example of this type of reasoning, consider the advertisement:

> If you brush your teeth with SCRUB, then you will have no cavities.

The advertiser would like you to assume, fallaciously, the inverse statement:

> If you do not brush your teeth with SCRUB, then you will have cavities.

The final form of valid reasoning that we shall consider here is based upon the tautology

$$[(p \rightarrow q) \land (q \rightarrow r)] \rightarrow (p \rightarrow r)$$

(§2-6, Exercise 9). Consider these specific examples:

> **1.** If you like this book, then you like mathematics.
> If you like mathematics, then you are intelligent.
> Therefore, if you like this book, then you are intelligent.

79

2. If a polygon is a square, then it has 3 sides.
If a polygon has 3 sides, then it has 4 angles.
Therefore, if a polygon is a square, then it has 4 angles.

3. If a polygon is a square, then it has 4 sides.
If a polygon has 4 sides, then it has 3 angles.
Therefore, if a polygon is a square, then it has 3 angles.

In the second example both premises are false, the conclusion is true, and the argument is valid. In the third example the first premise is true, the second premise is false, the conclusion is false, yet the argument is again valid, since it is based upon a tautology. Thus the truth of the conclusion does not enter in any way into a discussion of the validity of an argument. The *validity* of an argument depends solely upon its form; that is, whether or not it is based upon a tautology. An argument is valid if the conjunction of the premises implies the conclusion. The *truth* of a conclusion of a valid argument depends completely upon the truth of the premises.

It is important to emphasize the fact that validity has nothing to do with the question of whether the conclusion is true or false. The conclusion may be false, yet the argument may be valid if the chain of reasoning is correct. On the other hand, the conclusion may be true, yet the argument may not be valid if the reasoning is incorrect.

In each of the exercises that follow, bear in mind that the validity of each argument depends upon its form rather than the truth of the sentences used.

EXERCISES

Determine whether or not each argument is valid.

1. Given: If Elliot is a freshman, then Elliot takes mathematics.
Given: Elliot is a freshman.
Conclusion: Elliot takes mathematics.

2. Given: If you like dogs, then you will live to be 120 years old.
Given: You like dogs.
Conclusion: You will live to be 120 years old.

3. If the Yanks win the game, then they win the pennant.
They do not win the pennant.
Therefore, they did not win the game.

4. If you like mathematics, then you like this book.
You do not like mathematics.
Therefore, you do not like this book.

5. If you work hard, then you are a success.
You are not a success.
Therefore, you do not work hard.

6. If you are reading this book, then you like mathematics.
   You like mathematics.
   Therefore, you are reading this book.

7. If you are reading this book, then you like mathematics.
   You are not reading this book.
   Therefore, you do not like mathematics.

8. If you work hard, then you will pass the course.
   If you pass the course, then your teacher will praise you.
   Therefore, if you work hard, then your teacher will praise you.

9. If you like this book, then you like mathematics.
   If you like mathematics, then you are intelligent.
   Therefore, if you are intelligent, then you like this book.

10. If you are a blonde, then you are lucky.
    If you are lucky, then you will be rich.
    Therefore, if you do not become rich, then you are not a blonde.

Supply a conclusion so that each of the following arguments will be valid.

11. If you drink milk, then you will be healthy.
    You are not healthy.
    Therefore, ...

12. If you eat a lot, then you will gain weight.
    You eat a lot.
    Therefore, ...

13. If you like to fish, then you enjoy swimming.
    If you enjoy swimming, then you are a mathematician.
    Therefore, ...

14. If you do not work hard, then you will not get an A.
    If you do not get an A, then you will have to repeat the course.
    Therefore, ...

15. If you like this book, then you are not lazy,
    If you are not lazy, then you will become a mathematician.
    Therefore, ...

PEDAGOGICAL CONSIDERATIONS

1. Begin a collection of fallacies in reasoning that you find in newspapers or magazines, or that you hear on radio or television.

2. Examine a newspaper or magazine for if-then statements that are used in advertising. Then analyze the statement to see whether the converse or inverse of the statement is used.

3. The National Council of Teachers of Mathematics (N.C.T.M.) is a professional organization dedicated to the improvement of the teaching of mathematics at all grade levels. Among their many publications are *The Arithmetic Teacher,* of interest primarily to mathematics teachers in grades K through 8, and *The Mathematics Teacher,* primarily for grades 7 through 12. For further information, as well as for an application for student membership at reduced rates, write to the N.C.T.M. at 1201 Sixteenth Street, N.W., Washington, D.C., 20036.

The N.C.T.M. publishes yearbooks for teachers on a variety of different topics. Of particular interest to the reader of this text will be the twenty-seventh Yearbook, entitled *Enrichment Mathematics for the Grades.* This yearbook presents a collection of mathematical topics that may be used as enrichment for talented students. Obtain a copy of this book and, in particular, read Chapter 22 by Robert L. Swain entitled "Logic: For Teacher, For Pupil."

2-8

Euler diagrams

In the eighteenth century the Swiss mathematician Leonhard Euler used diagrams to present a visual approach to the study of the validity of arguments. The diagrams used in this section are an aid to reasoning but are not essential to the reasoning process. We use a region, usually a circular region, P to represent the situations in which a statement p is true. Then p is false in all situations represented by points of P'. Similarly, we use a region Q for a statement q, and so forth.

First consider the statement $p \rightarrow q$ (if p, then q). This statement is equivalent to saying "all p are q," that is, p is a subset of q; if you have p, then you must have q. We show this by drawing circles to represent p and q with p drawn as a subset of q. Consider the statements:

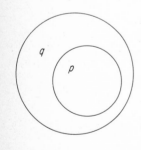

p: It is a lemon.
q: It is a piece of fruit.

Then the figure at the left may be used as a visualization of the statement "All lemons are pieces of fruit." Statements p and q are called **consistent statements** since they can both be true of the same object.

Next consider the statements:

p: It is a lemon.
s: It is an orange.

Statements p and s are called **contrary statements** since they cannot both be true of the same object. If we wish to draw a diagram to represent the statement "No lemon is an orange," then we may draw two circles whose intersection is the empty set:

Note that this same diagram can be used to represent either of the following equivalent statements:

> If it is a lemon, then it is not an orange.
> If it is an orange, then it is not a lemon.

We may also represent statements such as the following by means of Euler diagrams:

> p: It is a lemon.
> $\sim p$: It is not a lemon.

If we draw a circle to represent the statement p, then $\sim p$ may be represented by the set of points outside of p; that is, by the complement of p.

Statements such as p and $\sim p$ are called **contradictory statements.** Exactly one of the statements must be true; exactly one of the statements must be false. If either statement is true, then the other is false. If either statement is false, then the other is true.

Let us now return to the question of validity and explore a visual approach to the argument

$$[(p \rightarrow q) \wedge (q \rightarrow r)] \rightarrow (p \rightarrow r).$$

We have seen that the statement "if p, then q" is equivalent to saying "all p's are q's" and may be drawn with p as a subset of q. Similarly, the statement "if q, then r" can be written in the form "all q's are r's"; that is, q is a subset of r. To represent these two statements we construct circles in such a way that p is contained within q, and q is contained within r, as follows:

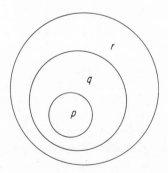

We have followed instructions carefully for our two premises. From the diagram it is clear that p must be a subset of r. That is, we conclude that "all *p*'s are *r*'s" or "if *p*, then *r*," which is what we set out to demonstrate as the conclusion of our argument.

From the final diagram it should also be clear that each of the following conclusions would *not* be valid:

$$\text{All } r\text{'s are } q\text{'s.} \qquad r \rightarrow q$$
$$\text{All } q\text{'s are } p\text{'s.} \qquad q \rightarrow p$$

Can you see that each of the following conclusions follows logically from the diagram drawn?

$$\text{If not } r, \text{ then not } p. \qquad {\sim}r \rightarrow {\sim}p$$
$$\text{If not } q, \text{ then not } p. \qquad {\sim}q \rightarrow {\sim}p$$

Euler diagrams may be used to test the validity of an argument. We draw one or more Euler diagrams, making use of the premises and avoiding the introduction of additional premises. The structure of the diagram indicates whether or not the argument is valid. As in §2-7 we disregard the truth values of the premises and conclusions.

Example 1: Test the validity of this argument:

All undergraduates are sophomores.
All sophomores are attractive.
Therefore, all undergraduates are attractive.

Solution: A diagram of this argument follows:

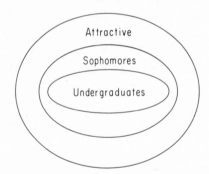

The argument is valid.

Notice in Example 1 that you may or may not agree with the conclusion or with the premises. The important thing is that you are *forced* to draw the diagram in this manner. The conclusion is an inescapable consequence of the given premises. You should not allow the everyday meaning of words to alter your thinking. Thus the preceding argument can be stated abstractly as in these three statements:

All *u*'s are *s*'s.

All *s*'s are *a*'s.
Therefore, all *u*'s are *a*'s.

Here we rely on logic alone rather than any preconceived notions about the meaning of such words as "undergraduate," "sophomore," and "attractive."

Example 2: Draw a diagram and test the validity of this argument:

All freshmen are clever.
All attractive people are clever.
Therefore, all freshmen are attractive.

Solution: The first premise tells us that the set of freshmen is a subset of the set of clever individuals. This is drawn as follows:

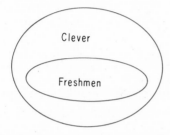

Next we need to draw a circle to represent the set of attractive people as a subset of the set of clever individuals. However, there are several possibilities here:

Each of the figures drawn represents a distinct possibility, but only one shows that all freshmen are attractive. Therefore, since we are not *forced* to arrive at this conclusion, the argument is said to be *not valid*. Each of the figures, on the other hand, forces you to arrive at the following valid conclusions:

Some clever people are attractive.
Some clever people are freshmen.

Consider next the statement:

Some freshmen are attractive.

This statement means that there exists *at least one* freshman who is attractive. It does not, however, preclude the possibility that all freshmen are attractive. Now consider the following argument whose validity we wish to check:

> Some freshmen are attractive.
> All girls are attractive.
> Therefore, some freshmen are girls.

To test the validity of this argument we draw Euler circles for the two premises and then see whether we are forced to accept the conclusion. Since some freshmen are attractive, we will draw the following but recognize that the set of freshmen could also be drawn as a proper subset of the set of attractive people.

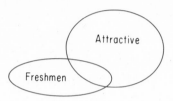

Next we draw a circle to represent the set of girls as a proper subset of the set of attractive people. Here are several possible ways to do this:

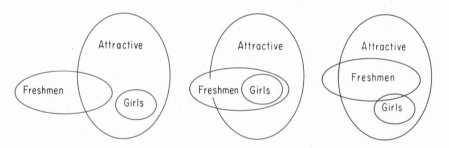

Notice that in two of the figures we find that some freshmen are girls. But we are not forced to adopt this conclusion, as seen in the first of the three figures drawn. Accordingly, we conclude that the argument is not valid.

Example 3: Test the validity of the following argument:

> All *a*'s are *b*'s.
> Some *a*'s are *c*'s.
> Therefore, some *c*'s are *b*'s.

Solution: Note that since some *a*'s are *c*'s, then there are also some *c*'s that are *a*'s. However, since all *a*'s are *b*'s, it follows that some *c*'s must also be *b*'s as in these diagrams:

  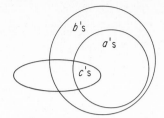

The argument is valid.

EXERCISES

Use each of the following diagrams and write at least five valid conclusions.

1.

2.

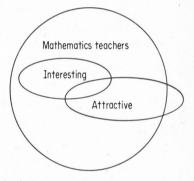

Draw a diagram and test the validity of each argument:

3. All students love mathematics.
   Harry is a student.
   Therefore, Harry loves mathematics.

4. All juniors are brilliant.
   All brilliant people love mathematics.
   Therefore, if you are a junior, then you are brilliant.

5. All girls are beautiful.
   All beautiful people like this book.
   Therefore, if you like this book, then you are a girl.

6. All mathematics teachers are dull.
   Some Ph.D.'s are dull.
   Therefore:
   (a) Some mathematics teachers are Ph.D.'s.
   (b) Some dull people have Ph.D.'s.

7. All juniors are clever.
   Some juniors are males.

   Therefore:

   (a) Some males are clever.
   (b) Some males are juniors.
   (c) Some clever people are males.
   (d) All males are juniors.

8. All boys are handsome.
   Some boys are athletes.
   Therefore, some athletes are handsome.

9. All mathematics teachers are interesting.
   All attractive individuals are interesting.
   Some mathematics teachers are kind.

   Therefore:

   (a) Some interesting people are kind.
   (b) Some mathematics teachers are attractive.
   (c) All mathematics teachers are attractive.
   (d) All mathematics teachers are kind.
   (e) Some kind individuals are attractive.
   (f) No mathematics teachers are attractive.
   (g) No attractive individuals are interesting.

10. All x's are y's.
    Some y's are z's.

    Therefore:

    (a) Some x's are z's.   (b) Some z's are y's.

11. All a's are b's.
    All b's are c's.
    Some d's are not c's.

    Therefore:

    (a) All a's are c's.   (b) Some d's are not b's.
    (c) Some d's are not a's.   (d) Some d's are c's.

12. All booms are mooms.
    Some booms are tooms.
    Some zooms are booms.

    Therefore:

    (a) Some mooms are booms.
    (b) Some zooms are mooms.
    (c) Some mooms are tooms.
    (d) No toom is a zoom.
    (e) Some zooms are moom-tooms.

PEDAGOGICAL CONSIDERATIONS

1. One elementary textbook uses Euler circles (diagrams) to show the use of the words "all," "some," and "no." Here is an example:

All counting numbers are integers.
Some integers are counting numbers.
No counting numbers are negative integers.

Draw a similar diagram to show this set of statements:

All triangles are polygons.
Some polygons are squares.
No square is a triangle.

Then prepare at least five other similar figures using arithmetic or geometric concepts that are found at the elementary school level.

2. Another interesting exercise for elementary school youngsters is to present an Euler diagram and have them draw inferences from it. In the following you are to complete each blank space with one of the words "all," "some," or "no."

———— squares are polygons.
———— squares are hexagons.
———— polygons are squares.

Prepare a collection of at least five other similar exercises, using concepts found at various grades in the elementary mathematics curriculum.

2-9

quantifiers Some statements are intended to be *universally* true for the people or things under consideration:

All sophomores are attractive.
All squares are rectangles.

Other statements are intended to assure us that there *exists* at least one case in which the statement is true.

Some teachers are intelligent.
Some parallelograms are squares.

These last two statements can also be written as:

There exists at least one teacher who is intelligent.
There exists at least one parallelogram that is a square.

The words "all" and "some" serve as quantifiers. "All" may be used to assert that a statement is universally true and is called a **universal quantifier.** "Some" may be used to assert that there exists a case for which the statement is true and is called an **existential quantifier.**

In mathematics, quantifiers are particularly helpful to express sentences so that they can be identified as true or false. For example,

$$x + 2 = 5$$

is true if $x = 3$ and is false otherwise. Thus we don't know whether $x + 2 = 5$ is true or false unless we know about the values of x. We know that the sentence is not true for all whole numbers; that is, it is false to say

$$x + 2 = 5 \quad \text{for all whole numbers } x.$$

We know that the sentence is true for at least one whole number. There exists a whole number x such that $x + 2 = 5$. We may write this in symbols for whole numbers x in this way:

$$\exists_x: \quad x + 2 = 5$$

We use $\exists_x$ as the symbol for the existential quantifier "there exists an x such that."

Now consider this sentence:

$$x + 2 = 2 + x \quad \text{for all real numbers.}$$

By the commutative property of addition we know that this is always true. We write this in symbols for real numbers x in this way:

$$\forall_x: \quad x + 2 = 2 + x$$

We use $\forall_x$ as the symbol for the universal quantifier "for all x."

Careful consideration of quantifiers is particularly important because of the common popular confusion of "all are not" and "not all are." The following statements are equivalent:

Not all students are sophomores.
Some students are not sophomores.

Similarly, these statements are equivalent:

All my students are not dishonest.
No one of my students is dishonest.
All of my students are honest.

Notice that to negate (deny) a universally quantified statement it is only necessary to find one case for which the statement is false, that is, one counterexample. Thus to prove that not all students are sophomores it is sufficient to find one student who is not a sophomore. However, to negate (deny) an existentially quantified statement it is necessary to prove that the statement is false for *all* students. Thus to prove that no one of my students is dishonest it is necessary to prove that all of my students are honest (not dishonest).

Example 1: State in words the negation of each statement:
(a) There is a rectangle that is not a square.
(b) All triangles are right triangles.

Solution: (a) All rectangles are square. (b) There exists a triangle that is not a right triangle.

Example 2: State in words the negation of the statement "Some books are worth reading."

Solution: It is often helpful to consider the denial of a statement by various rephrasings of the statement. Thus we can go from the given statement to "There is at least one book that is worth reading." The denial of this statement is then "There aren't any books worth reading," which can finally be stated in the form "All books are not worth reading."

EXERCISES

In Exercises 1 through 4 assign a value to the variable to make each sentence true.

1. $x - 2 = 7$ **2.** $2x = 12$

3. $2x + 3 = 15$ **4.** $x^2 = 9$

5. In Exercises 1 through 4 assign a value to the variable to make each sentence false.

6. In Exercises 1 through 4 use a quantifier to make each sentence true.

7. In Exercises 1 through 4 use a quantifier to make each sentence false.

State in words the negation of each statement.

8. All apples are pieces of fruit.

9. All numbers are whole numbers.

10. Any whole number is positive.

11. Any fraction represents a rational number.

12. There exist rational numbers that are not integers.

13. There exist real numbers that are not rational numbers.

14. Some complex numbers are real numbers.

15. Some rational numbers are integers.

PEDAGOGICAL CONSIDERATIONS

Elementary school teachers have a special need for understanding quantifiers since they, at least informally, are teaching students how to use the English language. Correct usage of quantifiers is essential for student understanding and for fair treatment of students.

1. Give at least three examples of sentences involving quantifiers that could be used with first graders.

2. Repeat Exercise 1 for fourth graders.

3. Describe three elementary school classroom situations in which negations of quantified sentences might arise.

CHAPTER TEST

1. Use p: John studies hard; q: Jim is lazy. Think of "is lazy" as "does not study hard" and write each of these statements in symbolic form:
  (a) Neither John nor Jim is lazy.
  (b) It is not true that John and Jim are both lazy.

2. Copy and complete the truth table:

p	q	$\sim$	$[(\sim p)$	$\vee$	$q]$
T	T				
T	F				
F	T				
F	F				
		(d)	(a)	(c)	(b)

3. Give the truth value of each statement:
   (a) If $6 + 2 = 10$, then $7 \times 2 = 11$.
   (b) If $7 \times 2 = 10$, then $6 \times 2 = 12$.

4. Write the converse, inverse, and contrapositive of the statement:

$$\text{If } x < 3, \text{ then } x \neq 3.$$

5. Write each statement in if-then form:
   (a) Right angles are congruent.
   (b) To like to swim it is necessary that you like water.

6. Write each statement in symbolic form using:

$$p: \quad \text{I wear a coat.}$$
$$q: \quad \text{It is snowing.}$$

   (a) For me to wear a coat it is necessary that it be snowing.
   (b) I wear a coat only if it is snowing.

7. Write each statement in symbolic form:
   (a) p is sufficient for q.
   (b) p is necessary and sufficient for $(\sim q)$.

8. Identify as true or false:
   (a) $3 \times 5 = 35$ only if $6 \times 5 = 70$.
   (b) For $2 \times x = 6$ it is sufficient that $2 \times 5 = 6$.

9. Test the validity of each argument:
   (a) All dogs love their masters.
       If an animal loves its master, it should be well cared for.
       Therefore, all dogs should be well cared for.
   (b) If x is a positive integer, then $x^2 > 0$.
       $x^2 \not> 0$.
       Therefore, x is not a positive number.

10. Use a truth table to determine whether or not the statement

$$[(p \rightarrow q) \wedge (\sim q)] \rightarrow (\sim p)$$

   is a tautology.

3

Mathematical Systems

Much of mathematics is concerned with the study of basic principles. In this chapter we shall explore some of these basic principles while studying a collection of abstract, but interesting, systems.

A **mathematical system** involves a set of elements (such as the set of counting numbers), one or more operations (such as addition, subtraction, multiplication, and division), one or more relations (such as the equality $2 \times 3 = 6$), and some axioms (rules) which the elements, operations, and relations satisfy. For example, we assume that every system involves the relation of equality and the axiom that $a = a$; that is, any quantity is equal to itself. Actually, you have been working with mathematical systems ever since you started school. We shall begin this chapter with the study of a specific mathematical system, and shall relate the properties of this system to those of familiar sets of numbers.

3-1

an abstract system

Let us define an abstract system composed of the set of elements:

$$\{*, \#, \Sigma, [\ \ ]\}$$

For ease of notation we shall call this set Y; that is,

$$Y = \{*, \#, \Sigma, [\ \ ]\}.$$

Next we define an operation which we call "multition" and denote this with the symbol ~. The operation is defined by means of the following table:

~	*	#	Σ	[ ]
*	*	#	Σ	[ ]
#	#	Σ	[ ]	*
Σ	Σ	[ ]	*	#
[ ]	[ ]	*	#	Σ

The operation ~ is a **binary operation,** since it associates a single element with any *two* elements of the given set.

To "multify" two elements of Y, we read the first of these elements in the vertical column at the left and the second in the horizontal row across the top. The answer is found where the column and the row intersect within the table. Verify from the table that each of the following statements is correct:

$$\# \sim \Sigma = [\ \ ]$$
$$[\ \ ] \sim \# = *$$
$$\Sigma \sim * = \Sigma$$

The task of major importance before us is to examine this system for its basic properties. The following are some of the more important ones.

1. Whenever any two elements of the set Y are combined by the process of multition, the result is a unique element of the original collection Y of elements. In other words, there is one and only one answer whenever two elements of Y are multified, and this answer is always one of the elements of Y. We say that *the set Y is closed under the operation of multition.*

In general a set S is said to be **closed** under a given binary operation if that operation associates a unique element of the set S with any two elements of the set.

Consider, in ordinary arithmetic, all the numbers that we use for counting:

$$C = \{1, 2, 3, 4, 5, \ldots\}$$

This set of counting numbers is closed under the operations of addition and multiplication. That is, the sum of any two counting numbers is again a counting number, and the product of any two counting numbers is a counting number. Is the set of counting numbers closed under subtraction? That is, is the difference of *any* two counting numbers always a counting number?

2. Consider the order in which we combine any two elements of the set Y. Does $\# \sim \Sigma = \Sigma \sim \#$? Does $[\ \ ] \sim * = * \sim [\ \ ]$? In general, if a and b represent any two elements of the set Y, does $a \sim b = b \sim a$? We find the answer to be in the affirmative in each case and summarize this property by saying that *the elements of set Y satisfy the commutative property for multition.*

In general, a set of elements is said to satisfy the **commutative property** for a particular operation if the result obtained by combining any two elements of the set under that operation does not depend upon the order in which these elements are combined.

The set of counting numbers is commutative under addition and multiplication. That is, the sum of any two counting numbers is the same, regardless of the order in which the elements are added. Thus $5 + 7 = 7 + 5$, $8 + 3 = 3 + 8$, etc. Similarly, the set of counting numbers satisfies the commutative property for multiplication, since the product of any two counting numbers is the same regardless of the order in which the elements are multiplied. Thus $5 \times 7 = 7 \times 5$, $8 \times 3 = 3 \times 8$, etc.

Example 1: Show that subtraction of counting numbers is not commutative.

Solution: A single *counterexample* is sufficient to show that a property does not hold. Since $7 - 3 = 4$ and $3 - 7 = -4$, $7 - 3 \neq 3 - 7$ and subtraction is not commutative. (The symbol $\neq$ is read "is not equal to.")

---

3. Next we combine three elements of the set Y. Can you find the answer for $\Sigma \sim [\ \ ] \sim \#$? Here we see that two possibilities exist. We could combine Σ and $[\ \ ]$ first and then combine the result with $\#$, or we could combine $[\ \ ]$ and $\#$ first and then combine Σ with the result obtained. Let us try both ways:

$$(\Sigma \sim [\ \ ]) \sim \# = \# \sim \# = \Sigma$$
$$\Sigma \sim ([\ \ ] \sim \#) = \Sigma \sim * = \Sigma$$

We obtain the same result in both cases and will do so for any choice of three members of the set Y. We call this the **associative property** for multition, and write, in symbols:

$$a \sim (b \sim c) = (a \sim b) \sim c$$

where a, b, and c may be replaced by any elements of the set Y.

The set of counting numbers is associative with respect to both addition and multiplication. That is, the sum or product of any three counting numbers is the same regardless of the order in which they are associated. For example,

$$3 + (5 + 7) = 3 + 12 = 15 \quad \text{and} \quad (3 + 5) + 7 = 8 + 7 = 15$$
$$3 \times (5 \times 7) = 3 \times 35 = 105 \quad \text{and} \quad (3 \times 5) \times 7 = 15 \times 7 = 105.$$

Thus

$$3 + (5 + 7) = (3 + 5) + 7 \quad \text{and} \quad 3 \times (5 \times 7) = (3 \times 5) \times 7.$$

Example 2: Show that the set of counting numbers is not associative with respect to division.

Solution: Again, one counterexample is sufficient.

$24 \div (6 \div 2) = 24 \div 3 = 8$, whereas $(24 \div 6) \div 2 = 4 \div 2 = 2$.

Thus $24 \div (6 \div 2) \neq (24 \div 6) \div 2$, and division of counting numbers is not associative.

---

4. The set Y contains a unique element, $*$, which leaves every other element unchanged with respect to multition. For example,

$$* \sim * = *, \quad \# \sim * = \#, \quad \Sigma \sim * = \Sigma, \quad \text{and} \quad [\ ] \sim * = [\ ].$$

The element with this property is called the **identity element** for multition.

In ordinary arithmetic, what is the identity element for addition? That is, what can we add to 7 to obtain 7? Does $9 + 0 = 9$? Do you see that 0 is the identity element for addition? However, 0 is not an element of the set of counting numbers. Therefore, we must say that the set of counting numbers does not contain an identity element with respect to addition. Does the set of counting numbers contain an identity element with respect to multiplication?

5. We examine one more property of the set Y. Replace each blank in the following by an element of the set Y to obtain a correct statement:

$$* \sim \text{——} = *$$
$$\# \sim \text{——} = *$$
$$\Sigma \sim \text{——} = *$$
$$[\ ] \sim \text{——} = *$$

The correct answers are, respectively, $*$, $[\ ]$, Σ, and $\#$. We find that for each element of the set Y there is another element of Y such that the first "multified" by the second produces the identity element $*$. Each of these second elements is called the **inverse with respect to multition** (or simply the **inverse**) of the first element of the statement in which it was used. Thus the inverse of $*$ is $*$, the inverse of $\#$ is $[\ ]$, the inverse of Σ is Σ, and the inverse of $[\ ]$ is $\#$. Note that $*$ and Σ serve as their own inverses.

The set of counting numbers does not contain the inverses for its elements with respect to either addition or multiplication. However, when we extend our system to include *all* integers, then every element has an inverse under addition. Thus the inverse of 5 is -5, in that $5 + (-5) = 0$. Similarly, the inverse of -3 is 3, in that $(-3) + 3 = 0$.

In later extensions of our number system we consider all fractions of the form $\dfrac{a}{b}$, where a and b are integers and $b \neq 0$. This set then contains the inverse for each of its nonzero elements under multiplication. Thus the inverse of 5 is $\frac{1}{5}$, in that $5 \times \frac{1}{5} = 1$. Similarly, the inverse of $\frac{2}{3}$ is $\frac{3}{2}$, in that $\frac{2}{3} \times \frac{3}{2} = 1$.

In summary, the set Y

1. Is closed with respect to multition.

2. Is commutative with respect to multition.

3. Is associative with respect to multition.

4. Contains an identity element with respect to multition.

5. Contains an inverse for each of its elements with respect to multition.

We may summarize all five of these properties by saying that the elements of the set Y form a **commutative group** under multition.

The set of counting numbers does not form a group under addition since the identity and inverses are missing. The set of counting numbers does not form a group under multiplication since the inverses of numbers different from 1 are missing. However, the set of counting numbers

1. Is closed with respect to addition.

2. Is commutative with respect to addition.

3. Is associative with respect to addition.

4. Is closed with respect to multiplication.

5. Is commutative with respect to multiplication.

6. Is associative with respect to multiplication.

7. Contains the identity with respect to multiplication.

EXERCISES

Find the answer to each of the following from the table given in this section.

1. # ~ *

2. * ~ Σ

3. [ ] ~ [ ]

4. Σ ~ #

Verify that each of the following is true and state the name of the property that each illustrates:

5. Σ ~ # = # ~ Σ

6. [ ] ~ (# ~ Σ) = ([ ] ~ #) ~ Σ

7. # ~ * = #

8. # ~ [ ] = *

9. (# ~ Σ) ~ * = # ~ (Σ ~ *)

10. [ ] ~ * = [ ]

11. [ ] ~ # = *

12. [ ] ~ * = * ~ [ ]

Answer Exercises 13 through 22 by using the following table, which defines an operation ⊙ for the elements of the set {△, □, Q}.

⊙	△	□	Q
△	Q	△	□
□	△	□	Q
Q	□	Q	△

13. Find △ ⊙ □.

14. Find Q ⊙ △.

15. Does Q ⊙ □ = □ ⊙ Q?

16. Does △ ⊙ Q = Q ⊙ △?

17. Does (Q ⊙ □) ⊙ △ = Q ⊙ (□ ⊙ △)?

18. Does □ ⊙ (Q ⊙ △) = (□ ⊙ Q) ⊙ △?

19. Does an affirmative answer to the two preceding exercises prove that the set satisfies the associative property with respect to ⊙? Explain your answer.

20. Is there an identity element for ⊙? If so, what is it?

21. Does each element have an inverse with respect to ⊙? If so, name the inverse of each element.

22. Summarize the properties of the set {△, □, Q} with respect to the operation ⊙.

Answer Exercises 23 through 26 by using the following table, which defines an operation ▱ for the set {•, [, !, $}.

▱	•	[	!	$
•	•	[	!	$
[	[	!	o	~
!	!	o	~	[
$	$	[	!	•

23. Is the set closed with respect to ▱? Justify your answer.

24. Does the set satisfy the commutative property for ▱? Explain.

25. Does the set contain an identity element for ▱? If so, what is it?

26. Which elements of the set have inverses with respect to ▱? Name each of these inverses.

1. Show that the set of counting numbers is not commutative with respect to division.

2. Show that the set of counting numbers is not associative with respect to subtraction.

3. What is the identity element with respect to multiplication for the set of counting numbers? Give several specific examples to justify your answer.

4. Show that the set of counting numbers does not form a group under addition.

5. Let $\#$ mean "Select the number if the two given numbers are equal, and the smaller of the two given numbers if they are unequal." Then $5 \# 5 = 5$ and $5 \# 3 = 3$. For the set of counting numbers, does $a \# b = b \# a$? Does $a \# (b \# c) = (a \# b) \# c$?

6. Let $\$$ mean "Select the first of two given numbers." Then $5 \$ 2 = 5$. For the set of counting numbers, does $a \$ b = b \$ a$? Does $a \$ (b \$ c) = (a \$ b) \$ c$?

*7. Use the definitions given in Exercises 5 and 6 and show that $a \$ (b \# c) = (a \$ b) \# (a \$ c)$ for all counting numbers a, b, and c.

*8. Referring to the definitions given in Exercises 5 and 6, does $a \# (b \$ c) = (a \# b) \$ (a \# c)$?

PEDAGOGICAL CONSIDERATIONS

Elementary school students use the properties of numbers to simplify arithmetic computations. For example, most students find 7×10 a much simpler problem than 10×7 until they recognize that these are simply different statements of the same problem. Make a worksheet of 20 problems to help elementary school students recognize and make effective use of each property:

1. Commutative property of multiplication.

2. Commutative property of addition.

3. Operations may be considered in terms of a **function machine** with three basic parts:

For the rule "add 5" complete this array:

Input	3	7	——	——	21	——
Output	——	——	8	11	——	21

4. Make a worksheet for elementary school students for the rule:
(a) subtract 1; **(b)** multiply by 5; **(c)** square the number; **(d)** square the number and add 2; **(e)** square the number and subtract 1.

5. The "add 5" rule was used in Exercise 3 above. Consider the "subtract 5" rule for each element in the output in Exercise 3 above and note that in each case the answer is the corresponding element of the input from Exercise 3. This relation between "add 5" and "subtract 5" rules is described by referring to these operations as **inverse operations**; each "undoes" the other.

The inverse operation may be obtained by thinking of the function machine as running backwards from output to input. Find the rule for the inverse operation for the operation:
(a) subtract 2;
(b) add 7;
(c) multiply by 3;
(d) divide by 2.

Operations in which each input consists of a single number are **unary operations**. The following unary operations are indicated with an arrow to show the correspondence or mapping of the given number (input) onto the new number (output). Each operation in the following exercises has been given a name to aid class discussions.

6. This is how we "star" a number.
$$6* \to 8 \qquad 7* \to 9 \qquad 0* \to 2$$
Can you "star" these numbers?
$$5* \to ? \qquad 9* \to ? \qquad 12* \to ?$$

7. This is how we "quote" a number.
$$"5" \to 26 \qquad "3" \to 10 \qquad "7" \to 50$$
Can you "quote" these numbers?
$$"6" \to ? \qquad "10" \to ? \qquad "9" \to ?$$

8. This is how we "bar" a number:
$$\overline{2} \to 7 \qquad \overline{5} \to 16 \qquad \overline{11} \to 34$$
Can you "bar" these numbers?
$$\overline{4} \to ? \qquad \overline{6} \to ? \qquad \overline{12} \to ?$$

Arrows are also often used for binary operations. Try to discover the meaning of the operation $\#$ from the examples given. The binary operation $\#$ has a different meaning in each exercise.

9. $3 \# 4 \to 7,$ $2 \# 0 \to 2,$ $5 \# 1 \to 6,$ $2 \# 2 \to 4.$
10. $2 \# 4 \to 3,$ $5 \# 7 \to 6,$ $8 \# 10 \to 9,$ $0 \# 2 \to 1.$
11. $5 \# 3 \to 9,$ $2 \# 4 \to 7,$ $0 \# 3 \to 4,$ $1 \# 5 \to 7.$
12. $2 \# 3 \to 7,$ $3 \# 4 \to 13,$ $1 \# 5 \to 6,$ $0 \# 3 \to 1.$
*13. $3 \# 4 \to 2,$ $4 \# 3 \to 5,$ $5 \# 1 \to 9,$ $1 \# 1 \to 1.$
*14. $3 \# 5 \to 2,$ $4 \# 5 \to 1,$ $2 \# 8 \to 0,$ $1 \# 6 \to 3.$

15. The talented elementary school student will enjoy exploring abstract operations and their properties. In so doing, he will obtain additional insight into many facets of mathematics. Consider, for example, the binary operation $\odot$ defined for any two counting numbers a and b in this way:

$$a \odot b = 10a + b$$

See how many properties you can find for this operation. For example, explore closure, commutativity, and associativity. What relationship, if any, does this operation have to our decimal system of numeration?

3-2
the distributive property

See if you can discover the meaning of the operations $\#$ and $\otimes$ from the following examples:

$$
\begin{array}{ll}
5 \# 3 = 3 & 8 \otimes 3 = 8 \\
2 \# 8 = 2 & 0 \otimes 3 = 0 \\
3 \# 7 = 3 & 6 \otimes 1 = 6 \\
4 \# 4 = 4 & 3 \otimes 3 = 3
\end{array}
$$

You should have discovered the following meanings of $\#$ and $\otimes$:

$\#$: Select the smaller number of the two if the numbers are different; select that number if both are the same.

$\otimes$: Select the first number of the two.

Thus $5 \# 9 = 5$ because 5 is smaller than 9, whereas $5 \otimes 9 = 5$ because 5 is the first of the two numbers in the given expression. Note that where both numbers were the same we wrote $4 \# 4 = 4$.

We may also combine both operations and determine the meaning of expressions such as

$$a \# (b \otimes c) \quad \text{and} \quad (a \# b) \otimes (a \# c)$$

where a, b, and c represent any of the numbers of ordinary arithmetic. For example, let $a = 3$, $b = 5$, $c = 1$; then the first expression becomes

$$3 \# (5 \otimes 1) = 3 \# 5 = 3.$$

Using the same values in the second expression, we have

$$(3 \# 5) \otimes (3 \# 1) = 3 \otimes 1 = 3.$$

Thus we find

$$3 \# (5 \otimes 1) = (3 \# 5) \otimes (3 \# 1).$$

In general, we may write

$$a \# (b \otimes c) = (a \# b) \otimes (a \# c)$$

for all replacements of *a*, *b*, and *c*. We state, formally, that # is **distributive with respect to** ⊗. This property, known as the **distributive property**, may be somewhat easier to visualize for the numbers of arithmetic.

Consider, for example, the expression $3(5 + 8)$. According to the distributive property, we may evaluate this expression in two different ways and obtain the same answer either way. Thus

$$3(5 + 8) = 3(13) = 39$$
$$(3)(5) + (3)(8) = 15 + 24 = 39$$

so that

$$3(5 + 8) = (3)(5) + (3)(8).$$

This example illustrates the fact that because of the distributive property we may either add first and then multiply, or find the two products first and then add. In general, we say that for all replacements of *a*, *b*, and *c*,

$$a(b + c) = ab + ac.$$

Formally we call this the **distributive property for multiplication with respect to addition,** or simply the *distributive property*. Note that addition is *not* distributive with respect to multiplication since, for example,

$$3 + (5 \times 8) \neq (3 + 5) \times (3 + 8),$$

that is,

$$3 + 40 \neq 8 \times 11.$$

It is the distributive property that allows us, in algebra, to make such statements as:

$$2(a + b) = 2a + 2b$$
$$3(x - y) = 3x - 3y$$

It is the distributive property that elementary school youngsters use in multiplication. Consider the problem 7×43. The distributive property is used by thinking of 7×43 as

$$7 \times (40 + 3)$$

as shown in the example.

$$
\begin{array}{r}
43 \\
\times 7 \\
\hline
21 = 7 \times 3 \\
280 = 7 \times 40 \\
\hline
301 = (7 \times 3) + (7 \times 40)
\end{array}
$$

The distributive property can also be used in developing shortcuts in multiplication. Thus the product 8×99 can be found quickly as follows:

$$8 \times 99 = 8(100 - 1) = 800 - 8 = 792$$

It has been said that the distributive property is one of the unifying themes of mathematics.

EXERCISES

For Exercises 1 through 6, let $ mean "select the larger of the two given numbers," and let $\otimes$ mean "select the first of two given numbers":

1. Find 3 $ 8, 8 $ 3, 3 $\otimes$ 7, 7 $\otimes$ 3.
2. Find 3 $ (5 $ 2), (3 $ 5) $ 2.
3. Find 7 $\otimes$ (2 $\otimes$ 3), (7 $\otimes$ 2) $\otimes$ 3.
4. Find 5 $ (1 $\otimes$ 3), (5 $ 1) $\otimes$ (5 $ 3).
5. Find 3 $\otimes$ (2 $ 7), (3 $\otimes$ 2) $ (3 $\otimes$ 7).
6. Show that $a \otimes (b \, \$ \, c) = (a \otimes b) \, \$ \, (a \otimes c) = a$ for all replacements of a, b, and c. Does $a \, \$ \, (b \otimes c) = (a \, \$ \, b) \otimes (a \, \$ \, c)$?
7. In ordinary arithmetic, is addition distributive with respect to addition? That is, does $a + (b + c) = (a + b) + (a + c)$ for all possible replacements of a, b, and c?
8. In ordinary arithmetic, is multiplication distributive with respect to multiplication? That is, does $a \times (b \times c) = (a \times b) \times (a \times c)$ for all possible replacements of a, b, and c?

Use the distributive property to find each of these products by means of a shortcut:

9. 7×79 10. 5×58

*   1 a replacement for n to make each sentence true:*

11. $8(2 + 3) = 8 \cdot 2 + 8 \cdot n$ 12. $5(3 + 4) = 5 \cdot 3 + n \cdot 4$
13. $7(4 + n) = 7 \cdot 4 + 7 \cdot 5$ 14. $n(5 + 7) = 3 \cdot 5 + 3 \cdot 7$
15. $(6 + 7)5 = 6 \cdot 5 + n \cdot 5$ 16. $(5 + n)3 = 5 \cdot 3 + 9 \cdot 3$
17. $3 \cdot 5 + 3 \cdot 7 = 3(5 + n)$ 18. $4 \cdot 6 + 4 \cdot n = 4(6 + 3)$
19. $3 \cdot n + 7 \cdot n = (3 + 7)n$ 20. $3(4 + n) = 3 \cdot 4 + 3 \cdot n$

PEDAGOGICAL CONSIDERATIONS

Elementary school students use the distributive property of multiplication with respect to addition in many ways. For example:

$$30 + 40 = (3 \times 10) + (4 \times 10) = (3 + 4) \times 10 = 7 \times 10 = 70$$
$$(17 \times 3) + (13 \times 3) = (17 + 13) \times 3 = 30 \times 3 = 90$$

Make a worksheet of 20 problems to help elementary school students recognize the usefulness of this distributive property.

Let us create another finite mathematical system to establish firmly the ideas presented thus far. If it is now 9 P.M. as you begin to read this section, what time will it be in 5 hours? (We hope it won't take you that long to complete your reading!) Do you see that the statement $9 + 5 = 2$ is a correct statement if we are talking about positions on a 12-hour clock?

Let us consider the numerals 1 through 12 on a 12-hour clock as the elements of a set T and consider addition on this clock to be based upon counting in a clockwise direction. Thus to find the sum $9 + 5$ we start at 9 and count 5 units in a clockwise direction to obtain the result 2.

Verify that each of the following is correct:

$$8 + 7 = 3 \qquad \text{(on a 12-hour clock)}$$
$$5 + 12 = 5 \qquad \text{(on a 12-hour clock)}$$
$$3 + 11 = 2 \qquad \text{(on a 12-hour clock)}$$

We may make a table of addition facts on a 12-hour clock as follows:

+	1	2	3	4	5	6	7	8	9	10	11	12
1	2	3	4	5	6	7	8	9	10	11	12	1
2	3	4	5	6	7	8	9	10	11	12	1	2
3	4	5	6	7	8	9	10	11	12	1	2	3
4	5	6	7	8	9	10	11	12	1	2	3	4
5	6	7	8	9	10	11	12	1	2	3	4	5
6	7	8	9	10	11	12	1	2	3	4	5	6
7	8	9	10	11	12	1	2	3	4	5	6	7
8	9	10	11	12	1	2	3	4	5	6	7	8
9	10	11	12	1	2	3	4	5	6	7	8	9
10	11	12	1	2	3	4	5	6	7	8	9	10
11	12	1	2	3	4	5	6	7	8	9	10	11
12	1	2	3	4	5	6	7	8	9	10	11	12

Note that regardless of where we start on the clock, we shall always be at the same place 12 hours later. Thus for any element t of set T we have

$$t + 12 = t \qquad \text{(on a 12-hour clock)}.$$

Let us attempt to define several other operations for this arithmetic on the 12-hour clock. What does multiplication mean? Multiplication by an integer may be considered as repeated addition. For example, 3×5 on the 12-hour clock is equivalent to $5 + 5 + 5$. Since $5 + 5 = 10$ and $10 + 5 = 3$, we know that $3 \times 5 = 3$ on the 12-hour clock.

Verify that each of the following is correct:

$$4 \times 5 = 8 \quad \text{(on the 12-hour clock)}$$
$$3 \times 9 = 3 \quad \text{(on the 12-hour clock)}$$
$$3 \times 7 = 9 \quad \text{(on the 12-hour clock)}$$

The following examples provide further illustrations.

Example 1: Solve the equation $t + 6 = 2$ for t, where t may be replaced by any one of the numerals on a 12-hour clock.

Solution: On the 12-hour clock $8 + 6 = 2$; therefore $t = 8$. Note that we may also solve for t by starting at 6 on the clock and counting, in a clockwise direction, until we reach 2.

Example 2: Using the numerals on a 12-hour clock, find a replacement for t such that $\frac{9}{7} = t$.

Solution: We can think of the statement $\frac{9}{7} = t$ as one of the four equivalent statements:

$$9 = 7 \times t \qquad 9 = t \times 7$$
$$\frac{9}{7} = t \qquad \frac{9}{t} = 7$$

Then use the sentence $9 = 7 \times t$ to solve the problem. How many groups of 7 are needed to produce 9 on the 12-hour clock? From a multiplication table, or by trial and error, we find that $7 \times 3 = 9$ (on the 12-hour clock); thus $t = 3$.

Example 3: What is $3 - 7$ on the 12-hour clock?

Solution: Let t represent a numeral on the 12-hour clock. The statement "$3 - 7 = t$" is equivalent to "$t + 7 = 3$." From the table of addition facts $8 + 7 = 3$ (on the 12-hour clock); thus $t = 8$.

From a slightly different point of view we may solve Example 3 by counting in a clockwise direction from 12 to 3, and then counting 7 units in a counterclockwise direction. We complete this process at 8. Thus $3 - 7 = 8$ (on the 12-hour clock).

EXERCISES

1. Is the set of numbers used in clock arithmetic closed (a) with respect to addition? (b) with respect to multiplication?

2. Make and complete a table of multiplication facts on the 12-hour clock.

3. Do the set of numbers on the clock contain identity elements with respect to addition and multiplication? If so, what are they?

4. List the numbers of clock arithmetic in a column. Beside each number list the inverse with respect to addition and the inverse with respect to multiplication, if one exists. (Thus the inverse of 3, with respect to addition, is 9, whereas 3 does not have an inverse with respect to multiplication.)

Solve each problem as on a 12-hour clock.

5. $8 + 7$ 6. $8 + 11$

7. $9 - 12$ 8. $5 - 9$

9. 3×9 10. 5×5

11. $1 \div 5$ 12. $2 \div 7$

Find all possible replacements for t for which each sentence is a true sentence for the numerals on a 12-hour clock.

13. $t + 9 = 5$ 14. $t - 3 = 11$

15. $8 + t = 2$ 16. $3 - t = 10$

17. $3 \times t = 3$ 18. $7 \times t = 11$

19. $\dfrac{t}{5} = 8$ 20. $\dfrac{t}{7} = 9$

*21. $\dfrac{2}{t} = 3$ *22. $2 + t = 2 - t$

*23. $t + 12 = t$ *24. $3 - t = 5 + t$

Solve each problem as on a 12-hour clock.

25. $9 + 9$ 26. $8 + 6$

27. 5×9 28. 3×8

29. 6×6 30. 8×7

PEDAGOGICAL CONSIDERATIONS

Elementary school students are often taught to consider number facts by families of related facts. For example, here is a family of related number facts:

$$5 - 2 = 3$$
$$5 = 2 + 3$$
$$5 = 3 + 2$$
$$5 - 3 = 2$$

Here is another family of related number facts:

$$6 \div 2 = 3$$
$$6 = 2 \times 3$$
$$6 = 3 \times 2$$
$$6 \div 3 = 2$$

Any two members of a family of related number facts are equivalent statements. Any statement may be used to "solve" or to "check" another statement in the same family.

Use families of number facts to explain your answers to Exercises 7, 8, 11, 12, 14, 16, 19, 20, and 21 of this section.

3-4

modular arithmetic Let us next consider a mathematical system based on a clock for five hours, numbered 0, 1, 2, 3, and 4 as in the accompanying figure.

Addition on this clock may be performed by counting as on an ordinary clock. However, it seems easier to think of addition as rotations in a clockwise direction. Thus $3 + 4$ indicates that we are to start at 0 and move 3 units, then 4 more. The result is 2.

We interpret 0 to mean no rotation as well as to designate a position on the clock. Then we have such facts as:

$$3 + 0 = 3 \quad \text{(on a five-hour clock)}$$
$$4 + 1 = 0 \quad \text{(on a five-hour clock)}$$

Verify that the following table of addition facts on a five-hour clock is correct:

+	0	1	2	3	4
0	0	1	2	3	4
1	1	2	3	4	0
2	2	3	4	0	1
3	3	4	0	1	2
4	4	0	1	2	3

Multiplication on the five-hour clock is defined as repeated addition, as in §3-3. For example, 3×4 on the five-hour clock is equivalent to $4 + 4 + 4$. Since $4 + 4 = 3$, and $3 + 4 = 2$, we see that $3 \times 4 = 2$ on the five-hour clock.

We define subtraction and division using equivalent statements as follows:

Subtraction: Since the statements "$a - b = x$" and "$a = b + x$" are equivalent, we define $a - b = x$ *if and only if* $b + x = a$.

Division: Since the statements "$a \div b = x$" and "$a = b \times x$" are equivalent, we define $a \div b = x$ *if and only if* $b \times x = a$.

Regardless of where we start on this five-hour clock, we shall always be at the same place five hours later. Usually the symbols of a mathematical system based on clock arithmetic begin with 0, and the system is called a **modular arithmetic.** In the illustration of this section we say that we have an arithmetic modulo 5, and we use the symbols 0, 1, 2, 3, 4. Formally, specific facts in this mathematical system are written as follows:

$$3 + 4 \equiv 2 \ (\text{mod } 5), \text{ read "}3 + 4 \text{ is congruent to 2, modulo 5"}$$
$$4 \times 2 \equiv 3 \ (\text{mod } 5), \text{ read "}4 \times 2 \text{ is congruent to 3, modulo 5"}$$

In general, two numbers are **congruent modulo 5** if and only if they differ by a multiple of 5. Thus 3, 8, 13, and 18 are all congruent to each other modulo 5. We may write, for example

$$18 \equiv 13 \ (\text{mod } 5)$$
$$8 \equiv 3 \ (\text{mod } 5)$$

Notice that every integer in ordinary arithmetic is congruent modulo 5 to exactly one element of the set F, where $F = \{0, 1, 2, 3, 4\}$. Indeed, the set F is the set of all possible remainders when any number is divided by 5. The elements of F are the elements of arithmetic modulo 5.

Example 1: Solve for x where x may be replaced by any element in arithmetic modulo 5: $2 - 3 = x$.

Solution: Note that the statement "$2 - 3 = x$" is equivalent to "$3 + x = 2$." Since $3 + 4 = 2$, $x = 4$.

Example 2: Solve for x in arithmetic modulo 5: $\frac{2}{3} = x$.

Solution: Note that the statement "$\frac{2}{3} = x$" is equivalent to "$3 \times x = 2$." Since $3 \times 4 = 2$, $x = 4$.

Let us explore the properties of the mathematical system based upon the set $F = \{0, 1, 2, 3, 4\}$ and the operation addition as defined in this section. The following properties are of special interest.

1. *The set F of elements in arithmetic modulo 5 is closed with respect to addition.* That is, for any pair of elements, there is a unique element which represents their sum and which is also a member of the original set. We note, for example, that there is one and only one entry in each place of the table given in this section, and that each entry is an element of the set F.

2. *The set F of elements in arithmetic modulo 5 satisfies the commutative property for addition.* That is,

$$a + b = b + a$$

where a and b are any elements of the set F. Specifically, we see that $3 + 4 = 4 + 3$, $2 + 3 = 3 + 2$, etc.

3. *The set F of elements in arithmetic modulo 5 satisfies the associative property for addition.* That is,

$$(a + b) + c = a + (b + c)$$

for all elements a, b, and c of the set F.

As a specific example we evaluate $3 + 4 + 2$ in two ways:

$$(3 + 4) + 2 \equiv 2 + 2 \text{ or } 4, \text{ modulo } 5$$
$$3 + (4 + 2) \equiv 3 + 1 \text{ or } 4, \text{ modulo } 5$$

4. *The set F of elements in arithmetic modulo 5 includes an identity element for addition.* That is, the set contains an element 0 such that the sum of any given element and 0 is the given element. That is,

$$0 + 0 = 0, \quad 1 + 0 = 1, \quad 2 + 0 = 2,$$
$$3 + 0 = 3, \quad 4 + 0 = 4.$$

5. *Each element in arithmetic modulo 5 has an inverse with respect to addition.* That is, for each element a of set F there exists an element a' of F such that $a + a' = 0$, the identity element. The element a' is said to be the inverse of a. Specifically, we have the following:

The inverse of 0 is 0,	$0 + 0 \equiv 0 \pmod 5$.
The inverse of 1 is 4,	$1 + 4 \equiv 0 \pmod 5$.
The inverse of 2 is 3,	$2 + 3 \equiv 0 \pmod 5$.
The inverse of 3 is 2,	$3 + 2 \equiv 0 \pmod 5$.
The inverse of 4 is 1,	$4 + 1 \equiv 0 \pmod 5$.

As in §3-1, we may summarize these five properties by saying that the set F of elements in arithmetic modulo 5 forms a commutative group under addition.

EXERCISES

Each of the following is based upon the set of elements in arithmetic modulo 5.

1. Make and complete a table of multiplication facts.

2. Verify that the commutative property for multiplication holds for at least two specific instances.

3. Verify that the associative property for multiplication holds for at least two specific instances.

4. What is the identity element with respect to multiplication?

5. Find the inverse of each element with respect to multiplication.

6. Verify for at least two specific instances that arithmetic modulo 5 satisfies the distributive property for multiplication with respect to addition.

Solve each of the following for x.

7. $2 + x \equiv 1 \pmod 5$

8. $x + 4 \equiv 2 \pmod 5$

9. $x + 3 \equiv 2 \pmod 5$

10. $4 + x \equiv 1 \pmod 5$

11. $1 - 3 \equiv x \pmod 5$

12. $1 - 4 \equiv x \pmod 5$

13. $2 - 4 \equiv x \pmod 5$

14. $3 - x \equiv 4 \pmod 5$

15. $2 \times x \equiv 3 \pmod 5$

16. $4 \times x \equiv 2 \pmod 5$

17. $4 \times x \equiv 1 \pmod 5$

18. $3 \times x \equiv 1 \pmod 5$

19. $\frac{1}{2} \equiv x \pmod 5$

20. $\frac{4}{3} \equiv x \pmod 5$

21. $\frac{3}{x} \equiv 2 \pmod 5$

22. $\frac{2}{x} \equiv 3 \pmod 5$

23. $x + 3 \equiv x \pmod 5$

24. $x + 1 \equiv 3 - x \pmod 5$

SUPPLEMENTARY EXERCISES

1. Consider arithmetic on a clock with four elements, (0, 1, 2, 3), as in the figure. Complete tables for the addition and multiplication facts for arithmetic modulo 4. Explore these systems and list as many properties as you can for each.

2. Use the results of Exercise 1 and tell whether the set of elements in arithmetic modulo 4 forms a commutative group with respect to (a) addition and (b) multiplication. If not, tell which properties are not satisfied under the given operation.

Find all possible replacements for x for which each sentence is a true statement.

3. $x + 5 \equiv 0 \pmod 7$

4. $x - 3 \equiv 2 \pmod 4$

5. $3x \equiv 1 \pmod 7$

6. $x \times x \equiv 1 \pmod 8$

7. $\frac{x}{4} \equiv 3 \pmod 9$

8. $\frac{2}{x} \equiv 3 \pmod 7$

9. $1 - x \equiv 4 \pmod 6$

10. $4 + x \equiv 1 \pmod 7$

11. $x + 5 \equiv 1 \pmod 8$

12. $2 - x \equiv 3 \pmod 6$

13. $2x \equiv 3 \pmod 6$

14. $\frac{x}{3} \equiv 4 \pmod 8$

15. $x - 5 \equiv 2 \pmod 7$

16. $7 + x \equiv 1 \pmod{12}$

17. We describe the property illustrated by the equation $3 \times 4 = 0$ in arithmetic modulo 12, where the product is zero but neither number

is zero, by saying that 3 and 4 are zero divisors. Are there other zero divisors in arithmetic modulo 12? If so, list them.

18. Consider a modulo 7 system where the days of the week correspond to numbers as follows: Monday—0; Tuesday—1; Wednesday—2; Thursday—3; Friday—4; Saturday—5; Sunday—6. Memorial Day, May 30, is the 150th day of the year and falls on a Thursday. In that same year, on what day of the week does July 4, the 185th day of the year, fall? On what day does Christmas, the 359th day of the year fall?

PEDAGOGICAL CONSIDERATIONS

Many of the topics considered in this chapter can be used directly with elementary school children.

1. Consider the mathematical system represented by the table.

×	a	b	c
a	a	a	a
b	a	b	c
c	a	c	b

(a) Is the system closed under multiplication?

(b) Is multiplication commutative?

(c) Identify the multiplicative identity.

(d) Identify the multiplicative inverse of each element that has a multiplicative inverse.

(e) Does the system form a group under multiplication?

2. Consider the mathematical system represented by the table.

+	a	b	c
a	a	b	c
b	b	c	a
c	c	a	b

(a) Is the system closed under addition?

(b) Is addition commutative?

(c) Identify the additive identity.

(d) Identify the additive inverse of each element that has an additive inverse.

(e) Check three of the twenty-seven possible applications of the associative property of addition.

(f) Does the system appear to form a commutative group under addition?

3. Consider the numerals 0, 1, 2, on a three-hour clock. For the arithmetic on a three-hour clock:

(a) Make an addition table.

(b) Does the system appear to form a group under addition?

(c) Make a multiplication table.

(d) Does the system appear to form a group under multiplication?

4. Think of the counterclockwise rotations on the plane of an equilateral triangle PQR about its center. Let R_1 represent a rotation of $0°$, R_2 a rotation of $120°$, and R_3 a rotation of $240°$ as shown in the figures.

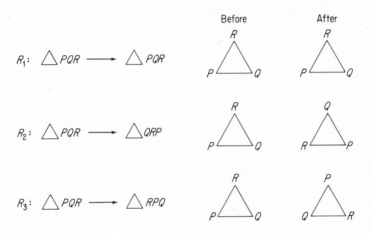

(a) Let multiplication $R_i \times R_k$ represent the successive application of the rotation R_i and then the rotation R_k and complete the following table.

$\times$	R_1	R_2	R_3
R_1	R_1	—	—
R_2	—	R_3	—
R_3	—	—	—

(b) Is the system closed under this operation?

(c) Identify the identity element for this operation.

(d) Compare the completed table with the table in Exercise 2.

(e) Does the system appear to form a group under the given operation?

5. Discuss the similarities of the tables considered in Exercises 2, 3(a), and 4.

6. Here is an exercise that may be used to provide a laboratory type experience for elementary school students. Take an index card and label it in each corner as shown. Also write these same letters in the corresponding corners on the reverse side of the card. Call this initial position *I*. Rotate the card 180° to obtain position *R*. Rotate the card about its horizontal axis, from the initial position, to obtain position *H*. Finally, rotate the card about its vertical axis, from the initial position, to obtain position *V*.

Use the symbol ∼ to mean "followed by." Thus *H* ∼ *R* means a rotation about the horizontal axis followed by a rotation of 180°. You should find *H* ∼ *R* = *V*. Verify also that *V* ∼ *H* = *R*.

Make and complete a table for this system whose horizontal and vertical headings are *I*, *R*, *H*, and *V*. List as many properties for this system as you can find.

7. This exercise may also be used to provide a laboratory type experience. Consider a soldier facing in a given direction. He is then given various commands, such as "right face," "left face," "about face," and "as you were." The last command tells him to retain whatever position he may be in at the time.

A few specific examples of his movements should clarify matters. If he is facing to the north and is given the command "about face," then in his new position he is facing to the south. If we let *R* represent the command "right face," *A* represent the command "about face," and *L* represent the command "left face,"then *R* followed by *A* is equivalent to the single command *L*.

Use the symbol ∘ to represent the operation "followed by" and verify that each of the following is correct:

$$L \circ L = A$$
$$L \circ A = R$$
$$A \circ R = L$$

Now suppose that our soldier makes a left turn and then we wish him to remain in that position. We make use of the command "as you were," represented by *E*. Thus *L* ∘ *E* = *L*; also *E* ∘ *L* = *L*.

(a) Make a table summarizing all possible movements, using the headings *E*, *R*, *L*, and *A*.

(b) Find $R \circ R, A \circ A, L \circ L$.

(c) Does $R \circ L = L \circ R$?

(d) Does $R \circ (L \circ A) = (R \circ L) \circ A$?

(e) Do the set of given commands appear to be commutative and associative with respect to the operation $\circ$?

(f) Find the identity element with respect to $\circ$.

(g) List the inverse of each of the elements of the set.

(h) Is the set of commands closed with respect to $\circ$?

CHAPTER TEST

Refer to the table below to answer Exercises 1 through 4:

$\otimes$	#	*	Σ	Δ
#	Σ	Δ	#	*
*	Δ	#	*	Σ
Σ	#	*	Σ	Δ
Δ	*	Σ	Δ	#

1. Which element is the identity element?

2. What is the inverse of Δ?

3. What is $(\# \otimes \Delta) \otimes *$?

4. For all replacements for a and b for the set $Y = \{\#, *, \Sigma, \Delta\}$, $a \otimes b$ is an element of Y. What property does this illustrate?

Solve as on a 12-hour clock:

5. (a) $9 + 7$ (b) 5×11

6. (a) $2 - 7$ (b) $2 \div 5$

Find all possible replacements for x for which each sentence is a true statement:

7. (a) $3 + x \equiv 1 \pmod 5$ (b) $2 - x \equiv 4 \pmod 5$

8. (a) $x + 2 \equiv 0 \pmod 7$ (b) $3 - x \equiv 5 \pmod 8$

9. (a) $3x \equiv 2 \pmod 6$ (b) $\dfrac{x}{3} \equiv 2 \pmod 5$

10. Use the distributive property to evaluate $3(2 + 4)$ in arithmetic modulo 5 in two different ways.

4

Systems
of
Numeration

We habitually take for granted the use of our system of numeration, as well as our computational procedures. However, these represent the creative work of man through the ages. We can gain a better appreciation of our system of numeration and methods of computation by examining other systems.

In this chapter we shall travel through the years and first examine the system of numeration used by the ancient Egyptians as found in their hieroglyphics. Then we shall move to contemporary times and study binary notation, the mathematical basis for the modern computer.

Throughout this exploration the reader should note the liberation in mankind's mathematical thinking that has taken place from early times to the present day, a liberation that matches his creative developments in other areas as well. Systems of numeration form an important part of each educated person's knowledge and provide the basis for many topics that are currently emphasized in elementary and secondary school mathematics curricula.

4-1

numbers without numerals

Our number concepts are based upon sets. We have considered equivalent sets as sets that have the same number of elements. We can also match sets to determine whether one set has more or less elements than another. Early man spent many years finding ways of indicating how many elements there were in a set. Undoubtedly just as you and I might hold up three fingers to indicate a set of three elements, early man used matching sets. For example, he may have used sets of pebbles to keep track of the number of sheep that he has.

There is evidence that members of the Bugilai tribe in British New Guinea indicated numbers by pointing in order to that number of parts of the body. The pointing is reported to have been done by the right index finger to these parts of the body in order:

left-hand little finger	1	tarangesa
left-hand ring finger	2	meta kina
left-hand middle finger	3	guigimeta
left-hand index finger	4	topea
left-hand thumb	5	manda
left wrist	6	gaben
left elbow	7	trankgimbe
left shoulder	8	podei
left breast	9	ngama
right breast	10	dala

Note that no words or numerals were needed for this sort of communication. Later when number words were developed in the Bugilai dialect, they had original meanings from the parts of body as indicated in the second and third columns of the array. Other tribes developed similar procedures and some tribes could designate numbers up to 31 using parts of the body. One tribe is reported to have been able to represent all numbers up to 100.

Parts of the body could be used to point out matching sets for small sets of elements. As soon as man started thinking of a part of the body as representing a set, such as a hand representing a set of five elements, he was starting to develop numerals. The use of a set of parts of the body to match one-to-one a specified set was a use of numbers without numerals. Only the matching was involved. There was a set of five fingers but man had not reached the stage of thinking of the set of five fingers as a single symbol, a hand, to represent five. For the rest of this chapter we shall be exploring ways of representing numbers by symbols, that is, systems of numeration.

4-2

**Egyptian
numeration**

A number is an abstract concept. A **numeral** is a name or symbol for a number. We write numerals; we can not write numbers. From a similar point of view, when you see the word "dog" on a piece of paper, you see the name of an animal rather than the animal itself. We shall make this distinction between a number and a numeral in this book only where it proves helpful in understanding some concept.

Many early people used strokes to represent numbers:

The number of strokes that a person can recognize as a distinct number varies from person to person. Somewhere around 5, 6, 7, or 8 we start seeing "many" strokes instead of a specific number of strokes. Several primitive tribes had number languages that consisted only of

one   two   many

or perhaps

one   two   three   many.

The early Egyptians used groupings of strokes so that they could recognize the number of strokes for 1 through 9.

I,   II,   III,   IIII,   III II,   III III,   III IIII,   IIIIIIII,   III III III

Other patterns were also used. Since strokes were either present or not present, they could be arranged horizontally, vertically, or in both rows and columns. For example, consider these groupings:

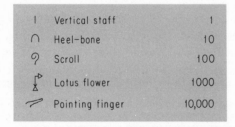

Probably because man has ten fingers a new symbol is often introduced for ten. We would write this symbol in our notation as 10. The ancient Egyptians used the symbol ∩. The ancient Babylonians used the symbol ❬. The Romans used X. All of these symbols—and there are others—are merely numerals; that is, they are different ways of representing (providing a name for) the same number.

Let us explore one system, the one used by the ancient Egyptians, in greater detail. They used a new symbol for each power of ten. Here are some of their symbols, a description of the physical objects they are supposed to represent, and the number they represent as expressed in our notation:

I	Vertical staff	1
∩	Heel-bone	10
⌐	Scroll	100
⌐	Lotus flower	1000
⌐	Pointing finger	10,000

This Egyptian system is said to have a **base** of ten, but has no place value. The "base" ten is due to the use of powers of ten. We call our system of numeration a **decimal system** to emphasize our use of powers of ten. The absence of a place value means that the position of the symbol does not affect the number represented. For example, in our decimals system of numeration, 23 and 32 represent different numbers. In the Egyptian system ∩| and |∩ are different ways of writing eleven, that is, different names or numerals for the same number. (The former notation is the one normally found in their hieroglyphics.) Here are some other comparisons of decimal and ancient Egyptian number symbols.

Computation in the ancient Egyptian system is possible, although tedious. For example, we use these steps to add 27 and 35.

Observe that, in this Egyptian system, an indicated collection of ten ones was replaced by a symbol for ten before the final computation took place. In our decimal system we mentally perform a similar exchange of ten ones for a ten when we express $(7 + 5)$ as one ten and two ones. We exchange kinds of units in a similar manner in subtraction.

The ancient Egyptians multiplied by a process of doubling. This process is based on the fact that any number may be represented as a sum of powers of two. For example, $19 = 1 + 2 + 16$. To find the product 19×25 we first double 25 as follows:

(1) x 25 = (25)

(2) x 25 = (50)

4 x 25 = 100

8 x 25 = 200

(16) x 25 = (400)

Then we use the distributive property and find the product 19×25 by adding the multiples of 25 that correspond to 1, 2, and 16:

$$19 = 1 + 2 + 16$$
$$19 \times 25 = (1 + 2 + 16) \times 25$$
$$= 25 + 50 + 400 = 475$$

Example 1: Use the Egyptian method of multiplication to find the product 23×41.

Solution:

$$23 = 1 + 2 + 4 + 16$$
$$23 \times 41 = (1 + 2 + 4 + 16) \times 41$$
$$= 41 + 82 + 164 + 656 = 943$$

Later the Egyptians adopted a more refined and automatic procedure for multiplication known as **duplation and mediation** that involves doubling one factor and halving the other. For example, to find the product 19×25 we may successively halve 19, discarding remainders at each step, and successively double 25 as follows:

Note: since all remainders are discarded, one-half of 19 is recorded as 9 and one-half of 9 is recorded as 4.

This process is complete when a 1 appears in the column of numbers which are being halved. Opposite each number in this column of halves there is a corresponding number in the column of numbers being doubled. The product 19×25 is found as the sum of the numbers that are opposite the *odd* numbers in the column of halves:

$$19 \times 25 = 25 + 50 + 400 = 475$$

Note that this process automatically selects the addends to be used in determining the product; one need not search for the appropriate powers of two to be used.

Example 2: Use the process of duplation and mediation to find the product 23 × 41.

Solution:

$$23 \longrightarrow \boxed{41}$$
$$11 \longrightarrow \boxed{82}$$
$$5 \longrightarrow \boxed{164} \qquad 23 \times 41 = 41 + 82 + 164 + 656 = 943$$
$$2 \qquad 328$$
$$1 \longrightarrow \boxed{656}$$

EXERCISES

Write in ancient Egyptian notation.

1. 25
2. 138
3. 1426
4. 40
5. 12,407
6. 5723

Write in decimal notation.

7. ∩∩ ı ı

9. 𝑓◁ 𝟫 ı ı

11. 𝑓◁ 𝟫𝟫𝟫 ∩∩ ı ı ı ı

8. 𝟫𝟫 ∩ ı

10. 𝑓◁𝑓◁ 𝟫𝟫 ∩ ı

12. 𝑓◁ 𝟫𝟫 ∩ ı ı

Write in ancient Egyptian notation and perform the indicated operation in that system.

13.   42
    +21

14.   153
    +62

15.   238
    +135

16.   431
    −213

17.   1243
    −137

18.   507
    −124

Use the Egyptian method of doubling to find these products.

19. 17 × 45
20. 15 × 35
21. 27 × 31
22. 31 × 19

Use the Egyptian method of duplation and mediation and find these products.

23. 19 × 33   *24.* 24 × 51

25. 21 × 52   *26.* 37 × 80

27. The Egyptians did not have a symbol for zero. Why do we need such a symbol in our system of numeration? Why did they find it unnecessary to invent such a symbol in order to have symbols for other numbers?

PEDAGOGICAL CONSIDERATIONS

1. Develop a short unit on Roman numerals for use with an elementary school class.

2. Select an elementary school grade level and for students in that grade describe briefly the system of numeration used by **(a)** the Babylonians about 2000 B.C. (using sixty); **(b)** the Greeks about 400 B.C. (using letters from their alphabet); **(c)** the Chinese about 200 B.C. (using sticks laid on a table); **(d)** the Mayans about 300 A.D.

3. Compare several numeration systems according to their use of addition (add the numbers represented by the symbols used), subtraction, multiplication, and place value.

4. The School Mathematics Study Group (S.M.S.G.) is a nonprofit curriculum group financed by the National Science Foundation. During the 1960's, S.M.S.G. exerted considerable influence on the development of the mathematics curriculum for grades K through 12 and published experimental textbooks at each of these grade levels. Try to locate and examine some of these books written for the elementary grades. In particular, write up a brief description of the use of hand numerals in their first grade text.

5. Elementary school textbooks vary greatly in their emphasis upon the distinction between numbers and numerals. The following questions may be used to identify some of the difficulties and ways of handling them. For example, a common practice is to use single quotes to refer to a numeral. Thus '2' refers to the numeral rather than the number, whereas 2 without quotes refers to the number.

   (a) Which is larger, 3 or 7 ?

   (b) Which is larger, '3' or '7' ?

   (c) Comment upon this statement: One half of 22 is 11, but one half of '22' is '2'.

There exist numerous examples of the ways in which ancient man performed his computations. Some of these may prove of interest to the reader. The method of multiplication that appeared in one of the first published arithmetic texts in Italy, the *Treviso Arithmetic* (1478), is an interesting one. Let us use it to find the product of 457 and 382. (We shall refer to this process here as "galley" multiplication, although it was called "Gelosia" multiplication in the original text.)

First prepare a "galley" with three rows and three columns, and draw the diagonals as in the figure. Our choice for the number of rows and columns is based on the fact that we are to multiply two three-digit numerals.

Place the digits 4, 5, and 7 in order from left to right at the top of the columns. Place the digits 3, 8, and 2 in order from top to bottom at the right of the rows. Then each product of a digit of 457 and a digit of 382 is called a **partial product** and is placed at the intersection of the column and row of the digits. The diagonal separates the digits of the partial product (tens digit above units digit). For example, $3 \times 7 = 21$, and this partial product is placed in the upper right-hand corner of the "galley"; $5 \times 8 = 40$, and this partial product is placed in the center of the galley; $4 \times 2 = 8$, and this partial product is entered as 08 in the lower left-hand corner of the galley. See if you can justify each of the entries in the completed array shown below.

After all partial products have been entered in the galley, we add along diagonals, starting in the lower right-hand corner and carrying to the next diagonal sum where necessary. The next diagram indicates this pattern.

The completed problem appears in the form shown below. We read the final answer, as indicated by the arrow in the figure, as 174,574. Note that we read the digits in the opposite order to that in which they were obtained.

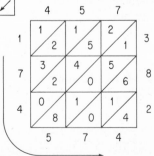

Example: Use galley multiplication and multiply 372 by 47.

Solution:

Answer: 17,484

This procedure works because we are really listing all partial products before we add. Compare the following two computations.

$$
\begin{array}{r}
872 \\
\times\,685 \\
\hline
10 \\
35 \\
40 \\
16 \\
56 \\
64 \\
12 \\
42 \\
48 \\
\hline
597{,}320
\end{array}
$$

$\left.\begin{array}{} \\ \end{array}\right\}$ 5 x 872

$\left.\begin{array}{} \\ \end{array}\right\}$ 80 x 872

$\left.\begin{array}{} \\ \end{array}\right\}$ 600 x 872

597,320 = 685 x 872

Note that the numerals along the diagonals correspond to those in the columns at the right.

The English mathematician, John Napier, made use of this system as he developed what proved to be one of the forerunners of the modern computing machines. His device is referred to as **Napier's rods,** or **Napier's bones,** named after the material on which he had numerals printed. Napier (1550–1617) is often spoken of as the inventor of logarithms.

To make a set of these rods we need to prepare a collection of strips of paper, or other material, with multiples of each of the digits listed. Study the set of rods shown on page 125 as Figure (A).

Note, for example, that the rod headed by the numeral 9 lists the multiples of 9:

$$9, 18, 27, 36, 45, 54, 63, 72, \text{ and } 81$$

We can use these rods to multiply two numbers. To multiply 7 × 483, place the rods headed by numerals 4, 8, and 3 alongside the index as shown in Figure (B).

124

Index	0	1	2	3	4	5	6	7	8	9
1	0/0	0/1	0/2	0/3	0/4	0/5	0/6	0/7	0/8	0/9
2	0/0	0/2	0/4	0/6	0/8	1/0	1/2	1/4	1/6	1/8
3	0/0	0/3	0/6	0/9	1/2	1/5	1/8	2/1	2/4	2/7
4	0/0	0/4	0/8	1/2	1/6	2/0	2/4	2/8	3/2	3/6
5	0/0	0/5	1/0	1/5	2/0	2/5	3/0	3/5	4/0	4/5
6	0/0	0/6	1/2	1/8	2/4	3/0	3/6	4/2	4/8	5/4
7	0/0	0/7	1/4	2/1	2/8	3/5	4/2	4/9	5/6	6/3
8	0/0	0/8	1/6	2/4	3/2	4/0	4/8	5/6	6/4	7/2
9	0/0	0/9	1/8	2/7	3/6	4/5	5/4	6/3	7/2	8/1

(A)

Index	4	8	3
1	0/4	0/8	0/3
2	0/8	1/6	0/6
3	1/2	2/4	0/9
4	1/6	3/2	1/2
5	2/0	4/0	1/5
6	2/4	4/8	1/8
7	2/8	5/6	2/1
8	3/2	6/4	2/4
9	3/6	7/2	2/7

(B)

Consider the row of numerals alongside 7 on the index.

Add along the diagonals, as in "galley multiplication."

7 x 483 = 3381

The same arrangement of rods may be used to read immediately the product of 483 and any other one-digit number. With practice one can develop skill in using these rods for rapid computation. For example, 8 × 483 = 3864 as follows:

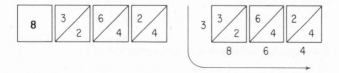

Combining the two previous results, we can find the product 87 × 483:

$$
\begin{array}{rl}
483 & 7 \times 483 = 3{,}381 \\
\underline{\times 87} & \underline{80 \times 483 = 38{,}640} \\
& 87 \times 483 = 42{,}021
\end{array}
$$

Note that we are able to read only products with a one-digit multiplier directly from the rods.

EXERCISES

Multiply, using the "galley" method.

1. 492
   ×37

2. 568
   ×429

3. 432
   ×276

4. 4876
   ×27

5. 7025
   ×398

6. 5081
   ×2376

Construct a set of Napier's rods and use them to find each of the following products.

7. 256
   ×8

8. 427
   ×9

9. 387
   ×5

10. 592
    ×7

11. 7256
    ×8

12. 427
    ×36

PEDAGOGICAL CONSIDERATIONS

1. Write up a lesson plan on galley multiplication for use in an elementary school class (specify the grade that you are considering).

2. Repeat Exercise 1 for multiplication using Napier's rods.

3. Prepare a demonstration set of Napier's rods for use in front of a class or prepare a set for use on an overhead projector.

We now turn our attention to the **decimal system of numeration**. It is a decimal system in that it is based on powers or groups of tens. Furthermore, it has place value in that the value of any digit used depends upon the position which it occupies. Thus, the two numerals 4 in 484 have quite different values.

To illustrate this latter concept we will write this number, 484, in what is known as **expanded notation**:

$$484 = 4 \text{ hundreds} + 8 \text{ tens} + 4 \text{ ones}$$
$$= (4 \times 100) + (8 \times 10) + (4 \times 1)$$

Note that one of the numerals 4 represents 4 hundreds, whereas the other 4 represents 4 units, that is, 4 ones.

It is convenient to use exponents when one is writing a number in expanded notation. An **exponent** is a number that tells how many times another number, called the base, is used as a factor in a product. For example, in the expression 7^2, the numeral 2 is the exponent and 7 is the base. Note that $7^2 = 7 \times 7 = 49$; also $7^3 = 7 \times 7 \times 7 = 343$.

Using exponents to write 484 in expanded notation, we have

$$484 = (4 \times 100) + (8 \times 10) + (4 \times 1)$$
$$= (4 \times 10^2) + (8 \times 10) + (4 \times 1).$$

Zero plays an important role in place value notation. Indeed, it is the use of the symbol 0 that allows us to write numbers as large as we wish in decimal notation, using only the ten digits. In Egyptian numeration, ∩ represents 10, ⌒ represents 100, ⌇ represents 1000, etc. Each new power of 10 requires a new symbol, whereas in our decimal system the symbol 0 allows us to write 10, 100, 1000, etc., using only the digits 0 and 1.

Example: Write 2306 in expanded notation.

Solution:

$$2306 = (2 \times 10^3) + (3 \times 10^2) + (0 \times 10) + (6 \times 1)$$

We may use the exponent 1 to indicate that a number is to be used as a factor only once; thus $10^1 = 10$. We define $10^0 = 1$. Using these exponents, we have

$$2306 = (2 \times 10^3) + (3 \times 10^2) + (0 \times 10^1) + (6 \times 10^0).$$

Any whole number N less than 10,000 can be expressed in the form

$$N = 1000T + 100h + 10t + u$$

where $\{T, h, t, u\} \subseteq \{0, 1, 2, 3, 4, 5, 6, 7, 8, 9\}$. Then

$$\frac{N}{2} = \frac{1000T}{2} + \frac{100h}{2} + \frac{10t}{2} + \frac{u}{2} = 500T + 50h + 5t + \frac{u}{2}.$$

Thus $N/2$ is a whole number if and only if $u/2$ is a whole number; that is, N is divisible by 2 if and only if u is divisible by 2.

$$\frac{N}{3} = \frac{(999+1)T}{3} + \frac{(99+1)h}{3} + \frac{(9+1)t}{3} + \frac{u}{3}$$

$$= 333T + 33h + 3t + \frac{T+h+t+u}{3}$$

Thus $N/3$ is a whole number if and only if $(T+h+t+u)/3$ is a whole number; that is, N is divisible by 3 if and only if the sum of its decimal digits is divisible by 3.

$$\frac{N}{4} = \frac{1000T}{4} + \frac{100h}{4} + \frac{10t}{4} + \frac{u}{4} = 250T + 25h + \frac{10t+u}{4}$$

Thus $N/4$ is a whole number if and only if $(10t+u)/4$ is a whole number; that is, N is divisible by 4 if and only if the number represented by its tens and units digits is divisible by 4.

According to these rules 1972 is divisible by 2 since its units digit 2 is divisible by 2; 1972 is not divisible by 3 since the sum of its digits is 19 and 19 is not divisible by 3, 1972 is divisible by 4 since 72 is divisible by 4. In the case of divisibility by 3 the process may be continued as follows: 1972 is not divisible by 3 since the sum of its digits is 19, the sum of the digits of 19 is 10, the sum of the digits of 10 is 1, and 1 is not divisible by 3.

Other explanations (proofs) and examples of divisibility rules are considered in the exercises.

EXERCISES

Write in expanded notation:

1. 237

2. 732

3. 4205

4. 2450

5. 125,200

6. 201,056

7. 600,100

8. 100,070

Write in decimal notation:

9. $(3 \times 10^3) + (2 \times 10^2) + (5 \times 10^1) + (3 \times 10^0)$

10. $(2 \times 10^5) + (4 \times 10^4) + (1 \times 10^3) + (3 \times 10^2)$
$$+ (5 \times 10^1) + (7 \times 10^0)$$

11. $(4 \times 10^5) + (3 \times 10^3) + (2 \times 10^2) + (5 \times 10^0)$

12. $(7 \times 10^6) + (8 \times 10^5) + (5 \times 10^4)$

13. $(9 \times 10^6) + (7 \times 10^0)$

14. (5×10^7)

15. (4×10^9)

16. (3×10^8)

Test each number for divisibility by (a) 2, (b) 3, (c) 4, (d) 5, (e) 6, (f) 8, (g) 9,
25, (i) 100:

17. 5280   18. 144   19. 1728

20. 16,527   21. 27,540   22. 19,678

23. 64,000   24. 37,600   25. 25,640

For decimal notation try to find a rule for divisibility by:

26. 5   27. 6   28. 8

29. 9   30. 25   31. 100

PEDAGOGICAL CONSIDERATIONS

Use expanded notation to explain the algorithm used for each exercise:

1. 125
   +371
   ────
    496

2. 753
   +271
   ────
   1024

3. 562
   −332
   ────
    230

4. 456
   −275
   ────
    181

5. (a) 231
       ×23
       ────
       693
      4620
      ────
      5313

   (b) 231
       ×23
       ────
       693
       462
       ────
       5313

Look at two or more series of elementary school textbooks and observe:

6. The grade levels at which decimal notation is introduced.

7. The manner in which decimal notation is developed.

4-5

**other
systems
of
numeration**

In our decimal system objects are grouped and counted in tens and powers of ten. For example, the diagram at the top of page 130 shows how one might group and count 134 items. We could, however, just as easily group sets of items in other ways.

$$134 = (1 \times 10^2) + (3 \times 10) + (4 \times 1)$$

In the next figure we see 23 asterisks grouped in three different ways.

2 tens + 3 ones   4 fives + 3 ones   2 nines + 5 ones

If we use a subscript to indicate our manner of grouping, we may write many different numerals (names) for the number of items in the same collection:

$$23_{\text{ten}} \qquad = \qquad 43_{\text{five}} \qquad = \qquad 25_{\text{nine}}$$

(2 *tens* + 3 ones)   (4 *fives* + 3 ones)   (2 *nines* + 5 ones)

43_{five} is read "four three, base five."

25_{nine} is read "two five, base nine."

Each of these numerals represents the number of asterisks in the same set of asterisks. Still another numeral for this number is 35_{six}:

$$35_{\text{six}} = 3 \ sixes + 5 \ \text{ones} = 18 + 5 = 23$$

We call our decimal system of numeration a **base ten** system; when no subscript is used, the numeral is understood to be expressed in base ten. When we group by fives, we have a **base five** system of numeration; that is, we name our system of numeration by the manner in which the grouping is accomplished.

Example 1: Draw a diagram for 18 objects and write the corresponding numeral (a) in base five and (b) in base eight notation.

Solution:

33_{five}

(3 fives + 3 ones)

22_{eight}

(2 eights + 2 ones)

Example 2: Write, in base six notation, a numeral for 27.

Solution: We note that $27 = 4$ sixes $+ 3$ ones; thus $27 = 43_{six}$.

EXERCISES

Write numerals for each of the following collections in the bases indicated by the manner of grouping.

1.
```
( * * * * )
( * * * * )
  * *
```

2.
```
( * * * * * * )
( * * * * * * )
( * * * * * * )
    * *
```

3.
```
( * * * )
( * * * )
( * * * )
```

4.
```
( * * * )
( * * * )
    *
```

Draw a diagram to show the meaning of each of the following.

5. 24_{five}

6. 23_{six}

7. 25_{seven}

8. 32_{nine}

9. 23_{four}

10. 12_{three}

Change to base ten notation.

11. 43_{five}

12. 24_{seven}

13. 32_{eight}

14. 51_{six}

15. 34_{five}

16. 32_{four}

17. Write, in base five notation, a numeral for 17.

18. Write, in base eight notation, a numeral for 43.

19. Write, in base six notation, a numeral for 20.

*20. Change 132_{five} to base ten notation.

PEDAGOGICAL CONSIDERATIONS

1. For base 2 notation try to find a rule for divisibility by **(a)** 10_2, that is, 2; **(b)** 100_2, that is, 4; **(c)** 1000_2, that is, 8.

2. For base 3 notation try to find a rule for divisibility by **(a)** 10_3, that is, 3; **(b)** 100_3, that is, 9; **(c)** 2.

131

3. For base 4 notation try to find a rule for divisibility by **(a)** 2; **(b)** 10_4, that is, 4; **(c)** 3.

4. Look at two or more series of elementary school textbooks and observe the uses that are made of other systems of notation.

5. Prepare a five-minute report that can be used to convince an inquiring parent of the value of teaching other systems of numeration to elementary school students.

4-6

**base
five
notation**

In this section we will explore, in some detail, the manner of writing numerals in another number base. We will work with a base five system since it is convenient to think of numbers written in base five notation in terms of hands and fingers. Thus 23_{five} may be thought of in terms of two hands and three fingers; that is, as 2 fives and 3 ones.

For convenience we will write all numbers in base five notation using the numeral 5 as a subscript, such as 23_5. As usual, numbers written without a subscript will be assumed to be in base ten notation. Thus it is correct to write

$$23_5 = 13.$$

Recall that these are merely two different names for the same number. (Two fives + three ones represent the same number as one ten + three ones.)

We can draw a diagram to show the meaning of 23_5 by drawing two groups of five and three ones as shown at the left.

In a similar manner we picture 123_5 as one group of 25, two groups of 5, and three ones. Note that the group of 25 represents five groups of 5, that is, one group of 5^2.

In the decimal system of notation we write numbers in terms of powers of ten using digits 0, 1, 2, 3, 4, 5, 6, 7, 8, and 9. For example,

$$234 = (2 \times 10^2) + (3 \times 10^1) + (4 \times 10^0).$$

In base five notation we write numbers in terms of powers of five using digits 0, 1, 2, 3, and 4. For example:

$$43_5 = (4 \times 5^1) + (3 \times 5^0)$$
$$324_5 = (3 \times 5^2) + (2 \times 5^1) + (4 \times 5^0)$$
$$2143_5 = (2 \times 5^3) + (1 \times 5^2) + (4 \times 5^1) + (3 \times 5^0)$$

Note that $5^0 = 1$.

The numbers 1 to 30 are written in the base five notation in the following table.

Base 10	Base 5	Base 10	Base 5	Base 10	Base 5
1	1	11	21_5	21	41_5
2	2	12	22_5	22	42_5
3	3	13	23_5	23	43_5
4	4	14	24_5	24	44_5
5	10_5	15	30_5	25	100_5
6	11_5	16	31_5	26	101_5
7	12_5	17	32_5	27	102_5
8	13_5	18	33_5	28	103_5
9	14_5	19	34_5	29	104_5
10	20_5	20	40_5	30	110_5

To translate a number that is not listed in the table from base 5 notation to base 10 notation, express the number in terms of powers of 5 and simplify.

Example 1: Write 3214_5 in base 10 notation.

Solution:

$$3214_5 = (3 \times 5^3) + (2 \times 5^2) + (1 \times 5^1) + (4 \times 5^0)$$
$$= (3 \times 125) + (2 \times 25) + (1 \times 5) + (4 \times 1)$$
$$= 434$$

To translate from base 10 to base 5, any one of several procedures may be adopted. Consider the problem

$$339 = (\qquad)_5.$$

When a number is expressed to the base 5, it is written in terms of powers of 5:

$$5^0 = 1, \quad 5^1 = 5, \quad 5^2 = 25, \quad 5^3 = 125, \quad 5^4 = 625, \quad \ldots$$

The highest power of 5 that is not greater than the given number is 5^3. This power of 5, namely $5^3 = 125$, can be subtracted from 339 twice. Then the remainder 89 is positive and less than 125.

$$
\begin{array}{r}
339 \\
-125 \\
\hline
214 \\
-125 \\
\hline
89
\end{array}
$$

Thus, we write 2×5^3 in the expansion of 339 to the base 5.

The next power of 5 is 5^2. This number can be subtracted from 89 three times to obtain a nonnegative remainder less than 25.

$$
\begin{array}{r}
89 \\
-25 \\
\hline
64 \\
-25 \\
\hline
39 \\
-25 \\
\hline
14
\end{array}
$$

Thus, we write 3×5^2 in the expansion.

Finally, we subtract 5 from 14 twice, write 2×5 in the expansion, and obtain 4 as a remainder.

$$
\begin{array}{r}
14 \\
-5 \\
\hline
9 \\
-5 \\
\hline
4
\end{array}
$$

$$
\begin{aligned}
339 &= 2(125) + 3(25) + 2(5) + 4 \\
&= (2 \times 5^3) + (3 \times 5^2) + (2 \times 5^1) + (4 \times 5^0) = 2324_5
\end{aligned}
$$

A group of 339 elements can be considered as two groups of 125 elements, three groups of 25 elements, two groups of 5 elements, and 4 elements.

An alternative procedure for changing 339 to the base 5 depends upon successive division by 5:

$$
\begin{aligned}
339 &= 67 \times 5 + 4 \\
67 &= 13 \times 5 + 2 \\
13 &= 2 \times 5 + 3
\end{aligned}
$$

Next, substitute from the third equation into the second. Then substitute from the second equation to the first, and simplify as follows:

$$
\begin{aligned}
13 &= 2 \times 5 + 3 \\
67 &= 13 \times 5 + 2 = (2 \times 5 + 3) \times 5 + 2 \\
339 &= 67 \times 5 + 4 = [(2 \times 5 + 3) \times 5 + 2] \times 5 + 4 \\
&= (2 \times 5^3) + (3 \times 5^2) + (2 \times 5^1) + (4 \times 5^0) = 2324_5
\end{aligned}
$$

The arithmetical steps involved in these computations can be performed as shown in the following array (often called an *algorithm*).

$$
\begin{array}{r|l}
5 & 339 \\
\cline{2-2}
5 & 67 \;-\!\!-\; 4 \\
\cline{2-2}
5 & 13 \;-\!\!-\; 2 \\
\cline{2-2}
5 & 2 \;-\!\!-\; 3 \\
\cline{2-2}
& 0 \;-\!\!-\; 2
\end{array}
$$

Read upward as 2324_5.

Note that the remainder is written after each division by 5. Then the remainders are read in reverse order to obtain the expression for the number to the base 5. This procedure works for integers only, not for fractional parts of a number.

Example 2:   Write 423 in base 5 notation.

Solution:

$$
\begin{array}{r|l}
5 & 423 \\
5 & 84 \;\text{——}\; 3 \\
5 & 16 \;\text{——}\; 4 \\
5 & 3 \;\text{——}\; 1 \\
& 0 \;\text{——}\; 3
\end{array}
$$

Answer: 3143_5

Check:

$$3143_5 = (3 \times 5^3) + (1 \times 5^2) + (4 \times 5^1) + (3 \times 5^0)$$
$$= 375 + 25 + 20 + 3 = 423$$

After computations in other bases have been considered, the method of successive division by the new base may be used in changing from one base to another. For example, we will be able to use this procedure to change from base 5 to base 10, successively dividing by 20_5. The computation must be done in base 5 notation. For the present we may change from base 5 to base 10 by expressing the number in terms of powers of 5.

EXERCISES

Write each number in decimal notation.

1. 423_5
2. 230_5
3. 444_5
4. 4230_5
5. 321_5
6. 3403_5
7. 1031_5
8. 342_5
9. 1314_5
10. 1423_5

Write each number in base 5 notation.

11. 382
12. 293
13. 782
14. 394
15. 625
16. 917
17. 137
18. 858
19. 507
20. 1000

Extend the concepts of this section and write each number in decimal notation.

*21. 437_8

*22. 2013_4

*23. 1011_2

*24. 321_{12}

*25. 132_{20}

*26. 312_6

*27. 214_{15}

*28. 1352_8

PEDAGOGICAL CONSIDERATIONS

1. For base 5 notation try to find rules for divisibility by (a) 10_5, that is, 5; (b) 4; (c) 2.
2. Prepare a brief introduction to base 5 notation using as many ways as possible to create interest.
3. Prepare a calendar for one month using base 5 numerals for the dates of the month.

4-7

other number bases

The base 5 system has been explored in detail merely for illustrative purposes; any other positive integer greater than 1 would have served just as well as a base. For each base N, the digits used are $0, 1, 2, 3, \ldots, N - 1$. For each base, powers of that base are used as place values for the digits. Consider, for example, the following numerals for the numbers 0 through 10 in several different bases.

Base 10	Base 3	Base 4	Base 5	Base 6	Base 7	Base 8
1	1	1	1	1	1	1
2	2	2	2	2	2	2
3	10_3	3	3	3	3	3
4	11_3	10_4	4	4	4	4
5	12_3	11_4	10_5	5	5	5
6	20_3	12_4	11_5	10_6	6	6
7	21_3	13_4	12_5	11_6	10_7	7
8	22_3	20_4	13_5	12_6	11_7	10_8
9	100_3	21_4	14_5	13_6	12_7	11_8
10	101_3	22_4	20_5	14_6	13_7	12_8

Note that for any positive integer N the numeral 10_N represents one group of N elements and no groups of one element. Thus, from the table:

$$10_3 = (1 \times 3) + (0 \times 1) = 3$$
$$10_4 = (1 \times 4) + (0 \times 1) = 4$$
$$10_5 = (1 \times 5) + (0 \times 1) = 5$$
$$10_6 = (1 \times 6) = (0 \times 1) = 6$$
$$10_7 = (1 \times 7) + (0 \times 1) = 7$$
$$10_8 = (1 \times 8) + (0 \times 1) = 8$$

To change from another base to base 10 notation, express the number in terms of powers of that base and simplify as in the following examples.

Example 1:   Change 324_8 to base 10.

Solution:

$$324_8 = (3 \times 8^2) + (2 \times 8^1) + (4 \times 8^0) = 212$$

Example 2:   Change 1231_4 to base 10.

Solution:

$$1231_4 = (1 \times 4^3) + (2 \times 4^2) + (3 \times 4^1) + (1 \times 4^0) = 109$$

Any number may be changed from base 10 to another base by dividing successively by the new base and using the remainders as in §4-6. This procedure is always performed in the notation of the old base. It has already been used for the base 5 and may be adapted for other bases as well.

Example 3:   Change 354 to base 8.

Solution:

$$
\begin{array}{r|l}
8 & 354 \\
\hline
8 & 44 \;\text{---}\; 2 \\
\hline
8 & 5 \;\text{---}\; 4 \\
\hline
& 0 \;\text{---}\; 5
\end{array}
$$

Answer: 542_8

Check:

$$542_8 = (5 \times 8^2) + (4 \times 8^1) + (2 \times 8^0)$$
$$= 320 + 32 + 2$$
$$= 354$$

Number bases greater than 10 are possible, but then new symbols must be introduced. Consider, for example, a base 12 system of numeration. The numeral 10_{12} represents $(1 \times 12) + (0 \times 1)$; that is, 12. Thus new symbols are necessary for 10 and 11. Let us use t for 10 and e for 11. Then we can count in base 12 as follows:

$$1, 2, 3, 4, 5, 6, 7, 8, 9, t, e, 10_{12}, 11_{12}, 12_{12}, \ldots$$

We can change a base 12 numeral to base 10 as in the following example.

Example 4: Change $2te5_{12}$ to base 10.

Solution:

$$2te5_{12} = (2 \times 12^3) + (10 \times 12^2) + (11 \times 12^1) + (5 \times 12^0)$$
$$= 3456 + 1440 + 132 + 5 = 5033$$

The reader should note that we make use of a base 12 system in many of our everyday uses of measurements. For example, linear measurement in terms of feet and inches is based on units of 12. Thus a person who is 5 feet 8 inches tall would be listed as having a height of 58_{12} in base 12 notation. Similarly, the use of dozens and gross is based on units of 12. An order for 3 gross, 5 dozen, 7 pencils could be written as 357_{12}:

$$357_{12} = (3 \times 12^2) + (5 \times 12) + (7 \times 12^0) = 499$$

EXERCISES

Change to base 10.

1. 472_8

2. 3213_4

3. 4552_6

4. 347_{12}

5. 1212_3

6. 523_7

7. 472_{12}

8. $3t5_{12}$

9. $15et_{12}$

10. 1101_2

Write each of the following as a base 12 numeral.

11. $(3 \times 12^3) + (5 \times 12^2) + (9 \times 12^1) + (8 \times 12^0)$

12. $(5 \times 12^3) + (10 \times 12^2) + (6 \times 12^1) + (5 \times 12^0)$

13. $(8 \times 12^3) + (7 \times 12^2) + (11 \times 12^1) + (0 \times 12^0)$

14. $(7 \times 12^3) + (11 \times 12^2) + (10 \times 12^1) + (8 \times 12^0)$

Change to the stated base.

15. $324 = (\quad)_4$

16. $576 = (\quad)_8$

17. $427 = (\quad)_6$

18. $114 = (\quad)_3$

19. $798 = (\quad)_5$

20. $257 = (\quad)_7$

21. $536 = (\quad)_9$

22. $182 = (\quad)_{12}$

23. $247 = (\quad)_{12}$

24. $2033 = (\quad)_{12}$

*25. $283 = ($ $)_{12}$ *26. $27 = ($ $)_2$

*27. $47 = ($ $)_2$ *28. $175 = ($ $)_2$

PEDAGOGICAL CONSIDERATIONS

1. In base 12 count by 1's from 1 to 50 and from 100 to 150.
2. Repeat Exercise 1 for base **(a)** 8; **(b)** 5; **(c)** 3.
3. In base 2 count by 1's from 1 to 50 and by 2's from 100 to 150.
4. In base 8 count by 2's from 1 to 99.
5. Repeat Exercise 4 for base 5.
6. In base 12 count by 3's from 3 to 99.
7. Repeat Exercise 6 for base 5.
8. For base 6 notation try to find rules for divisibility by **(a)** 2; **(b)** 3; **(c)** 4; **(d)** 5.
9. For base 12 notation try to find rules for divisibility by **(a)** 2; **(b)** 3; **(c)** 4; **(d)** 6; **(e)** 8; **(f)** 9; **(g)** 11.
10. Prepare a five-minute report on some of the recommendations of the Duodecimal Society of America (Office: 20 Carlton Place, Staten Island, New York 10304).
11. Compare base 12 notation and the arithmetic on a 12-hour clock. Note that all whole numbers can be represented in base 12 notation. The arithmetic on a 12-hour clock (§3-3) is actually a system of remainders with each whole number represented by a number from 0 through 11 or from 1 through 12. Find 5×8 in base 12 notation and also in the arithmetic on a 12-hour clock. Then find other numbers such as 7×9 to provide further examples of the comparison.

4-8
computation in base five notation

We can form an addition table for the numbers in base 5 notation very easily. Consider, for example, the problem $4_5 + 3_5$. This may be written as $4_5 + 1_5 + 2_5$. Now, $(4_5 + 1_5)$ is one group of 5, or 10_5. We then add 2_5 to obtain 12_5. This is equivalent to

$$(1 \times 5^1) + (2 \times 5^0) = (1 \times 5^1) + (2 \times 1) = 5 + 2 = 7.$$

In a similar manner, $4_5 + 4_5 = 4_5 + 1_5 + 3_5$. Now $(4_5 + 1_5) = 10_5$. We then add 3_5 to obtain the sum 13_5. This result could also have been obtained by grouping: $4_5 + 4_5$ can be represented as (****) + (****), which can be regrouped as (*****) + (***); that is, 13_5.

Shown below is a table of the number facts needed for addition problems in base 5. (You should verify each entry.)

+	0	1	2	3	4
0	0	1	2	3	4
1	1	2	3	4	10_5
2	2	3	4	10_5	11_5
3	3	4	10_5	11_5	12_5
4	4	10_5	11_5	12_5	13_5

The facts in this table may be used in finding sums of numbers, as illustrated in the following example.

Example 1: Find the sum of 432_5 and 243_5. Then check in base 10 notation.

Solution:

$$
\begin{array}{r}
432_5 \\
+243_5 \\
\hline
1230_5
\end{array}
$$

Check:

$$
\begin{aligned}
432_5 &= (4 \times 5^2) + (3 \times 5^1) + (2 \times 5^0) & = 117 \\
+243_5 &= (2 \times 5^2) + (4 \times 5^1) + (3 \times 5^0) & = 73 \\
\hline
1230_5 &= (1 \times 5^3) + (2 \times 5^2) + (3 \times 5^1) + (0 \times 5^0) & = 190
\end{aligned}
$$

Here are the steps used in Example 1. In each case the familiar symbol 5 has been used in place of 10_{five} to help the reader recognize that powers of 5 are involved. This convention will be followed throughout this chapter.

First add the column of 1's.

$$
\begin{array}{r}
(4 \times 5^2) + (3 \times 5^1) + (2 \times 5^0) \\
(2 \times 5^2) + (4 \times 5^1) + (3 \times 5^0) \\
\hline
10_5
\end{array}
$$

Next write the sum 10_5 of the 1's as 1×5^1 in the 5's column, and add the column of 5's.

$$
\begin{array}{r}
1 \times 5^1 \\
(4 \times 5^2) + (3 \times 5^1) + (2 \times 5^0) \\
(2 \times 5^2) + (4 \times 5^1) + (3 \times 5^0) \\
\hline
(13_5 \times 5^1) + (0 \times 5^0)
\end{array}
$$

141

sec. 4-8
computation
in
base
five
notation

Then write $10_5 \times 5^1$ from the previous sum in the 5^2 column and add the column of 5^2 entries.

$$
\begin{array}{l}
1 \times 5^2 \\
(4 \times 5^2) + (3 \times 5^1) + (2 \times 5^0) \\
\underline{(2 \times 5^2) + (4 \times 5^1) + (3 \times 5^0)} \\
(12_5 \times 5^2) + (3 \times 5^1) + (0 \times 5^0) = 1230_5
\end{array}
$$

Note that we "carry" groups of five in base 5 computation, just as we "carry" groups of ten in decimal computation.

Subtraction is not difficult if it is thought of as the inverse of addition. The table of addition facts in base 5 may again be used.

Example 2: Subtract in base 5 and check in base 10: $211_5 - 142_5$.

Solution: Think of 211_5 as

$$(2 \times 5^2) + (1 \times 5^1) + (1 \times 5^0)$$

then as

$$(1 \times 5^2) + (11_5 \times 5^1) + (1 \times 5^0)$$

then as

$$(1 \times 5^2) + (10_5 \times 5^1) + (11_5 \times 5^0).$$

Thus:

$$
\begin{array}{rl}
211_5 = & (1 \times 5^2) + (10_5 \times 5^1) + (11_5 \times 5^0) \\
-142_5 = & \underline{(1 \times 5^2) + \quad (4 \times 5^1) + \quad (2 \times 5^0)} \\
14_5 & \qquad\qquad\quad (1 \times 5^1) + \quad (4 \times 5^0)
\end{array}
$$

Check:

$$
\begin{array}{rl}
211_5 = & 56 \\
-142_5 = & 47 \\
\hline
14_5 = & \ 9
\end{array}
$$

This problem can be solved by borrowing and thinking in base 5 in the following steps:

$$
\begin{array}{lll}
\text{(a)} \quad 211_5 & \text{(b)} \quad 20^11_5 & \text{(c)} \quad 1^10^11_5 \\
\quad \underline{-142_5} & \quad \underline{-14\ 2_5} & \quad \underline{-1\ 4\ 2_5} \\
& \qquad\quad 4_5 & \qquad 1\ 4_5
\end{array}
$$

Each raised "1" can be interpreted to show that the number 211_5 has not been changed but rather expressed in a different form:

$$
\begin{array}{l}
211_5 = (2 \times 5^2) + (1 \times 5^1) + (1 \times 5^0) \\
20^11_5 = (2 \times 5^2) + (0 \times 5^1) + (1 \times 5^1) + (1 \times 5^0) = 211_5 \\
1^10^11_5 = (1 + 5^2) + (1 \times 5^2) + (0 \times 5^1) + (1 \times 5^1) \\
\qquad\qquad\qquad\qquad\qquad\qquad + (1 \times 5^0) = 211_5
\end{array}
$$

Problems in multiplication can be performed as repeated additions or as in base 10.

Example 3:   Find the product of 243_5 and 4_5.

Solution: We need the following facts:

$$4 \times 3 = 12_{10} = (2 \times 5^1) + (2 \times 5^0) = 22_5$$
$$4 \times 4 = 16_{10} = (3 \times 5^1) + (1 \times 5^0) = 31_5$$
$$4 \times 2 = 8_{10} = (1 \times 5^1) + (3 \times 5^0) = 13_5$$

The pattern for the computation when multiplying and "carrying" in base 5 notation is then precisely the same as that used in base 10 computation. For example, partial products are indented to represent the powers of 5 involved. In the long form, no attempt is made to "carry" mentally from one place to the next.

$$
\begin{array}{r}
243_5 \\
\times 4_5 \\
\hline
22_5 \\
31_5 \\
13_5 \\
\hline
2132_5
\end{array}
$$

Condensed form:

$$
\begin{array}{r}
243_5 \\
\times 4_5 \\
\hline
2132_5
\end{array}
$$

Multiplication by more than a one-digit multiplier is possible in other bases. Again, the pattern for computation is similar to that used in base 10 computation. Consider, for example, the product $34_5 \times 243_5$:

$$
\begin{array}{rl}
243_5 & \\
34_5 & \\
\hline
2132 & = \quad 4_5 \times 243_5 \\
13340 & = \quad 30_5 \times 243_5 \\
\hline
21022_5 &
\end{array}
$$

As in base 10 computation, the product $30_5 \times 243_5$ can be written without the final numeral 0 since the place value is taken care of by indenting.

The pattern used for division in base 10 can also be used for division in base 5. Consider, for example, the problem $121_5 \div 4_5$. As a first step it is helpful to make a table of multiples of 4_5 in base 5:

$$1_5 \times 4_5 = 4_5$$
$$2_5 \times 4_5 = 13_5$$
$$3_5 \times 4_5 = 22_5$$
$$4_5 \times 4_5 = 31_5$$

142

143

sec. 4-8
computation
in
base
five
notation

Since 12_5 is greater than 4_5 and less than 13_5, the first numeral in the quotient is 1.

$$\begin{array}{r} 1 \\ 4_5 \overline{)121_5} \\ 4 \\ \hline 31 \end{array}$$

As in base 10 computation, we multiply $1_5 \times 4_5$ and subtract this from 12_5 in the dividend. The next digit is then brought down. Note that $12_5 = 7$ and thus $12_5 - 4 = 3$.

Next we must divide 31_5 by 4_5. Note that $4_5 \times 4_5 = 31_5$. Thus we can complete the division, and there is no remainder.

$$\begin{array}{r} 14_5 \\ 4_5 \overline{)121_5} \\ 4 \\ \hline 31 \\ 31 \\ \hline \end{array}$$

Check:
$$\begin{array}{r} 14_5 \\ \times 4_5 \\ \hline 121_5 \end{array}$$

Example 4: Divide 232_5 by 3_5 and check by multiplication in base 5.

Solution:

$$\begin{aligned} 1_5 \times 3_5 &= 3_5 \\ 2_5 \times 3_5 &= 11_5 \\ 3_5 \times 3_5 &= 14_5 \\ 4_5 \times 3_5 &= 22_5 \end{aligned}$$

$$\begin{array}{r} 42_5 \\ 3_5 \overline{)232_5} \\ 22 \\ \hline 12 \\ 11 \\ \hline 1 \end{array}$$

Check:
$$\begin{array}{r} 42_5 \\ \times 3_5 \\ \hline 231_5 \\ +1_5 \\ \hline 232_5 \end{array}$$

The quotient is 42_5 and the remainder is 1_5. Note that in the check we multiply the quotient by the divisor, and add the remainder to obtain the dividend.

EXERCISES

Add in base 5 and check in base 10.

1. $\begin{array}{r} 32_5 \\ +24_5 \end{array}$

2. $\begin{array}{r} 24_5 \\ +34_5 \end{array}$

3. $\begin{array}{r} 143_5 \\ +234_5 \end{array}$

4. $\begin{array}{r} 432_5 \\ +443_5 \end{array}$

5. $\begin{array}{r} 2341_5 \\ +3421_5 \end{array}$

6. $\begin{array}{r} 4324_5 \\ +1442_5 \end{array}$

Subtract in base 5 and check in base 10.

7. 143_5
-31_5

8. 43_5
-24_5

9. 312_5
-123_5

10. 421_5
-234_5

11. 1321_5
-403_5

12. 3204_5
-1342_5

13. 400_5
-123_5

*14. 2003_5
-1344_5

Multiply in base 5 and check in base 10.

15. 342_5
$\times 4_5$

16. 243_5
$\times 3_5$

17. 1423_5
$\times 2_5$

18. 2304_5
$\times 3_5$

19. 24_5
$\times 32_5$

20. 32_5
$\times 43_5$

21. 243_5
$\times 34_5$

*22. 3243_5
$\times 324_5$

Divide in base 5 and check in base 10.

23. $4_5 \overline{)143_5}$

24. $3_5 \overline{)121_5}$

25. $3_5 \overline{)1234_5}$

26. $4_5 \overline{)3042_5}$

27. $11_5 \overline{)143_5}$

28. $32_5 \overline{)3031_5}$

29. $23_5 \overline{)2043_5}$

30. $41_5 \overline{)43201_5}$

31. Complete a table showing the basic multiplication facts for base 5.

*32. Complete the following tables of addition and multiplication facts for base 4.

+	0	1	2	3
0				
1				
2				
3				

x	0	1	2	3
0				
1				
2				
3				

*33. Complete the following tables of addition and multiplication facts for base 8.

+	0	1	2	3	4	5	6	7
0								
1								
2								
3								
4								
5								
6								
7								

x	0	1	2	3	4	5	6	7
0								
1								
2								
3								
4								
5								
6								
7								

*34. Using t for 10 and e for 11, complete the following tables of addition and multiplication facts for base 12.

+	0	1	2	3	4	5	6	7	8	9	t	e
0												
1												
2												
3												
4												
5												
6												
7												
8												
9												
t												
e												

x	0	1	2	3	4	5	6	7	8	9	t	e
0												
1												
2												
3												
4												
5												
6												
7												
8												
9												
t												
e												

Complete each of the indicated operations.

*35. $\begin{array}{r} 2453_6 \\ +3425_6 \end{array}$

*36. $\begin{array}{r} 13201_4 \\ +12323_4 \end{array}$

*37. $\begin{array}{r} 374_8 \\ -146_8 \end{array}$

*38. $\begin{array}{r} 5206_7 \\ -2354_7 \end{array}$

*39. $\begin{array}{r} 256_8 \\ \times 45_8 \end{array}$

*40. $\begin{array}{r} 1302_4 \\ \times 213_4 \end{array}$

*41. $3_4\overline{)1231_4}$

*42. $25_6\overline{)3245_6}$

PEDAGOGICAL CONSIDERATIONS

All teachers, and especially teachers of elementary mathematics, need to be concerned with students of varying levels of abilities and to make special provisions for such students. Of special concern in recent years are those slow learners who come from disadvantaged areas and have been culturally deprived. The National Council of Teachers of Mathematics has been interested in such problems, and has published materials in these areas. For example, *Experiences In Mathematical Ideas* consists of two volumes of learning activities for students in grades 5 through 8 who have had little prior success in mathematics. On the other hand, the twenty-seventh Yearbook, *Enrichment Mathematics for the Grades*, consists of numerous suggestions for supplementary teaching units for the academically talented student in the elementary grades.

1. Read and report on at least one of the units described in each of the afore-mentioned publications.

2. Read and report on at least three different articles found in professional journals dealing with teaching the slow learner.

3. Repeat Exercise 2 for the gifted student.

4. Prepare a list of ten characteristics of the slow learner.

5. Repeat Exercise 4 for the gifted student.

6. Numerous suggestions can be found in the literature for teaching techniques suitable for the slow learner. For example, it is claimed that they learn best through an activity-oriented curriculum centered about laboratory type experiences. Prepare a list of 10 specific suggestions for teaching mathematics to the slow learner in the elementary school.

7. Repeat Exercise 6 for the gifted student.

8. Visit an elementary mathematics class and report on what provisions, if any, are made for students of different levels of ability.

4-9

binary notation

Numbers written to the base 2 are of special interest because of their application in modern electronic computers. This system of notation is called **binary notation,** and makes use of only two digits, 0 and 1.

Recent spacecraft that took pictures of the planet Mars transmitted the data back to earth in binary notation. On earth, with the aid of computers, this data was converted into pictures of the surface of the planet that appeared in many local newspapers.

Although there is some evidence that the basic concepts of binary notation were known to the ancient Chinese about 2000 B.C., it is only within recent years that it has been widely applied in computer mathematics, card sorting operations, and other electronic devices. Some of the place values in the binary system are as follows:

$$2^7 \quad 2^6 \quad 2^5 \quad 2^4 \quad 2^3 \quad 2^2 \quad 2^1 \quad 2^0$$
$$128 \quad 64 \quad 32 \quad 16 \quad 8 \quad 4 \quad 2 \quad 1$$

Thus, in binary notation, $11011101_2 = 215$ as shown in the following example.

Example 1: Write 11011101_2 in base 10 notation.

Solution:

$$11011101_2 = (1 \times 2^7) + (1 \times 2^6) + (0 \times 2^5) + (1 \times 2^4)$$
$$+ (1 \times 2^3) + (1 \times 2^2) + (0 \times 2^1) + (1 \times 2^0)$$
$$= 128 + 64 + 16 + 8 + 4 + 1 = 221$$

Here are the first 16 counting numbers written in binary notation:

Base 10	Base 2	Base 10	Base 2
1	1	9	1001_2
2	10_2	10	1010_2
3	11_2	11	1011_2
4	100_2	12	1100_2
5	101_2	13	1101_2
6	110_2	14	1110_2
7	111_2	15	1111_2
8	1000_2	16	10000_2

Tables for addition and multiplication in binary notation are easy to complete.

+	0	1
0	0	1
1	1	10_2

x	0	1
0	0	0
1	0	1

Example 2: Multiply: $101_2 \times 1101_2$.

Solution:

$$
\begin{array}{r}
1101_2 \\
101_2 \\
\hline
1101 \\
11010 \\
\hline
1000001_2
\end{array}
$$

Check:

$$1101_2 = 13$$
$$101_2 = \underline{5}$$
$$65$$
$$1000001_2 = (1 \times 2^6) + (1 \times 2^0)$$
$$= 64 + 1 = 65$$

The seven-digit numeral 1000001 is the name for the letter A when that letter is sent over teletype to a computer. The American Standard Code for Information Interchange (ASCII) was adopted in 1967 and is used to convert all letters, numerals, and other symbols into binary notation.

Binary numerals can be shown by means of electric lights flashing on and off. If a light is on, the digit 1 is represented; whereas if it is off, the digit 0 is shown. Similarly the digits 0 and 1 can be used in a card sorting operation. The reader can gain an appreciation of the process used by construction of a small set of punched cards. First prepare a set of sixteen index cards with four holes punched in each and a corner notched as in the figure.

Next represent the numbers 0 through 15 on these cards in binary notation. Cut out the space above each hole to represent 1; leave the hole untouched to represent 0. Several cards are shown in the figure.

After all the cards have been completed in this manner, shuffle them thoroughly and align them, making certain that they remain "face up." (The notched corners will help indicate when the cards are right side up.) Then, going from right to left, perform the following operation: Stick a pencil or other similar object through the first hole and lift up. Some of the cards will come up, namely, those in which the holes have not been cut through to the edge of the card (that is, those cards representing numbers whose units digit in binary notation is 0).

Place the cards that lift up in front of the other cards and repeat the same operation for the remaining holes in order from right to left. When you have finished, the cards should be in numerical order, 0 through 15.

Note that only four operations are needed to arrange the sixteen cards. As the number of cards is doubled, only one additional operation will be needed each time to place them in order. That is, 32 cards may be placed in numerical order with five of the described card sorting operations; 64 cards may be arranged with six operations; 128 cards with seven operations; and so forth. Thus a large number of cards may be arranged in order with a relatively small number of operations. For example, over one billion cards may be placed in numerical order with only thirty sortings.

EXERCISES

Write each number in binary notation.

1. 28   *2.* 45

3. 19   4. 128

5. 152   6. 325

Change each number to decimal notation.

7. 1111_2 8. 10101_2

9. 110111_2 10. 101010_2

11. 100110_2 12. 11011011_2

Perform the indicated operation in binary notation and check in base 10.

13. $\begin{array}{r} 1101_2 \\ +1011_2 \\ \hline \end{array}$ 14. $\begin{array}{r} 10011_2 \\ +10101_2 \\ \hline \end{array}$

15. $\begin{array}{r} 11011_2 \\ -10110_2 \\ \hline \end{array}$ 16. $\begin{array}{r} 110101_2 \\ -10111_2 \\ \hline \end{array}$

17. $\begin{array}{r} 1101_2 \\ \times 11_2 \\ \hline \end{array}$ 18. $\begin{array}{r} 10110_2 \\ \times 101_2 \\ \hline \end{array}$

19. In the ASCII code the numbers for B, D, and G are 66, 68, and 71, respectively. Find the binary representation of (a) B; (b) D; (c) G.

20. In the ASCII code capital letters A, B, C, ... in alphabetical order are assigned numbers 65, 66, 67, ... What is the word transmitted by the code 1010010, 1010101, 1001110?

21. Write the number 214 in base 8 and then in base 2 notation. Can you discover a relationship between these two bases?

*Base 8 notation is often called **octal notation**. Write in*

22. $101,111,001,010_2$ 23. $11,101,011,001_2$

24. $1,001,101,110,010_2$ 25. $10,010,101,011,011_2$

26. $111,110,101,101,011,001_2$ 27. $1,101,011,001,000,100,110_2$

Write in binary notation.

28. 435_8 29. 6023_8

30. 4257_8 31. 3624_8

32. 12345_8 33. 76543_8

*34. Complete a set of 32 punched cards to represent the numbers 0 through 31. Here five holes are needed on each card and five "lifting" operations are necessary to place the set in numerical order.

PEDAGOGICAL CONSIDERATIONS

Many recreational items are based on the binary system of notation. Consider, for example, the boxes shown, within which the numbers 1 to 15 are placed according to the following scheme.

D	C	B	A
8	4	2	1
9	5	3	3
10	6	6	5
11	7	7	7
12	12	10	9
13	13	11	11
14	14	14	13
15	15	15	15

In box *A* place all numbers that have a 1 in the units place when written in binary notation. In box *B* place those with a 1 in the second position from the right in binary notation. In *C* and *D* are those numbers with a 1 in the third and fourth positions, respectively.

Next, ask someone to think of a number and tell you in which box or boxes it appears. You then tell that person his number by finding the sum of the first number in each box he mentions. Thus, if his number is 11, he lists boxes *A*, *B*, and *D*. You then find the sum $1 + 2 + 8$ as the number under discussion.

1. Explain why the method given for finding a number after knowing the boxes in which it appears works as it does.

2. Extend the boxes to include all the numbers through 31. (A fifth column, *E*, will be necessary.) Then explain for the set of five boxes how to find a number if one knows in which boxes it appears.

3. You can use binary notations to identify any one of eight numbers by means of three questions that can be answered as "yes" of "no." Thus consider the numbers from 0 through 7 written in binary notation. The place values are identified by columns *A*, *B*, and *C*.

	C	B	A
0	0	0	0
1	0	0	1
2	0	1	0
3	0	1	1
4	1	0	0
5	1	0	1
6	1	1	0
7	1	1	1

The three questions to be asked are: "Does the number have a 1 in position *A*?" "Does the number have a 1 in position *B*?" "Does the number have a 1 in position *C*?" Suppose the answers are "Yes, No, Yes." Then the number is identified as 101_2; that is, 5. Extend this process to show how you can guess a number that someone is thinking of from 0 through 15 by means of four "yes–no" questions.

4. One can count on one's fingers in base 2 by considering a finger held up as 1 and a finger down as 0. Here are several representations for the first five counting numbers.

1_2	10_2	11_2	100_2	101_2
1	2	3	4	5

Show how to count through 31 in binary notation using the fingers on one hand.

5. Find a reference to the game of Nim and study its relationship to the binary system of notation.

CHAPTER TEST

1. Use the Egyptian method of duplation and mediation to find the product 17×31.

2. Multiply, using the "galley" method: 325×437.

3. Write in expanded notation: 23,457.

4. Write a numeral for 58 in: **(a)** base 8 notation; **(b)** base 5 notation.

5. Change to base 10:
   (a) 234_5 **(b)** 125_6

6. Change to base 10:
   (a) $3te_{12}$ **(b)** 1011_2

7. Change to the stated base:
   (a) $243 = (\quad)_5$ **(b)** $47 = (\quad)_4$

8. Compute in base 5:
   (a) $\begin{array}{r} 243_5 \\ +132_5 \\ \hline \end{array}$ **(b)** $\begin{array}{r} 3021_5 \\ -1314_5 \\ \hline \end{array}$

9. Compute in base 5:
   (a) $\begin{array}{r} 1342_5 \\ \times\ 3_5 \\ \hline \end{array}$ **(b)** $4_5 \overline{)3212_5}$

10. Compute in binary notation:
   (a) $\begin{array}{r} 1111_2 \\ +1101_2 \\ \hline \end{array}$ **(b)** $\begin{array}{r} 1011_2 \\ \times 11_2 \\ \hline \end{array}$

5

The Set of Integers

You have been concerned with sets of numbers and operations on numbers since you first started your study of mathematics. In this and the following two chapters we shall present a brief development of various collections of numbers and their properties. We are not so much concerned with routine operations on these numbers as we are with key concepts that form the basis for the structure of our number system. Although many of these ideas may be familiar, they will be presented in a way that should help to develop an appreciation of this structure and the reason for its development.

Elementary school teachers introduce and extensively use properties of the integers. The set of integers is the first major extension of the set of counting numbers and their properties. This extension is made so that subtraction will always be possible; that is, so that numbers can be used in representing situations such as temperatures above and below freezing and in solving a wider range of problems.

The famous mathematician Leopold Kronecker (1823–1891) once said "God created the natural numbers; everything else is man's handiwork." Thus we shall begin our study of the number system with this set of natural numbers, now generally referred to as the set of counting numbers.

5-1
counting numbers

A number is associated with a set of elements whenever the elements of the set are counted. The set of counting numbers is often represented as the set C:

$$C = \{1, 2, 3, 4, 5, \ldots\}$$

Sums and products of counting numbers may be introduced using cardinal numbers of sets. If A and B are two sets such that $A \cap B = \varnothing$, then

$$n(A) + n(B) = n(A \cup B).$$

Multiplication may be introduced in terms of successive addition. For example,

$$2 \times 3 = 3 + 3, \qquad 3 \times 2 = 2 + 2 + 2.$$

Multiplication may also be introduced using equivalent sets that have no common elements, that is, disjoint equivalent sets. For example, let $A = \{a, b\}$, $B = \{p, q\}$, and $C = \{x, y\}$. Then

$$n(A) = n(B) = n(C) = 2$$
$$A \cap B = \varnothing, \qquad A \cap C = \varnothing, \qquad B \cap C = \varnothing$$

and

$$n(A \cup B \cup C) = 3 \times 2.$$

The use of equivalent sets to represent multiplication can be extended to obtain a new operation on sets. Let $P = \{1, 2, 3\}$ and $Q = \{a, b\}$. A new set with cardinal number 3 may be formed by taking each element of P with a given element of Q:

$$n\{(1, a), (2, a), (3, a)\} = 3$$
$$n\{(1, b), (2, b), (3, b)\} = 3$$

Notice that each of these sets has pairs of elements such as $(2, b)$ as its elements. Each pair has been purposely written with an element of P first and an element of Q second. Any pair with one element considered as the first element and the other element considered as the second element is an **ordered pair**.

For any two given sets A and B the set of all possible ordered pairs with an element of A as the first element and an element of B as the second element is the **Cartesian product** of A and B, written $A \times B$. For $P = \{1, 2, 3\}$ and $Q = \{a, b\}$,

$$P \times Q = \{(1, a), (2, a), (3, a), (1, b), (2, b), (3, b)\}.$$

Notice that for $n(P) = 3$ and $n(Q) = 2$ the set $P \times Q$ may be considered either as two sets of three elements each (one set of pairs including a and one set of pairs including b) or as three sets (one for each element of P) of two elements each. Thus

$$n(P) \times n(Q) = n(P \times Q) = n(Q) \times n(P)$$
$$3 \quad \times \quad 2 \quad = \quad\quad 6 \quad\quad = \quad 2 \quad \times \quad 3.$$

Commutativity may also be shown from $n(P \times Q) = n(Q \times P)$; that is,

$$n\{(1, a), (2, a), (3, a), (1, b), (2, b), (3, b)\}$$
$$= n\{(a, 1), (a, 2), (a, 3), (b, 1), (b, 2), (b, 3)\}.$$

Example 1: Use a Cartesian product of sets to illustrate $2 \times 4 = 8$.

Let $A = \{a, b\}$ and $B = \{p, q, r, s\}$. Then $n(A) = 2$, $n(B) = 4$, $A \times B = \{(a, p), (b, p), (a, q), (b, q), (a, r), (b, r), (a, s), (b, s)\}$, and $2 \times 4 = n(A) \times n(B) = n(A \times B) = 8$.

s	(a, s)	(b, s)
r	(a, r)	(b, r)
q	(a, q)	(b, q)
p	(a, p)	(h, p)
	a	b

The elements of a Cartesian product are often represented in an array for the sets A and B of Example 1 as shown at the left.

The properties of the set of counting numbers under addition and multiplication may be described using the terminology introduced in Chapter 3. For any counting numbers a, b, c we have:

Closure, $+$: There is one and only one counting number $a + b$.

Closure, $\times$: There is one and only one counting number $a \times b$.

Commutative, $+$: $a + b = b + a$; the sum of any two counting numbers is the same regardless of the order in which the numbers are added.

Commutative, $\times$: $a \times b = b \times a$; the product of any two counting numbers is the same regardless of the order in which the numbers are multiplied.

Associative, $+$: $(a + b) + c = a + (b + c)$.

Associative, $\times$: $(a \times b) \times c = a \times (b \times c)$.

Identity, $\times$: $a \times 1 = 1 \times a = a$; the counting number 1 is the **multiplicative identity** and is also called the **identity element for multiplication.** This property is often referred to as the **multiplication property of one.**

Distributive Property, $\times$ with respect to $+$:

$$a \times (b + c) = (a \times b) + (a \times c).$$

The set of counting numbers does not include an identity for addition since zero is not a counting number. Also the set of counting numbers is *not* closed under subtraction or under division. For example, $2 - 5$ and $12 \div 5$ are impossible if you are restricted to using counting numbers. These two counterexamples are sufficient to show that the properties do not hold.

Example 2: Apply properties of counting numbers to the left member of the equation $25 \times (11 \times 4) = (25 \times 4) \times 11$ to obtain the right member of the equation. Show each step and name each property.

Solution:

$$\begin{aligned}
25 \times (11 \times 4) &= 25 \times (4 \times 11) &\quad &\text{Commutative, } \times. \\
&= (25 \times 4) \times 11 &\quad &\text{Associative, } \times.
\end{aligned}$$

EXERCISES

1. Find $A \times B$ for $A = \{1, 2\}$ and $B = \{1, 2, 3, 4\}$.
2. Find $P \times Q$ for $P = \{1, 2, 3\}$ and $Q = \{1, 2, 3\}$.
3. Show that subtraction of counting numbers is not commutative.
4. Show that division of counting numbers is not commutative.
5. Does $8 - (3 - 2) = (8 - 3) - 2$? Is subtraction of counting numbers associative?
6. Does $12 \div (6 \div 2) = (12 \div 6) \div 2$? Is division of counting numbers associative?
7. Show that addition of counting numbers is not distributive with respect to multiplication.

For each arithmetic statement name the property of counting numbers that is illustrated.

8. $3 \times 2 = 2 \times 3$
9. $6 + 5 = 5 + 6$
10. $2 + (3 \times 5) = 2 + (5 \times 3)$
11. $2 \times (3 \times 5) = (2 \times 3) \times 5$
12. $1 \times 4580 = 4580$
13. $(25 \times 14) + (25 \times 26) = 25 \times (14 + 26)$
14. $17 \times (15 + 21) = 17 \times (21 + 15)$
15. $48 + (19 + 7) = (19 + 7) + 48$
16. $27 \times (5 + 3) = (5 + 3) \times 27$
17. $17 \times (10 + 1) = (17 \times 10) + (17 \times 1)$
18. $5280 \times (7 + 13) = 5280 \times (13 + 7)$

Identify the property of counting numbers that is used for each step in these applications of our usual addition and multiplication algorithms.

19.
$$
\begin{aligned}
12 + 25 &= (10 + 2) + (20 + 5) &&\text{Numeration system.} \\
&= 10 + [2 + (20 + 5)] &&\text{(a)} \\
&= 10 + [(20 + 5) + 2] &&\text{(b)} \\
&= 10 + [20 + (5 + 2)] &&\text{(c)} \\
&= 10 + (20 + 7) &&\text{Number fact.} \\
&= (10 + 20) + 7 &&\text{(d)} \\
&= [(1 \times 10) + (2 \times 10)] + 7 &&\text{Numeration system.} \\
&= [(1 + 2) \times 10] + 7 &&\text{(e)} \\
&= (3 \times 10) + 7 &&\text{Number fact.} \\
&= 37 &&\text{Numeration system.}
\end{aligned}
$$

20.
$$
\begin{aligned}
12 \times 6 &= (10 + 2) \times 6 &&\text{Numeration system.} \\
&= (10 \times 6) + (2 \times 6) &&\text{(a)} \\
&= 60 + 12 &&\text{Number facts.}
\end{aligned}
$$

$= 60 + (10 + 2)$	Numeration system.
$= (60 + 10) + 2$	(b)
$= [(6 \times 10) + 10] + 2$	Number fact.
$= [(6 \times 10) + (1 \times 10)] + 2$	(c)
$= (6 + 1) \times 10 + 2$	(d)
$= 7 \times 10 + 2$	Number facts.
$= 72$	Numeration system.

Apply properties of counting numbers to the left member of each equation to obtain the right member. Show each step and name each property.

21. $92 + (50 + 8) = (92 + 8) + 50$

22. $(25 \times 17) \times 4 = 17 \times 100$

23. $37 \times (1 + 100) = 3700 + 37$

24. $(2 + 3) + (8 + 7) = (2 + 8) + (3 + 7)$

25. $(73 + 19) + (7 + 1) = (73 + 7) + (19 + 1)$

26. $(26 \times 1) \times (10 \times 2) = 10 \times (26 \times 2)$

Explain your answer for each question:

27. Note that $6 \div (3 \div 1) = (6 \div 3) \div 1$. Is division of counting numbers associative?

28. Note that $4^2 = 2^4$. Is the operation of raising to a power commutative for counting numbers?

PEDAGOGICAL CONSIDERATIONS

The properties of counting numbers are used to simplify arithmetic computations throughout the elementary school grades. For example, the number fact $7 \times 10 = 70$ is easily learned. However, the problem 10×7 often leads an elementary school student to computations 1×7, 2×7, 3×7, ..., 10×7 and is rather hard *until* the student realizes that 7×10 and 10×7 are two ways of stating the same problem. Many educators feel that students *should not be told* but rather *should be helped to discover* for themselves such useful properties of numbers. In other words, many matched pairs of problems (7×10 and 10×7, 2×15 and 15×2, and so forth) should be given until the student observes the pattern of the commutativity of multiplication. Such experiences of discovery and the recognition of the labor-saving effect of the pattern that is discovered should be developed whenever possible.

1. Two rows of three blocks may be used to represent 2×3. Describe how you could represent 3×2 and illustrate $2 \times 3 = 3 \times 2$.

2. Repeat Exercise 1 for 4×5.

3. Extend Exercise 1 to show that $(2 \times 3) \times 4 = 2 \times (3 \times 4)$.

4. Repeat Exercise 3 for $(3 \times 4) \times 5 = 3 \times (4 \times 5)$.

5. Extend Exercise 1 to show that $2 \times (3 + 4) = (2 \times 3) + (2 \times 4)$.

6. Extend Exercise 1 to show that $3 \times (2 + 5) = (3 \times 2) + (3 \times 5)$.

As in Exercises 19 and 20 of the preceding set of exercises, write out and explain each step for the usual algorithm for:

7. 3×15 **8.** 8×57 **9.** 21×17

10. $21 + 17$ **11.** $32 + 45$ **12.** $65 + 27$

13. In the introduction to this chapter there appears a quotation from the German mathematician, Kronecker. Start a collection of such famous sayings that might be of interest in mathematics classes. For example, Archimedes is supposed to have said, "Give me a place to stand and a lever long enough and I will move the earth."

5-2

prime numbers

Before extending the set of counting numbers to other sets we shall explore an interesting and useful classification of the counting numbers according to several of its subsets. First, however, certain definitions are needed.

The counting number 6 is divisible by 2, since there is a counting number 3 such that $6 = 2 \times 3$; the counting number 7 is not divisible by 2, since there exists no counting number b such that $7 = 2 \times b$. In general, a counting number n is **divisible by** a counting number t if and only if there is a counting number k such that $n = t \times k$. If n is divisible by t, then n is a **multiple of** t and t is a **factor of** n. For example, 6 is a multiple of 2 and 2 is a factor of 6.

The counting numbers are often considered in terms of the numbers by which they are divisible. The set

$$A = \{2, 4, 6, 8, 10, 12, \ldots\}$$

consists of the numbers that are divisible by 2—that is, the numbers expressible in the form $2k$, where k stands for a counting number. The set

$$B = \{3, 6, 9, 12, 15, 18, \ldots\}$$

consists of the numbers divisible by 3; the set

$$C = \{4, 8, 12, 16, 20, 24, \ldots\}$$

consists of the numbers divisible by 4; the set

$$D = \{5, 10, 15, 20, 25, 30, \ldots\}$$

consists of the numbers divisible by 5; the set

$$E = \{6, 12, 18, 24, 30, 36, \ldots\}$$

consists of the numbers divisible by 6; and so forth.

Notice that $C \subset A$; in other words, any number that is divisible by 4 is also divisible by 2. Notice also that $A \cap B = E$; in other words, the set of numbers that are divisible by both 2 and 3 is the set of numbers that are divisible by 6.

The number 1 divides every counting number since $k = 1 \times k$ for every counting number k. Accordingly, the number 1 is called a **unit**. Since the set A does not include all counting numbers, divisibility by 2 is a special property of the elements of the set A.

The number 2 is a member of the set A and is not a member of any other set; that is, 2 is not divisible by any counting number except itself and 1. Any counting number greater than 1 that is divisible only by itself and 1 is called a **prime number**. Thus 2 is prime, 3 is prime, 4 is not prime since 4 is divisible by 2 (that is, 4 is a member of the set A as well as the set C), 5 is prime, and 6 is not prime since 6 is divisible by both 2 and 3. The counting numbers that are greater than 1 and are not prime are called **composite numbers**. Note that every counting number greater than 1 is either prime or composite. The number 1 is neither prime nor composite.

We could extend the list of sets A, B, C, D, E, F to identify other prime numbers, that is, elements that belong to one and only one of these sets. However, the method used to select these sets may be applied to the entire set of counting numbers. We shall illustrate this for the set $\{1, 2, 3, \ldots, 100\}$ and thereby illustrate a method for finding prime numbers that was discovered by a Greek mathematician named Eratosthenes over two thousand years ago. The method is known as the **Sieve of Eratosthenes**. First we prepare a table of the counting numbers through 100. Then selected members of this set are excluded as follows.

Cross out 1, since we know that it is not classified as a prime number. Draw a circle around 2, the smallest prime number. Then cross out every following multiple of 2, since each one is divisible by 2 and thus is not prime. That is, cross out the numbers in the set $\{4, 6, 8, \ldots, 100\}$.

Draw a circle around 3, the next prime number in our list. Then cross out each succeeding multiple of 3. Some of these numbers, such as 6 and 12, will already have been crossed out because they are multiples of 2. That is, they are members of both sets A and B.

The number 5 is prime and is circled, and we exclude each fifth number after 5. The next prime number is 7, and we exclude each seventh number after 7. Note that 49 is the first multiple of 7 that has not already been excluded as being a member of another set of multiples. The next prime number is 11. Since all multiples of 11 in this set have already been excluded, the remaining numbers that have not been excluded are prime numbers and may be circled.

Notice that 49 is the first number that is divisible by 7 and is not also divisible by a prime number less than 7. In other words, each composite number less than 7^2 has at least one of its factors less than 7. Similarly, we might have observed that each composite number less than 5^2 has at least one factor less than 5. In general, *for any prime number p each composite number less than p^2 has a prime number less than p as a factor.*

We use this property to tell us when we have excluded all composite numbers from a set. In the set of numbers $\{1, 2, \ldots, 100\}$ we have considered the primes 2, 3, 5, and 7. The next prime is 11. Thus by our method we have already excluded all composite numbers up to but not including 11^2; that is, 121. In particular, we have identified the set of prime numbers less than or equal to 100.

Example 1:   List the set of prime numbers less than 70.

Solution: From the chart we identify this set as

$\{2, 3, 5, 7, 11, 13, 17, 19, 23, 29, 31, 37, 41, 43, 47, 53, 59, 61, 67\}$.

We have seen that every counting number greater than 1 is either a prime number or a composite number. Now we shall find that every counting number greater than 1 can be expressed in terms of its prime factors in essentially only one way.

Consider the various ways of factoring 24:

$$24 = 1 \times 24$$
$$24 = 2 \times 12$$
$$24 = 3 \times 8$$
$$24 = 4 \times 6$$
$$24 = 2 \times 2 \times 6$$
$$24 = 2 \times 3 \times 4$$
$$24 = 2 \times 2 \times 2 \times 3 = 2^3 \times 3$$

The last factorization in terms of the prime numbers 2 and 3 could be written as $2 \times 3 \times 2^2$ and in other ways. However, these ways are equivalent, since the order of the factors does not affect the product. Thus 24 can be expressed in terms of its prime factors in one and only one way.

One of the easiest ways to find the prime factors of a number is to consider the prime numbers

$$2, 3, 5, 7, 11, 13, 17, 19, 23, 29, 31, \ldots$$

in order and use each one as a factor as many times as possible. Then for 24 we would have

$$24 = 2 \times 12$$
$$= 2 \times 2 \times 6$$
$$= 2 \times 2 \times 2 \times 3.$$

Some people prefer to write these steps using division:

$$\begin{array}{r|l} 2 & 24 \\ \hline 2 & 12 \\ \hline 2 & 6 \\ \hline & 3 \end{array}$$

Since 3 is a prime number, no further steps are needed and $24 = 2^3 \times 3$.

Example 2: Express 3850 in terms of its prime factors.

Solution:

$$\begin{array}{r|l} 2 & 3850 \\ \hline 5 & 1925 \\ \hline 5 & 385 \\ \hline 7 & 77 \\ \hline & 11 \end{array} \qquad 3850 = 2 \times 5^2 \times 7 \times 11$$

In general, if a counting number n is greater than 1, then n has a prime number p_1 as a factor. Suppose

$$n = p_1 n_1.$$

Then if n is a prime number, $n = p_1$ and $n_1 = 1$. If n is not a prime number, then n_1 is a counting number greater than 1. In this case n_1 is either a prime number or a composite number. Suppose

$$n_1 = p_2 n_2 \quad \text{and thus} \quad n = p_1 p_2 n_2$$

where p_2 is a prime number. As before, if $n_2 \neq 1$, then

$$n_2 = p_3 n_3 \quad \text{and thus} \quad n = p_1 p_2 p_3 n_3$$

where p_3 is a prime number, and so forth. We may continue this process until some $n_k = 1$, since there are only a finite number of counting numbers less than n and

$$n > n_1 > n_2 > n_3 > \cdots > n_k = 1.$$

Then we have an expression for n as a product of prime numbers:

$$n = p_1 p_2 p_3 \cdots p_k$$

We call this the **prime factorization** of n, that is, the factorization of n into its prime factors. Except for the order of the factors, the prime factorization of any counting number greater than 1 is unique; that is, *any counting number greater than 1 may be expressed as a product of prime numbers in one and only one way.* As in the examples, we usually write the prime factorization as a product of powers of prime numbers.

Example 3: Find the prime factorization of 5280.

Solution:

2	5280
2	2640
2	1320
2	660
2	330
3	165
5	55
	11

$5280 = 2^5 \times 3 \times 5 \times 11$

EXERCISES

Let A = the set of numbers divisible by 2, B = the set of numbers divisible by 3, D = the set of numbers divisible by 5, F = the set of numbers divisible by 12, and H = the set of numbers divisible by 15. Restate each of the statements in Exercises 1 through 6 in terms of divisibility.

1. $H \subset B$
2. $H \subset D$
3. $B \cap D = H$
4. $F \subset A$
5. $F \subset B$
6. $F \subset (A \cap B)$

7. List the composite numbers between 20 and 40.

8. Adapt the Sieve of Eratosthenes to find the prime numbers less than or equal to 200.

9. Is every odd number a prime number? Is every prime number an odd number?

10. Exhibit a pair of prime numbers that differ by 1 and show that there is only one such pair possible.

11. Here is a famous theorem that has not yet been proved: Every even number greater than 2 is expressible as the sum of two prime numbers. (This theorem is often called **Goldbach's conjecture.**) Express each even number from 4 to 40 inclusive as a sum of two prime numbers.

12. Here is another famous theorem. Two prime numbers such as 17 and 19 that differ by 2 are called *twin primes*. It is believed but has not yet been proved that there are infinitely many twin primes. Find a pair of twin primes that are between **(a)** 25 and 35; **(b)** 55 and 65; **(c)** 95 and 105.

13. A set of three prime numbers that differ by 2 is called a *prime triplet*. Exhibit a prime triplet and explain why it is the only possible triplet of primes.

14. It has been conjectured but not proved that every odd number greater than 5 is expressible as the sum of three prime numbers. Verify this for the numbers 7, 9, 11, 13, and 15.

*15. What is the largest prime that you need to consider to be sure that you have excluded all composite numbers less than or equal to **(a)** 200; **(b)** 500; **(c)** 1000?

Express each number as the product of two counting numbers in as many different ways as possible.

16. 12

17. 15

18. 18

19. 20

20. 13

21. 29

Find the prime factorization of each number.

22. 76

23. 68

24. 215

25. 123

26. 738

27. 1425

28. 341

29. 818

PEDAGOGICAL CONSIDERATIONS

1. Examine several elementary school textbook series to determine at which grade level the concept of prime numbers is first introduced in the curriculum.

2. Many elementary school texts use the idea of a branching tree to help students think about factors. If the branches terminate with prime numbers, then we have a **prime-factor tree.** Two such trees are shown on the next page.

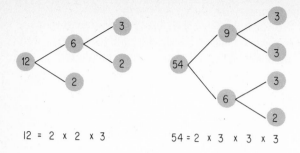

12 = 2 x 2 x 3   54 = 2 x 3 x 3 x 3

Draw prime-factor trees for 30, 60, and 96.

3. No one has ever been able to find a formula that will produce only prime numbers. At one time it was thought that the expression $n^2 - n + 41$ would give only prime numbers for the set of counting numbers as replacements for n. Show that prime numbers are obtained when n is replaced by 1, 2, 3, 4, and 5. Show that the formula fails for $n = 41$.

4. Euclid proved that the set of prime numbers is infinite. Refer to a textbook or a book on the history of mathematics and study this simple, yet elegant, proof.

5. At the time of the writing of this book the largest known prime was $2^{19,937} - 1$, a number with 6,002 digits in decimal notation. Look for and read a reference to man's search for large prime numbers through the use of computers.

5-3

applications of prime factorizations

We can make effective use of the concept of prime factorization in many arithmetic situations. Let us first consider the various factors of counting numbers. Recall that a counting number t is said to be a factor of another counting number n if and only if there is a counting number k such that $n = t \times k$. Now consider the set of factors of 12 and the set of factors of 18:

12: {1, 2, 3, 4, 6, 12}
18: {1, 2, 3, 6, 9, 18}

The set of **common factors** of 12 and 18 consists of those numbers that are factors of both 12 and 18, that is,

{1, 2, 3, 6}.

The largest member of this set, 6, is called the (G.C.F.) of the two numbers. In general, the **greatest common factor** of two or more counting numbers is the largest counting number that is a factor of each of the given numbers.

We may use the prime factorization of two counting numbers to find their greatest common factor. First express each number by its prime factorization. Then consider the prime numbers that are factors of both of the given numbers, and take the product of those prime numbers with each raised to the highest power that is a factor of *both* of the given numbers. For $12 = 2^2 \times 3$ and $18 = 2 \times 3^2$ we have G.C.F. $= 2 \times 3$, that is, 6.

Example 1:   Find the greatest common factor of 60 and 5280.

Solution:

$$
\begin{array}{r|l}
2 & 60 \\ \hline
2 & 30 \\ \hline
3 & 15 \\ \hline
  & 5
\end{array}
\qquad 60 = 2^2 \times 3 \times 5
$$

As in Example 3 of §5-2, we have $5280 = 2^5 \times 3 \times 5 \times 11$. Then the highest power of 2 that is a common factor of 60 and 5280 is 2^2; 3 is a common factor; 5 is a common factor; 11 is not a common factor. The greatest common factor of 60 and 5280 is $2^2 \times 3 \times 5$, that is, 60.

Example 2:   Find the greatest common factor of 3850 and 5280.

Solution:

$$3850 = 2 \times 5^2 \times 7 \times 11$$
$$5280 = 2^5 \times 3 \times 5 \times 11$$

The greatest common factor of 3850 and 5280 is $2 \times 5 \times 11$, that is, 110.

Example 3:   Find the G.C.F. of 12, 36, and 60.

Solution: First write the prime factorization of each number:

$$12 = 2^2 \times 3$$
$$36 = 2^2 \times 3^2$$
$$60 = 2^2 \times 3 \times 5$$

The G.C.F. is $2^2 \times 3$, that is, 12.

We may use the greatest common factor when we reduce (simplify) a fraction. For example,

$$\frac{60}{4880} = \frac{(2^2 \times 5) \times 3}{(2^2 \times 5) \times (2^2 \times 61)} = \frac{3}{2^2 \times 61} = \frac{3}{244}.$$

Since 3 is the only prime factor of the numerator and 3 is not a factor of the denominator, the fraction $\frac{3}{244}$ is in **lowest terms**. The numerator and the denominator do not have any common prime factors and are said to be **relatively prime**.

165

Example 4: Reduce the fraction $\frac{60}{168}$ to lowest terms.

Solution:

$$60 = 2^2 \times 3 \times 5$$
$$168 = 2^3 \times 3 \times 7$$

The greatest common factor is $2^2 \times 3$,

$$\frac{60}{168} = \frac{(2^2 \times 3) \times 5}{(2^2 \times 3) \times 2 \times 7} = \frac{5}{14}.$$

Let us now turn our attention to the concept of a multiple of a number. Recall that if a counting number t is a factor of a counting number n, then n is said to be a multiple of t. Consider the set of multiples of 12 and the set of multiples of 18:

12: $\{12, 24, 36, 48, 60, 72, 84, 96, 108, 120, \ldots\}$
18: $\{18, 36, 54, 72, 90, 108, 126, \ldots\}$

The set of **common multiples** of 12 and 18 consists of those numbers that are multiples of both 12 and 18, that is,

$$\{36, 72, 108, \ldots\}.$$

The smallest member of this set, 36, is called the (L.C.M.) of the two numbers. In general, the **least common multiple** of two or more counting numbers is the smallest number that is a multiple of each of the given numbers.

We may use the prime factorization of two numbers to find their lowest common multiple. First express each number by its prime factorization. Then consider the prime factors that are factors of either of the given numbers, and take the product of these prime numbers with each raised to the highest power that occurs in *either* of the prime factorizations. For $12 = 2^2 \times 3$ and $18 = 2 \times 3^2$, we have L.C.M. $= 2^2 \times 3^2$, that is, 36.

Example 5: Find the lowest common multiple of 3850 and 5280.

Solution: As in Examples 1 and 2,

$$3850 = 2 \times 5^2 \times 7 \times 11$$
$$5280 = 2^5 \times 3 \times 5 \times 11.$$

The lowest common multiple of 3850 and 5280 is

$$2^5 \times 3 \times 5^2 \times 7 \times 11$$

that is, 184,800.

Example 6: Find the L.C.M. of 12, 18, and 20.

Solution: First write the prime factorization of each number:

$$12 = 2^2 \times 3$$
$$18 = 2 \times 3^2$$
$$20 = 2^2 \times 5$$

The L.C.M. is $2^2 \times 3^2 \times 5$, that is, 180.

We use the lowest common multiple of the denominators of two fractions when we add or subtract fractions. For example, the lowest common multiple of 12 and 18 is 36:

$$\frac{7}{12} + \frac{5}{18} = \frac{21}{36} + \frac{10}{36} = \frac{31}{36}$$

The answer is in reduced form, since 31 and 36 are relatively prime.

Example 7: Simplify: $\frac{37}{5280} - \frac{19}{3850}$.

Solution: We use the lowest common multiple as found in Example 5:

$$\frac{37}{5280} - \frac{19}{3850} = \frac{37}{2^5 \times 3 \times 5 \times 11} - \frac{19}{2 \times 5^2 \times 7 \times 11}$$

$$= \frac{37 \times 5 \times 7}{2^5 \times 3 \times 5^2 \times 7 \times 11} - \frac{19 \times 2^4 \times 3}{2^5 \times 3 \times 5^2 \times 7 \times 11}$$

$$= \frac{1295 - 912}{2^5 \times 3 \times 5^2 \times 7 \times 11} = \frac{383}{184,800}$$

The instruction "simplify" is used as in Example 7 to mean "perform the indicated operations and express the answer in simplest form." In the case of fractions, "express in simplest form" means "reduce to lowest terms."

EXERCISES

List the factors of:

1. 20   2. 24   3. 9   4. 16
5. 28   6. 48   7. 60   8. 72

Write the prime factorizations and find the G.C.F. of:

9. 68 and 76   10. 123 and 215
11. 76 and 1425   12. 123 and 1425
13. 215 and 1425   14. 68 and 738

15. 12, 15, and 20   16. 18, 24, and 40

17. 12, 18, and 30   18. 15, 45, and 60

List the first five elements in the set of multiples of:

19. 7   20. 8

21. 15   22. 20

Write the prime factorizations and find the L.C.M. of:

23. 68 and 76   24. 123 and 215

25. 76 and 1425   26. 123 and 1425

27. 215 and 1425   28. 68 and 738

29. 12, 15, and 20   30. 18, 24, and 40

31. 12, 18, and 30   32. 15, 45, and 60

Simplify:

33. $\frac{123}{215}$ 34. $\frac{76}{1425}$

35. $\frac{3}{8} + \frac{5}{12}$ 36. $\frac{3}{4} - \frac{1}{6}$

37. $\frac{7}{10} - \frac{4}{15}$ 38. $\frac{11}{12} + \frac{3}{16}$

39. $\frac{11}{12} - \frac{9}{18}$ 40. $\frac{5}{68} + \frac{11}{76}$

41. $\frac{7}{123} - \frac{2}{215}$ 42. $\frac{41}{215} + \frac{19}{1425}$

*43. What is the *least* common factor of any two counting numbers?

*44. What is the *greatest* common multiple of any two counting numbers?

PEDAGOGICAL CONSIDERATIONS

Observe the uses of prime factorizations that have been made in this section. Then describe several of the uses of prime factorizations in elementary school mathematics. Check your expectations by looking at two or more series of elementary school textbooks.

5-4

equivalence and order relations

Equality is the most common relation among numbers. We have used equalities to state that two symbols or expressions are names for the same number. For example, $2 + 3 = 5$.

The equality relation has the following basic properties, and is therefore called an *equivalence relation:*

169

sec. 5-4
equivalence
and
order
relations

Reflexive: $a = a.$
Symmetric: If $a = b$, then $b = a.$
Transitive: If $a = b$ and $b = c$, then $a = c.$

In general, consider a relation R and a set S. Let a, b, and c represent arbitrary elements of S. The relation R is said to be an **equivalence relation** if these three properties are satisfied:

Reflexive: a R a for all elements a in S.
Symmetric: If a R b, then b R a for all a and b in S.
Transitive: If a R b and b R c, then a R c for all a, b, and c in S.

We illustrate these properties for various binary relations in the following examples. A **binary relation** is one that associates a single element with any two given elements of a set.

Example 1: Test the relation "is perpendicular to" for lines in a plane to see if this relation is (a) reflexive; (b) symmetric; (c) transitive.

Solution: (a) The relation is not reflexive since a line is not considered to be perpendicular to itself.

(b) If line m is perpendicular to line n, then line n is perpendicular to line m. The relation is symmetric.

(c) In the figure below, we have $m \perp n$, $n \perp s$, but m is not perpendicular to s. The relation is not transitive.

$m \perp n$
$n \perp s$
($\perp$ is read "is perpendicular to")

Example 2: Test the relation "weighs within five pounds of" to determine whether or not it is an equivalence relation.

Solution: The relation is reflexive, since a person weighs within five pounds of his own weight. Furthermore if A weighs within five pounds of B, then B weighs within five pounds of A; the relation is symmetric. However, one counterexample is sufficient to show that the relation is not transitive. Thus assume the weights of A, B, and C are 145, 149, and 153, respectively. Then A weighs within five pounds of B, B weighs within five pounds of C, but A does not weigh within five pounds of C. The relation is therefore not an equivalence relation.

Relations are widely used for many types of elements. For example, "being of the same age" is an equivalence relation among people. Consider three students Don, John, and Bill. Note that even though you do not know their ages, you do know that Don is the same age that he is. If Don and John are the same age, then John and Don are the same age. Also, if Don and John are the same age and John and Bill are the same age, then Don and Bill are the same age.

The relation of "being younger than" is not an equivalence relation since, for example, Don is not younger than himself (not reflexive). Also if Don is younger than Sue, then Sue is not younger than Don (not symmetric). However, if Don is younger than Sue and Sue is younger than Debbie, then Don is younger than Debbie. The relation "is younger than" is transitive.

Any relation may be considered as a subset of a Cartesian product. Suppose that Don, John, and Bill are each 19 years old, Sue is 20, and Debbie is 21. In the array the dots indicate the elements of $S \times S$, where $S = \{$Don, John, Bill, Sue, Debbie$\}$. For each element (a, b) of $S \times S$ the dot in the first array is circled if a is the same age as b.

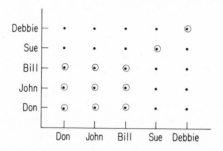

In the next array the dot for (a, b) is circled if a is younger than b.

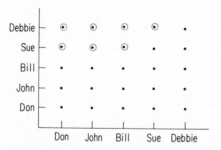

Other arrays could be made for other relations such as "is at least as old as," "is not the same age as," "is older than," "is taller than," "is a brother or sister of," "has the same color eyes as," and so forth.

For numbers the **order relations** *is less than* and *is greater than* have special importance because of their numerous interpretations such as *is younger than* and *is older than*. You may think of one number such as 2 as less than 5 because the graph of 2 is on the left of the graph of 5 on a num-

171

sec. 5-4
equivalence
and
order
relations

ber line such as the one below:

Formally, 2 is less than 5 because there is a counting number 3 that can be added to 2 to obtain 5, that is,

$$2 < 5 \quad \text{because} \quad 2 + 3 = 5.$$

We write $5 > 2$, 5 is greater than 2, if and only if $2 < 5$.

Example 3: Show the relation $<$ on the Cartesian product $A \times A$, where $A = \{1, 2, 3, 4, 5\}$.

Solution:

EXERCISES

1. Let $A = \{1\}$, $B = \{1, 2\}$, and $C = \{1, 2, 3\}$. Consider the statements
   (**a**) $A \, R \, A$;   (**b**) $A \, R \, B$;   (**c**) $A \, R \, C$;
   (**d**) $B \, R \, A$;   (**e**) $B \, R \, B$;   (**f**) $B \, R \, C$;
   (**g**) $C \, R \, A$;   (**h**) $C \, R \, B$;   (**i**) $C \, R \, C$.

   Which of these statements is true for the relation $\subseteq$, is a subset of?

2. As in Exercise 1, which of the statements is true for the relation $\subset$, is a proper subset of?

3. As in Exercise 1, which of the statements is true for $\not\subset$, is not a proper subset of?

*For Exercises 4 through 12 tell whether each relation is (**a**) reflexive, (**b**) symmetric, (**c**) transitive:*

4. For people, is a daughter of.

5. For students, is a classmate of.

6. For numbers, is less than, $<$.

7. For numbers, is less than or equal to, $\leq$.

8. For books, is heavier than.

9. For meals, is less expensive than.

10. For lines in a plane, is parallel to, $\parallel$.

11. For numbers, is divisible by.

12. For sets, is a subset of, $\subseteq$.

13. Consider the three relations $<$, $=$, $>$ and tell which relations can be used in place of R in the statement:
   (a) 5 R 4
   (b) 25 R 42
   (c) $(10 + 3)$ R $(3 + 10)$
   (d) $[2 + (3 \times 5)]$ R $[8 + (6 + 2)]$
   (e) $[(5 \times 2) + 1]$ R $[(5 \times 2) + (5 \times 1)]$

14. In Exercise 13 exactly one relation could be used in each case. This is a general property of counting numbers called the **trichotomy law**; for any counting numbers a, b exactly one of the relations must hold $a < b$, $a = b$, $a > b$. State which of the three relations hold for:
   (a) $5 + (7 \times 6)$ and $6 + (7 \times 5)$
   (b) b and $b + 1$
   (c) $b + d$ and $a + d$ if $b < a$
   (d) $b \times d$ and $a \times d$ if $b > a$

15. Explain each statement:
   (a) Either $a = b$ or $a \neq b$.
   (b) If $a \neq b$, then either $a < b$ or $a > b$.

PEDAGOGICAL CONSIDERATIONS

The three properties of an equivalence relation should be satisfied in all cases in which the symbol $=$ is used. Other symbols may be used for equivalence relations. In many classes it is worth while to use a variety of examples (age, height, nationality, shape, size, cardinal number, and so forth) to illustrate that being alike in any respect is an equivalence relation. Then, since there are so many ways in which elements can be alike, it often seems worth while to use different symbols to show the manner in which the elements are alike. For example, for geometric figures we use:

$=$ for two names for the same figure.

$=$ for the measures of two figures of the same size.

$\sim$	for two figures of the same shape.
$\cong$	for two figures of the same shape and size.

Order relations are used with young children in many ways. For example, they may be asked to arrange a set of blocks in order from smallest to largest, to line up in order from shortest to tallest, and so forth.

1. Give three examples of common uses of equivalence relations in first grade classes.

2. Repeat Exercise 1 for order relations.

The Secondary School Mathematics Curriculum Improvement Study is a federally financed group that is involved with writing experimental materials for gifted mathematics students in grades 7 through 12. Their headquarters is located at Teachers College, Columbia University, New York City and they operate under the direction of Dr. Howard F. Fehr. In their seventh grade materials they use arrow diagrams to show relations in the following way:

a is related to b b is related to a a is related to a

$_aR_b$ $_bR_a$ $_aR_a$

If a relation is reflexive, there is a loop at every element of the given set under discussion. If a relation is symmetric, then all arrows that go from an element a to an element b must also have an arrow that goes from b to a. A relation is transitive if whenever there is an arrow from a to b, and one from b to c, then there is also an arrow from a to c.

3. Here is a graph for the relation "divides" for the set $M = \{2, 3, 6, 12\}$. From the graph, show that the relation is reflexive and transitive but not symmetric.

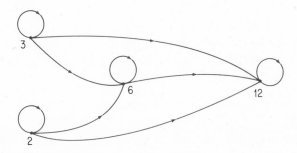

Draw arrow graphs for each of the following relations on the sets given:

4. Is greater than for $A = \{2, 5, 6, 9, 10\}$.

5. Is greater than or equal to for $B = \{3, 8, 9, 12\}$.

6. Is within three pounds of for the set of weights $C = \{130, 132, 134\}$.

whole
numbers

The counting numbers serve man well and are actually all that he needs for many purposes. However, with just this collection of numbers at his disposal he is unable to ask for *half* of a piece of pie, or to say that he has *no* money in his pocket. We need to expand our number system and shall do so by first introducing a numeral 0 to represent the number zero of elements in the empty set. Zero is the cardinal number of any empty set.

We also need the number zero to solve sentences such as $5 + n = 5$. If $5 + n = 5$, then $n = 0$ and $5 + 0 = 5$.

> **Identity, +:** $a + 0 = a = 0 + a$; the number zero is the additive identity and is also called the **identity element for addition**. This property is often referred to as the **addition property of zero**.

When we include 0 with the set of counting numbers, we form a new set W of **whole numbers**:

$$W = \{0, 1, 2, 3, 4, 5, \ldots\}$$

Note that the only difference between the set of counting numbers and the set of whole numbers is the element 0. Often we say that the set of whole numbers consists of the **union** of the set of counting numbers and zero.

We can locate this set of numbers on a number line and identify points on the line by numbers. The numbers are the **coordinates** of the points; the points are the **graphs** of the numbers. First draw a line and select any point of the line as the **origin** with coordinate 0, and any other point as the unit point with coordinate 1. Usually the number line is considered in a horizontal position with the unit point on the right of the origin.

The length of the line segment with the origin and the unit point as endpoints is the **unit distance** for marking off a scale on the line. The point representing any given counting number may be obtained by marking off successive units to the right of the origin.

We can now **graph** different sets of numbers on a number line as in the following examples. We shall use the verb "graph" to mean "draw the graph of."

Example 1: Graph the set of whole numbers less than 3 on a number line.

Solution:

Example 2: Graph the set of whole numbers between 1 and 5 on a number line.

Solution: The set to be graphed is {2, 3, 4}.

$$0 \quad 1 \quad 2 \quad 3 \quad 4 \quad 5 \quad 6$$

The set of whole numbers has all of the properties that were described for the set of counting numbers, as well as having an identity element for addition. In summary these properties are:

Closure, +	Closure, ×
Commutative, +	Commutative, ×
Associative, +	Associative, ×
Identity, +	Identity, ×
Distributive, × with respect to +	

The distinguishing feature between the set of counting numbers and the set of whole numbers is the presence of 0 as a whole number. The number 0 not only serves as the additive identity, but also has the property:

Zero, ✗: $a \times 0 = 0 \times a = 0$; the product of any whole number a and 0 is 0. This property is frequently referred to as the **multiplication property of zero.**

EXERCISES

In Exercises 1 through 10 graph each set of numbers on a number line.

1. {1, 2, 3}
2. {1, 5, 7}
3. {0, 2, 4, 6}
4. {0}
5. The set of counting numbers less than 5.
6. The set of whole numbers less than 5.
7. The set of whole numbers between 0 and 5.
8. The set of counting numbers between 1 and 3.
9. The set of counting numbers between 0 and 1.
10. The set of whole numbers greater than or equal to 3 and less than 7.
11. Show that there is a one-to-one correspondence between the elements of the set of whole numbers less than 5 and the elements of the set of counting numbers less than 6.

175

12. If a and b represent whole numbers and $a \times b = 0$, what can you conclude about a and b?

13. Repeat Exercise 12 if $a \times b = 1$; if $a \times b = 2$; if $a \times b = 3$; if $a \times b = 4$.

14. Does every point on the number line have a whole number as its coordinate? Is every whole number the coordinate of some point on the number line?

15. Let n represent any whole number. How many whole numbers are there between n and $n + 1$?

16. If n represents any whole number, then $n + 1$ is called the **successor** of n. Does every whole number have a successor? Is every whole number the successor of some other whole number?

17. Show that there is a one-to-one correspondence between the elements of the set of whole numbers and the elements of the set of counting numbers.

18. Consider $A \times \varnothing$ and $\varnothing \times A$ for any set A. Identify these sets and the associated property of whole numbers.

PEDAGOGICAL CONSIDERATIONS

1. Look at two or more series of elementary school textbooks and summarize the uses of whole numbers that you find.

2. Many interesting games and puzzles that involve counting numbers only can be used in the classroom. One such puzzle involves using four fours, and any operation with which the student is familiar. The objective is to represent as many counting numbers as possible in this way. Here are several examples:

$$1 = \tfrac{44}{44}, \qquad\qquad 2 = (4 \div 4) + (4 \div 4)$$
$$3 = (4 + 4 + 4) \div 4, \qquad 4 = 4 + [(4 - 4) \div 4]$$

See how far you can extend this list.

3. Another interesting puzzle is to represent the set of counting numbers using the digits of a year *in the order in which they appear.* Here are several examples, using the year 1975:

$$1 = 1^{975}, \qquad 3 = 1 + [(\sqrt{9} + 7) \div 5]$$

How many others can you find?

4. "Pretzel charts" such as the following are popular with elementary school students. The input numbers, the numbers used at the start, may be varied for additional practice. Make up several similar charts of your own.

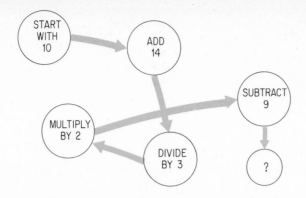

5. Flow charts are also popular with young students. For the following, find the output numbers for each of the given inputs. Then make up several other similar examples of your own.

INPUT	OUTPUT
4	?
5	?
0	?
11	?
12	?

Cross-number puzzles are very popular with elementary school students and serve as a novel and interesting way to provide practice with fundamental skills. The first figure on the next page shows such a puzzle for addition; the student is to find the sum of each row, column, and diagonal. In the second figure the sums are shown for each row and column. These partial sums, 15 + 15 and 17 + 13, are then added across and down to give the sum 30 shown in the box in the lower right hand corner. This serves as a check in that both sums should be the same. The third figure shows a final check on the student's work. He is asked to add along the diagonals, 7 + 9 and 8 + 6. These diagonal sums are placed in the circles

177

and their sum, $14 + 16$, should give the number found in the box, 30.

6. Prepare a similar cross-number puzzle for multiplication.

7. Prepare cross-number puzzles for subtraction and division. In these cases the diagonal check does not seem to work. See if you can suggest a modification of the procedures used for addition and multiplication so that the diagonals may be used as checks.

8. Elementary school students have fun with puzzles that have missing entries. Complete the following and make up similar ones of your own for use in class.

(a)   **(b)**

5-6

integers Once again we extend our number system. With just the set of whole numbers at our disposal, we are still unable to find replacements for n that make these sentences true:

$$5 + n = 2$$
$$n + 4 = 0$$
$$2 - 7 = n$$

That is, the solution set for each of these sentences is the empty set if only the set of whole numbers may be used as the set of possible replacements for the **variable n.**

We begin our extension of the number system by considering points on the number line to the left of the origin 0. For each point on the number line that is the coordinate of a whole number n, locate a point to the left of the origin that is the same distance from 0 as the point with coordinate n. We call the coordinate of this new point **negative n,** written $-n$, and refer to the number that $-n$ represents as the **opposite** of n.

In the figure the point with coordinate negative 3 is located three units to the left of 0. We say that the opposite of 3 is -3; also, the opposite of -3 is 3. Similarly, the point with coordinate -5 is located five units to the left of 0, and we say that 5 and -5 are opposites of each other. Finally we agree that the opposite of 0 is 0; that is, $-0 = 0$.

We call the set of whole numbers together with the set of the opposites of the whole numbers, the set I of **integers**:

$$I = \{\ldots, -3, -2, -1, 0, 1, 2, 3, \ldots\}$$

Note that the graph of the set of integers continues indefinitely both in the positive direction (to the right of 0) and in the negative direction (to the left of 0).

The set of integers may be considered as the union of these sets of numbers:

Positive integers: $\{1, 2, 3, 4, 5, \ldots\}$
Zero: $\{0\}$
Negative integers: $\{\ldots, -5, -4, -3, -2, -1\}$

Example 1: Graph the set of integers between -3 and 2.

Solution:

Inverse, $+$: $n + (-n) = 0$; every integer n has an opposite $-n$ such that $n + (-n) = 0$. This property is sometimes referred to as the **addition property of opposites.**

Each of the integers n and $-n$ is said to be the **additive inverse** or opposite of the other. That is, the additive inverse of 3 is -3, the additive inverse of -3 is 3, and $3 + (-3) = 0$. Indeed, the distinguishing feature between the set of integers and the set of whole numbers is that every integer has an additive inverse *in the set*.

We are now able to solve the sentences such as those given at the beginning of this section.

Sentence	Solution
$5 + n = 2$	$n = -3$
$n + 4 = 0$	$n = -4$
$2 - 7 = n$	$n = -5$

In the following section we shall carefully review the operations of addition, subtraction, multiplication, and division with integers. However, based on past experience and intuition, the reader should see that we have the following properties of the set of integers under addition:

Closure, $+$: The sum of any two integers is an integer.

Associative, $+$: $a + (b + c) = (a + b) + c$ for all integers a, b, and c.

Identity, $+$: The set of integers contains an element 0 such that $a + 0 = 0 + a = a$ for every integer a.

Inverse, $+$: For each integer a, the set of integers contains an element $-a$ such that $a + (-a) = 0$.

Commutative, $+$: $a + b = b + a$ for all integers a and b.

We indicate that the set of integers has the first four properties by saying that the set of integers forms a **group under addition**. Furthermore, since addition of integers is also commutative, we say that the set of integers forms a **commutative group** with respect to addition. (See §3-1.)

Example 2: Show that subtraction of integers is not commutative.

Solution: We can show this by a single counterexample such as $5 - 2 = 3$ and $2 - 5 = -3$; $5 - 2 \neq 2 - 5$.

Example 3: Show that the set of integers does *not* form a group with respect to multiplication.

Solution: Multiplication of integers is closed and associative, and there exists an identity element 1 since $n \times 1 = 1 \times n = n$ for each integer n. However, the set of integers does not contain the multiplicative inverse for each of its elements. For example, there is no integer n such that $3 \times n = 1$.

In general, we have the following properties of the set of integers under multiplication:

Closure, $\times$: The product of any two integers is an integer.

Associative, $\times$: $a \times (b \times c) = (a \times b) \times c$ for all integers a, b, and c.

Identity, $\times$: The set of integers contains an element 1 such that $a \times 1 = 1 \times a = a$ for every integer a.

Commutative, $\times$: $a \times b = b \times a$ for all integers a and b.

There are a number of interesting proofs that we are now ready to explore which illustrate the type of reasoning that mathematicians use. First we need several definitions; then we shall consider several proofs and offer others in the exercises.

An integer is **even** if it is a multiple of 2, that is, if it may be expressed as $2k$, where k stands for an integer. Then the set of even integers is

$$\{\ldots, -6, -4, -2, 0, 2, 4, 6, \ldots\}.$$

An integer that is not even is said to be **odd**. Each odd integer may be expressed in the form $2k + 1$, where k stands for an integer. Then the set of odd integers is

$$\{\ldots, -7, -5, -3, -1, 1, 3, 5, 7, \ldots\}.$$

Example 4: Prove that the sum of any two even integers is an even integer.

Proof: Any two even integers m and n may be expressed as $2k$ and $2r$, where k and r stand for integers. Then

$$m + n = 2k + 2r = 2(k + r)$$

where $k + r$ stands for an integer, since the sum of any two integers is an integer. Therefore, $m + n$ is an even integer.

Example 5: Prove that the square of any even integer is an even integer.

Proof: Any even integer may be expressed as $2k$, where k stands for an integer. Then the square of the integer may be expressed as $(2k)^2$, where

$$(2k)^2 = (2k)(2k) = 2(2k^2).$$

Since k, k^2, and $2k^2$ all stand for integers, $(2k)^2$ stands for an even integer.

EXERCISES

1. Show that there is a one-to-one correspondence between the elements of the set of positive integers and the elements of the set of negative integers.

2. Show that the set of whole numbers does not form a group under addition.

Graph each set of numbers on a number line.

3. The set of integers between -5 and 3.

4. The set of integers -5 through 3 inclusive.

5. The set of integers that are the opposites of the first five counting numbers.

6. The set of integers that are the opposites of the members of the set $M = \{-3, -2, -1\}$.

Classify each of the statements in Exercises 7 through 15 as true or false.

7. Every counting number is an integer.

8. Every whole number is an integer.

9. Every integer is a whole number.

10. Every integer is either positive or negative.

11. Every integer is the opposite of some integer.

12. The set of integers is the same as the set of the opposites of the integers.

13. The set of negative integers is the same as the set of the opposites of the whole numbers.

14. The set of integers is closed under multiplication.

15. The set of integers is closed under division.

16. What is the intersection of the set of positive integers and the set of negative integers?

17. Is the union of the set of positive integers and the set of negative integers equal to the set of integers? Explain your answer.

18. Show that there is a one-to-one correspondence between the elements of the set of whole numbers and the elements of the set of integers.

19. Show that there is a one-to-one correspondence between the set of integers and the set of counting numbers.

Prove each statement.

20. The sum of any two odd integers is an even integer.

21. The product of any two even integers is an even integer.

22. The square of any odd integer is an odd integer.

23. If the square of an integer is odd, the integer is odd; if the square of an integer is even, the integer is even.

PEDAGOGICAL CONSIDERATIONS

1. Explore several elementary textbook series to determine the grade level at which the set of integers is first introduced. Report on how negative integers are introduced to elementary school students.

2. An interesting approach to addition of integers is to use the concept of positive and negative arrows. For example, consider the following:

These three arrows together illustrate −3:

These two arrows together illustrate +2:

Positive and negative arrows are then combined to give zero pairs as follows:

We have one negative arrow left. Thus $(-3) + (+2) = -1$.

Prepare a short unit introducing addition of integers in this way.

3. Here is a nomograph that can be used to find sums of integers. Just connect the point representing one addend on the A scale, with the corresponding point for the other addend on the B scale. The point where the line crosses the S scale will give the sum. The figure shows the sum $(+4) + (-6) = -2$. Make your own nomograph using graph paper and use it to find various sums of integers.

4. See if you can use your knowledge of elementary geometry to explain why the nomograph shown works as it does.

5. Try to determine how the nomograph shown can be used for subtraction of integers.

6. Prepare a nomograph that will enable you to find $a + 2b$ directly for any two integers a and b.

7. Read Booklet 2 (The Whole Numbers) of the twenty-ninth yearbook of the National Council of Teachers of Mathematics, *Topics in Mathematics for Elementary School Teachers.*

8. Read Booklet 9 (The System of Integers) of the thirtieth yearbook of the National Council of Teachers of Mathematics, *More Topics in Mathematics for Elementary School Teachers.*

5-7

operations with integers

It is assumed that the reader has had past experience with the four fundamental operations with integers, and some practice on these skills is offered in the exercises as a means of review. It may be enlightening, however, to review these operations in some detail.

The number line can be used quite effectively to help us add any two integers. Represent the first addend by an arrow that starts at 0, the origin. From the end of the first arrow, draw a second arrow to show the second addend. Positive numbers are shown by arrows that go to the right, and negative numbers by arrows that go to the left. Finally, the sum is found as the coordinate of the point at the end of the second arrow. These examples illustrate this number-line method for addition of integers.

Example 1: Illustrate on a number line: $(+3) + (-7)$.

Solution: Start at 0 and draw an arrow that goes three units to the right. Then show an arrow that goes seven units to the left. The second arrow ends at -4. Thus, $(+3) + (-7) = -4$.

Example 2: Illustrate on a number line: $(-2) + (-3)$.

Solution:

From the figure we see that $(-2) + (-3) = -5$.

Since subtraction is the opposite operation of addition, we can do every subtraction problem by transforming it into an equivalent addition problem. Thus compare these examples:

Subtraction	Addition
$7 - 3 = 4$	$7 + (-3) = 4$
$8 - 2 = 6$	$8 + (-2) = 6$
$12 - 7 = 5$	$12 + (-7) = 5$

In general we define the subtraction $a - b$ of two integers a and b as follows:

$$a - b = a + (-b)$$

Note that $-b$ can be thought of as $(-1) \times b$. Thus we may also define subtraction in this way:

$$a - b = a + (-1) \times b$$

Example 3: Illustrate on a number line: $(-2) - (+3)$.

Solution: By the definition of subtraction this can be thought of as $-2 + (-3)$; that is, -5. On the number line we think of this as a movement of two units from the origin in the negative direction, followed by a movement of three units in the negative direction, as in Example 2.

All too often the rules for multiplication of integers have been learned by rote without any attempt to see or to understand their development. Thus most people know that the product of two negative numbers is a positive number. They know that the product of a positive number and a negative number is a negative number. However, they are unable to give any justification for these rules. Informally we might say that if you lose $5 a day for 2 days, then you have lost $10, so $2 \times (-5) = -10$. We can justify such products by thinking of multiplication as repeated addition. For example:

$$2 \times (-5) = (-5) + (-5) = -10$$

Let us examine a mathematical justification for this product based upon the use of the distributive property.

Notice that under the usual properties of addition and multiplication we have the following:

$$5 + (-5) = 0$$
$$2 \times [5 + (-5)] = 2 \times 0 = 0$$
$$(2 \times 5) + [2 \times (-5)] = 0 \qquad \text{(by the distributive property)}$$

Since the sum of 2×5 and $2 \times (-5)$ is zero, each must be the opposite of the other. Then $2 \times (-5) = -10$, since $2 \times 5 = 10$. Furthermore, since $2 \times (-5) = (-5) \times 2$, we have $(-5) \times 2 = -10$. In general, for any positive integers b and c:

$$b + (-b) = 0$$
$$c \times [b + (-b)] = c \times 0 = 0$$
$$(c \times b) + [c \times (-b)] = 0$$

Thus $c \times b$ and $c \times (-b)$ are each the opposite of the other; $c \times b = cb$ and $c \times (-b) = -cb$. Since $cb = bc$ and $c \times (-b) = (-b) \times c$, we have $(-b) \times c = -bc$. Thus the product of a positive integer and a negative integer is a negative integer.

Justifying the product of two negative numbers is somewhat more difficult. Informally, we might say that if you lose \$5 a day for two days, then two days *ago* you had \$10 more than you have today. Therefore, $(-2) \times (-5) = +10$. Again, let us consider a mathematical justification for this product through the use of the distributive property as follows:

$$5 + (-5) = 0$$
$$(-2) \times [5 + (-5)] = (-2) \times 0 = 0$$
$$[(-2) \times 5] + [(-2) \times (-5)] = 0 \qquad \text{(by the distributive property)}$$

Since the sum of $(-2) \times 5$ and $(-2) \times (-5)$ is zero, each must be the opposite of the other. Furthermore, since $(-2) \times 5 = -10$, we have $(-2) \times (-5) = 10$. In general, for any positive integers b and c:

$$b + (-b) = 0$$
$$(-c) \times [b + (-b)] = (-c) \times 0 = 0$$
$$[(-c) \times b] + [(-c) \times (-b)] = 0$$

Thus $(-c) \times b$ and $(-c) \times (-b)$ are each the opposite of the other. Then, since $(-c) \times b = -cb$, we have $(-c) \times (-b) = cb$. Thus the product of two negative integers is a positive integer.

Just as subtraction is the opposite operation of addition, so also is division the opposite operation of multiplication. In general we define the quotient $a \div b$ of two integers a and b as follows:

$$a \div b = a \times \frac{1}{b}$$

Example 4: Find the quotient $(-12) \div (+3)$.

Solution:

$$-12 \div 3 = -12 \times \frac{1}{3} = -4$$

Through similar examples we see that the quotient of two positive integers is positive, the quotient of two negative integers is positive, and the quotient of a positive and a negative integer is negative. The set of integers is *not* closed with respect to division, as noted in the following example. The desire to make division always possible will lead to further extensions of the number system in the next chapter.

Example 5: Show that the set of integers is not closed with respect to division.

Solution: We can show this by a single counterexample: $8 \div 3$ cannot be named by an integer. That is, there is no integer n such that $3 \times n = 8$.

EXERCISES

Find and illustrate each of the following sums on a number line.

1. $(+8) + (+3)$ 2. $(+8) + (-3)$ 3. $(-8) + (+3)$
4. $(-8) + (-3)$ 5. $(-9) + (+12)$ 6. $(+7) + (-7)$
7. $(-5) + (-9)$ 8. $(+2) + (-6)$

Write each subtraction problem as an equivalent addition and find the difference.

9. $(+8) - (+3)$ 10. $(+3) - (+8)$ 11. $(-8) - (+3)$
12. $(-3) - (+8)$ 13. $(-3) - (-8)$ 14. $(-8) - (-3)$
15. $(-8) - (+8)$ 16. $(-8) - (-8)$

Find each product.

17. $(+5) \times (+4)$ 18. $(-5) \times (+4)$ 19. $(+5) \times (-4)$
20. $(-5) \times (-4)$ 21. $(-5) \times 0$ 22. $(-8) \times (+5)$
23. $(-7) \times (-6)$ 24. $(-6) \times (+9)$

Find each quotient.

25. $(+12) \div (+3)$ 26. $(+12) \div (-3)$ 27. $(-12) \div (+3)$
28. $(-12) \div (-3)$ 29. $0 \div (-3)$ 30. $(-8) \div (-8)$
31. $(+42) \div (-6)$ 32. $(-54) \div (-9)$

Perform the indicated operations.

33. $(-3) + [(-5) + (-7)]$ 34. $[(-3) + (-5)] + (-7)$
35. $(-2) \times [4 \times (-3)]$ 36. $[(-2) \times (4)] \times (-3)$
37. $(12) \div [6 \div (-2)]$ 38. $(12 \div 6) \div (-2)$

39. Use the results of Exercises 37 and 38 to form a conjecture about the associativity of integers with respect to division. Then find another example to help confirm your conjecture.

40. Use a counterexample to show that the set of integers is not commutative with respect to subtraction.

1. Many elementary textbooks introduce the set of negative integers by using raised negative signs. Thus the set N of negative integers would be written in this way:

$$\{\ldots, {}^{-}5, {}^{-}4, {}^{-}3, {}^{-}2, {}^{-}1\}$$

Discuss the pedagogical value for introducing the set in this manner' thus reserving the minus sign, $-$, for subtraction. Refer to a first year algebra text to see how the transition is made from raised negative signs to lowered ones.

2. Explore the School Mathematics Study Group junior high school texts to see their manner of introducing the rules for multiplying integers by means of discovery of patterns. Thus, for example, they ask their students to complete the following multiplication table using the pattern shown by the positive integers and zero:

X	−3	−2	−1	0	1	2	3
−3				0			
−2				0			
−1				0			
0	0	0	0	0	0	0	0
1				0	1	2	3
2				0	2	4	6
3				0	3	6	9

5-8

absolute value   In order to give a formal definition of addition and multiplication of integers there is one more concept that we need to introduce. On the number line you should see that there are two points that are at a distance of three units from the origin; they have coordinates 3 and -3. We use the symbol $|x|$, read "the *absolute value* of x," to represent the distance of a point from the origin on the number line.

Thus for the diagram shown we may write

$$|3| = 3 \quad \text{and} \quad |-3| = 3.$$

In general, we define the **absolute value** of any integer k as follows:

$$|k| = k, \qquad \text{if } k \text{ is positive}$$
$$|k| = -k, \qquad \text{if } k \text{ is negative}$$
$$|k| = 0, \qquad \text{if } k = 0$$

Notice in the second case that if k represents a negative number, then $-k$ represents a positive number. For example, if $x = -3, |-3| = -(-3) = 3$.

Example 1: Evaluate: $|-7| + |-2| + |+5|$.

Solution: First we note the following:

$$|-7| = 7, \qquad |-2| = 2, \qquad |+5| = 5$$

Thus

$$|-7| + |-2| + |+5| = 7 + 2 + 5 = 14.$$

Example 2: Evaluate: $|3 - 7|$.

Solution: First we evaluate the expression $3 - 7$ as $3 + (-7) = -4$. Thus $|3 - 7| = |-4| = 4$.

We are now able to define the sum of two integers a and b under these four separate cases:

Case 1: $a + b = |a| + |b|$, when $a \geq 0$ and $b \geq 0$.
For example, let $a = +3$ and $b = +5$:

$$(+3) + (+5) = |+3| + |+5| = 3 + 5 = 8$$

Case 2: $a + b = -(|a| + |b|)$, when $a \leq 0$ and $b \leq 0$.
For example, let $a = -3$ and $b = -5$:

$$(-3) + (-5) = -(|-3| + |-5|) = -(3 + 5) = -8$$

Case 3: $a + b = |a| - |b|$, when $a > 0$, $b < 0$, and $|a| > |b|$.
For example, let $a = +5$ and $b = -3$:

$$(+5) + (-3) = |+5| - |-3| = 5 - 3 = 2$$

Case 4: $a + b = -(|b| - |a|)$, when $a > 0$, $b < 0$, and $|b| > |a|$.
For example, let $a = +3$ and $b = -5$:

$$(-5) + (+3) = -(|-5| - |+3|) = -(5 - 3) = -2$$

Usually we find the sum of two integers by informal methods, as in the preceding section. However it is important to note that we can offer formal definitions for finding such sums. In a similar manner we can formally define the product of two integers a and b under the following cases:

Case 1: $a \times b = |a| \times |b|$, when $a > 0$ and $b > 0$.
For example, let $a = +3$ and $b = +5$:

$$(+3) \times (+5) = |+3| \times |+5| = 3 \times 5 = 15$$

Case 2: $a \times b = |a| \times |b|$, when $a < 0$ and $b < 0$.
For example, let $a = -3$ and $b = -5$:

$$(-3) \times (-5) = |-3| \times |-5| = 3 \times 5 = 15$$

Case 3: $a \times b = -(|a| \times |b|)$, when $a > 0$ and $b < 0$, or when $a < 0$ and $b > 0$.
For example, let $a = +3$ and $b = -5$:

$$(+3) \times (-5) = -(|+3| \times |-5|) = -(3 \times 5) = -15$$

For another example, let $a = -3$ and $b = +5$:

$$(-3) \times (+5) = -(|-3| \times |+5|) = -(3 \times 5) = -15$$

Note also that the definition does not include any mention of multiplication by 0. This case is covered by the multiplication property of zero, which states that for any integer n, $n \times 0 = 0$.

The various cases for sums and products of positive and negative integers may be summarized as follows:

The sum of two positive integers is a positive integer.
The sum of two negative integers is a negative integer.
The sum of a positive integer and a negative integer may
be either a positive integer or a negative integer.
The product of two positive integers is a positive integer.
The product of two negative integers is a positive integer.
The product of a positive integer and a negative integer is
a negative integer.

EXERCISES

Evaluate each of the following.

1. $|-7|$ 2. $|+15|$ 3. $-|+6|$ 4. $-|-6|$

5. $|+3| + |+5|$ 6. $|(+3) + (+5)|$

7. $|-3| + |-5|$ 8. $|(-3) + (-5)|$

9. $|-3| + |+5|$ 10. $|(-3) + (+5)|$

11. $-(|+3| + |-5|)$ 12. $-|(+3) + (-5)|$

13. $-(|-3| + |-5|)$ 14. $-|(-3) + (-5)|$

Use the definition of addition to find each sum.

15. $(+9) + (+5)$ 16. $(-9) + (-5)$

17. $(+9) + (-5)$ **18.** $(+5) + (-9)$

the definition of multiplication to find each product.

19. $(+7) \times (+6)$ **20.** $(-7) \times (-6)$

21. $(+7) \times (-6)$ **22.** $(-7) \times (+6)$

23. The absolute value of a number n is sometimes defined as being the greater of the number n or its opposite. Show with several examples that this definition is equivalent to the one given in this section.

24. Does $|a| + |b| = |a + b|$? Answer this question for each of these three cases: **(a)** $a > 0, b > 0$; **(b)** $a < 0, b < 0$; **(c)** $a > 0, b < 0$.

Find two replacements for n to make each of these sentences true.

25. $|n + 2| = 7$ **26.** $|n + 7| = 2$

27. $|n - 2| = 7$ **28.** $|n - 7| = 2$

PEDAGOGICAL CONSIDERATIONS

1. The University of Illinois Committee on School Mathematics has an interesting physical model to help explain the rules for multiplying integers. Read about this in their Algebra I course and report on it to class.

2. In many modern programs flow charts are used extensively to help explain steps in algorithms. Use various replacements for *a* and *b* and verify that the following flow chart gives the correct sum in each case.

3. Make a flow chart for finding the difference of two integers *a* and *b*.
4. Make a flow chart for finding the product of two integers *a* and *b*.
5. Make a flow chart for finding the quotient of two integers *a* and *b*.

CHAPTER TEST

1. Show a one-to-one correspondence between the set of whole numbers and the set of negative integers.

2. List the set of prime numbers that are less than 15.

3. Write the prime factorization of 300.

4. Find the greatest common factor of 120 and 140.

5. Find the least common multiple of 90 and 1500.

6. Classify each statement as true or false:
   (a) Every counting number is a whole number.
   (b) The opposite of every whole number is a negative integer.
   (c) The set of whole numbers is a subset of the set of integers.
   (d) The absolute value of every integer is a whole number.
   (e) The set of counting numbers is a proper subset of the set of whole numbers.

7. Tell whether each relation is reflexive, symmetric, and/or transitive:
   (a) For students in your class, "is taller than."
   (b) For the set of whole numbers, "is less than or is equal to."

8. Name the property illustrated by each statement for integers x, y, and z:
   (a) $(x + y) + z = x + (y + z)$ (b) $x(y + z) = xy + xz$

9. Perform the indicated operations:
   (a) $(+5) + (-9)$ (b) $(-3) - (-7)$
   (c) $(-8) \times (+3)$ (d) $(-45) \div (-5)$

10. Evaluate each of the following:
    (a) $|-5| + |+3|$ (b) $|(-5) + (+3)|$
    (c) $|-3| \times |+5|$ (d) $|(-3) \times (+5)|$

192

6

The Set of Rational Numbers

In the preceding chapter we extended the number system to include the set of integers. However, using only the integers as permissible replacements for the variable, we are still unable to solve such equations as $2n = 5$. That is, there is no integer n such that two times that integer is equal to five.

Using integers only, we are unable to solve many practical problems. The housewife can not use a recipe that calls for three cups of flour if she wishes to use only one-half the quantity. Thus the problem $3 \div 2$ has no solution in the system of integers. A child who wishes to divide a candy bar into two equal parts has no way to represent one of these parts. There is no integer to represent the solution to the problem $1 \div 2$.

Formally we say that the set of integers is not closed with respect to division. The quotient of two integers is not always an integer. Thus we need a further extension of the number system to make division always possible, except division by zero. In this chapter we make such an extension and call our new set of numbers the **set of rational numbers**.

6-1
symbols and definitions

A **rational number** is a number that can be expressed in the form

$$\frac{a}{b}$$

(often written a/b) where a is an integer and b is a counting number. Then

$$\frac{2}{3}, \frac{1}{2}, \frac{-7}{3}, \frac{0}{1}, \quad \text{and} \quad \frac{215}{524}$$

are examples of rational numbers. Many elementary school textbooks refer to positive numbers represented by such fractions as whole numbers and **fractional numbers**. We use rational numbers to include whole numbers, fractional numbers, and their opposites.

Example 1: Show that every integer is a rational number.

Solution: Every integer n can be written in the form $n/1$, the quotient of the integer and 1, and thus is a rational number. For example, 7 can be written as $\frac{7}{1}$.

Statements of equality such as

$$n = \frac{n}{1}$$

are used when two symbols represent the same number. Each rational number may be represented in many ways. For example,

$$\tfrac{6}{2}, \tfrac{15}{5}, \tfrac{30}{10}, \quad \text{and} \quad \tfrac{150}{50}$$

each represent the number $\frac{3}{1}$. In general, the equality of rational numbers a/b is *defined* as follows:

$$\frac{a}{b} = \frac{c}{d} \quad \text{if and only if} \quad ad = bc$$

Then

$$\tfrac{6}{2} = \tfrac{15}{5} \quad \text{since} \quad 6 \times 5 = 2 \times 15.$$

Order relations ($<$, $>$) among rational numbers are defined in a similar way:

$$\frac{a}{b} < \frac{c}{d} \quad \text{if and only if} \quad ad < bc$$

$$\frac{a}{b} > \frac{c}{d} \quad \text{if and only if} \quad ad > bc$$

Then

$$\tfrac{5}{8} < \tfrac{2}{3} \quad \text{since } 5 \times 3 < 8 \times 2$$

$$\tfrac{5}{7} > \tfrac{2}{3} \quad \text{since } 5 \times 3 < 7 \times 2.$$

Operations ($+$, $-$, $\times$, $\div$) upon rational numbers are also defined in terms of operations on integers. Definitions are given here for multiplication and addition. Later subtraction will be defined in terms of addition, and division will be defined in terms of multiplication.

The **product** of two rational numbers is defined as

$$\frac{a}{b} \times \frac{c}{d} = \frac{ac}{bd}.$$

For example,

$$\frac{3}{4} \times \frac{1}{5} = \frac{3 \times 1}{4 \times 5} = \frac{3}{20}.$$

The **sum** of two rational numbers is defined as

$$\frac{a}{b} + \frac{c}{d} = \frac{ad + bc}{bd}$$

since

$$\frac{a}{b} + \frac{c}{d} = \frac{a}{b} \times \frac{d}{d} + \frac{b}{b} \times \frac{c}{d} = \frac{ad}{bd} + \frac{bc}{bd}.$$

For example,

$$\frac{3}{4} + \frac{1}{5} = \frac{(3 \times 5) + (4 \times 1)}{4 \times 5} = \frac{15 + 4}{20} = \frac{19}{20}.$$

Note that the definition of a rational number prohibits the denominator from being 0, since 0 is not a counting number. Division by 0 is not permitted and it is important that we understand why this is so. To do so we should first see the relationship between the operations of division and multiplication.

$$a \div b \quad \text{is the number } n \text{ such that} \quad b \times n = a.$$

For example, $12 \div 3$ is the number n such that $3 \times n = 12$. Indeed, this is the way that elementary school children check their division problems.

$$\begin{array}{r} 4 \\ 3\overline{)12} \end{array} \qquad \textit{Check: } 3 \times 4 = 12$$

Now consider the problem $7 \div 0$.

$$\begin{array}{r} n \\ 0\overline{)7} \end{array} \quad 7 \div 0 \text{ is the number } n \text{ such that } 0 \times n = 7.$$

But $0 \times n = 0$ for all n, and can never be equal to 7.

In the preceding discussion we could have used any number other than 7 with similar results. Therefore we say that division by 0 is not possible.

Example 2: Show that $0 \div 0$ has no meaning.

Solution: First we note that $0 \div 0$ is the number n such that $0 \times n = 0$. But $0 \times n = 0$ for *all* n. Therefore $0 \div 0$ has no unique solution and is often said to be *indeterminate*.

These definitions of equality, order relations, multiplication, and addition arise from the representations of rational numbers on a number line and as fractions. The following exercises are intended to provide familiarity with the symbols through practice with the basic definitions before considering various interpretations and applications.

EXERCISES

Select the relation R ($=$, $<$, $>$) to make the statement true:

1. $\frac{1}{2}$ R $\frac{2}{3}$
2. $\frac{2}{3}$ R $\frac{5}{7}$
3. $\frac{1}{3}$ R $\frac{2}{7}$
4. $\frac{3}{4}$ R $\frac{7}{10}$
5. $\frac{5}{11}$ R $\frac{6}{13}$
6. $\frac{1}{8}$ R $\frac{2}{17}$

*Each symbol a/b has a **numerator** a and a **denominator***

7. Identify the numerator of (a) $\frac{2}{3}$; (b) $\frac{1}{5}$; (c) $\frac{2}{7}$; (d) $\frac{9}{7}$.
8. Identify the denominator of each symbol in Exercise 7.
9. Fractions with like denominators are used for addition and subtraction. Express each sum as a sum of fractions with denominator 12:

   (a) $\frac{1}{4} + \frac{2}{3}$ (b) $\frac{3}{4} + \frac{1}{3}$ (c) $\frac{5}{6} + \frac{7}{4}$

10. Any fraction a/b is equal to ak/bk for $k \neq 0$. Select a value of k that can be used in this way to obtain a fraction that is equal to the first fraction and that has the same denominator as the second:

    (a) $\frac{1}{2}, \frac{3}{4}$ (b) $\frac{2}{3}, \frac{5}{9}$ (c) $\frac{1}{2}, \frac{3}{10}$

    (d) $\frac{3}{4}, \frac{7}{20}$ (e) $\frac{3}{8}, \frac{7}{40}$ (f) $\frac{2}{50}, \frac{7}{1000}$

Use the definitions to find:

11. $\frac{2}{3} \times \frac{1}{5}$ 12. $\frac{3}{4} \times \frac{5}{2}$ 13. $\frac{2}{11} \times \frac{7}{3}$

14. $\frac{1}{3} \times \frac{5}{2}$ 15. $\frac{2}{5} \times \frac{3}{7}$ 16. $\frac{5}{3} \times \frac{2}{7}$

17. $\frac{1}{2} + \frac{1}{3}$ 18. $\frac{2}{3} + \frac{1}{2}$ 19. $\frac{3}{4} + \frac{1}{3}$

20. $\frac{3}{4} + \frac{2}{3}$ 21. $\frac{2}{5} + \frac{1}{3}$ 22. $\frac{5}{4} + \frac{3}{5}$

PEDAGOGICAL CONSIDERATIONS

1. Some elementary texts define a rational number in this way:

   A rational number is one that can be written in the form a/b, where a and b are integers and $b \neq 0$.

   Show that this definition is equivalent to the one given in this section.

2. An interesting way to introduce the need for fractional numbers is to present a class with a set of questions and ask whether a fractional number or a whole number is needed to answer the question. For example, consider such questions as these:

How many students are in this class?

A pound of butter is usually packed as four bars. Each bar is what part of a pound?

Prepare a set of five similar questions that illustrate this difference between counting numbers and fractional numbers.

3. By definition, a fraction is a numeral for a fractional number. Thus technically we may add and multiply fractional numbers, but not fractions. This language tends to become cumbersome and thus in this text we shall usually refer to operations with rational numbers. Select an elementary textbook for grades 5 or 6 and explore the manner in which they use the words "fraction," "fractional number," and "rational number."

6-2
rational numbers and the number line

Consider a number line and locate a point midway between the points with coordinates 0 and 1. The coordinate of the midpoint is called $\frac{1}{2}$. Then we may locate all points that correspond to "halves," such as $\frac{3}{2} = 1\frac{1}{2}$, $\frac{5}{2} = 2\frac{1}{2}$, $\frac{7}{2} = 3\frac{1}{2}$, and so forth. Also we may locate points with coordinates $-\frac{1}{2}$, $-1\frac{1}{2}$, $-2\frac{1}{2}$, $-3\frac{1}{2}$ and so forth. The graph of any rational number $n/2$, where n is an integer, may be found in this way.

The graphs of rational numbers $n/3$ may be found by dividing the unit interval from 0 to 1 into three congruent parts. The points of division are the graphs of $\frac{1}{3}$ and $\frac{2}{3}$. Then the points

$$0 = \frac{0}{3} \quad \frac{1}{3} \quad \frac{2}{3} \quad \frac{3}{3} = 1$$

corresponding to $\frac{4}{3}$, $\frac{5}{3}$, $\frac{6}{3} = 2$, $\frac{7}{3}$, ... and also the points corresponding to $-\frac{1}{3}$, $-\frac{2}{3}$, $-\frac{3}{3} = -1$, $-\frac{4}{3}$, ... may be located. Similarly, the points corresponding to multiples of $\frac{1}{4}$, $\frac{1}{5}$, $\frac{1}{6}$, and so forth may be located on a number line. Each rational number has a graph on the number line. Note that the set of rational numbers includes the set of integers as a subset.

Our number line has now become dense with points and "resembles" a complete line. The word "resembles" is used inasmuch as we shall find later that there are still "gaps" in the number line—that is, there are points on the number line that do not have rational numbers as their coordinates. However, the set of rational numbers is said to be dense, since between any two elements of the set, there is always another element of the set.

Example 1: Locate a rational number between $\frac{17}{19}$ and $\frac{18}{19}$.

Solution: Change each rational number to one with a denominator of 38.

$$\tfrac{17}{19} \times \tfrac{2}{2} = \tfrac{34}{38} \qquad \tfrac{18}{19} \times \tfrac{2}{2} = \tfrac{36}{38}$$

Clearly, $\frac{35}{38}$ lies between the two given rational numbers. Note also that

$$\tfrac{35}{38} = \tfrac{1}{2}(\tfrac{17}{19} + \tfrac{18}{19}).$$

Example 2: Locate a rational number between $\frac{17}{19}$ and $\frac{35}{38}$.

Solution: Write each rational number as one with a denominator of 76.

$$\tfrac{17}{19} \times \tfrac{4}{4} = \tfrac{68}{76} \qquad \tfrac{35}{38} \times \tfrac{2}{2} = \tfrac{70}{76}$$

Note that $\frac{69}{76}$ lies between the two given rational numbers. Also,

$$\tfrac{69}{76} = \tfrac{1}{2}(\tfrac{17}{19} + \tfrac{35}{38}).$$

We can summarize the results of the two preceding examples in this manner:

$$\tfrac{17}{19} < \tfrac{35}{38} < \tfrac{18}{19}$$
$$\tfrac{17}{19} < \tfrac{69}{76} < \tfrac{35}{38}$$

This process can be extended indefinitely. For example, to locate a rational number between $\frac{17}{19}$ and $\frac{69}{76}$, write each with a denominator of 152. Then find one-half the sum of the two numbers. This ability to locate a rational number between any two given rational numbers is what is meant by the *density property.*

Addition and subtraction of integers were considered on a number line in §5-7. The same procedures may be used for addition and subtraction of rational numbers. Multiplication of rational numbers is usually considered in terms of the definition in §6-1, that is,

$$\frac{a}{b} \times \frac{c}{d} = \frac{ac}{bd}.$$

Then if $a \neq 0$,

$$\frac{a}{b} \times \frac{b}{a} = \frac{ab}{ab} = 1$$

and $\dfrac{b}{a}$ is the **multiplicative inverse,** or **reciprocal,** of $\dfrac{a}{b}$, $a \neq 0$. Each rational number different from 0 has a rational number as its multiplicative inverse. For example, the multiplicative inverse of 2 is $\dfrac{1}{2}$, of $\dfrac{2}{3}$ is $\dfrac{3}{2}$, and of $\dfrac{-3}{4}$ is $\dfrac{4}{-3}$.

EXERCISES

Classify each statement as true or false.

1. Every integer is a rational number.
2. Every rational number is an integer.
3. The reciprocal of every rational number except 0 is a rational number.
4. The number 0 is not a rational number.
5. The multiplicative inverse of a positive rational number is negative.

Copy the following table. Use "✓" to show that the number listed at the top of a column is a member of the set listed at the side. Use "✗" if the number is not a member of the set.

Set	6. 2	7. −5	8. 0	9. $\frac{2}{3}$	10. $-\frac{3}{4}$
Counting numbers					
Whole numbers					
Integers					
Rational numbers					

11. Name two properties of the set of rational numbers that are not properties of the set of integers.
12. Find three rational numbers between $\frac{7}{9}$ and $\frac{8}{9}$.
13. Find three rational numbers between 0 and $\frac{1}{100}$.
14. Name a rational number between the rational numbers a/b and c/d.
15. Is there a next rational number after 0? Use a property of the set of rational numbers to explain your answer.

Find a replacement for n that will make each of the sentences below true where n must be a member of the set named at the top of the column. If there is no such replacement, write "none."

	Sentence	Counting numbers	Whole numbers	Integers	Rational numbers
16.	$n - 3 = 0$				
17.	$n + 3 = 3$				
18.	$n + 3 = 0$				
19.	$3n = 5$				
20.	$n \times n = 2$				

PEDAGOGICAL CONSIDERATIONS

1. Construct a number line, graph the necessary rational numbers, and represent on the line:

   (a) $\frac{3}{4} + \frac{5}{4}$ (b) $\frac{3}{4} - \frac{5}{4}$ (c) $2\frac{1}{4} + 1\frac{1}{2}$

   (d) $3\frac{3}{4} - 5\frac{1}{4}$ (e) $-2\frac{1}{4} + 3$ (f) $-1\frac{3}{4} - \frac{1}{2}$

2. The Greek philosopher Zeno proposed several paradoxes in the fifth century B.C. that were based on the density property. Prepare a report on these paradoxes, using a history of mathematics book as your reference. This report should be appropriate for upper elementary and junior high school students.

3. Take a small strip of paper about 12 inches long and label it 0 at the left end and 1 at the right end. Then fold the paper once to obtain the midpoint of the strip. Label this point as $\frac{1}{2}$. Continue with this paper-folding experiment to construct points representing the fourths and points representing the eighths.

4. A number line such as the following can be very helpful in giving young students an intuitive concept of order:

   $\frac{1}{2}$ is to the left of $\frac{7}{8}$; therefore $\frac{1}{2} < \frac{7}{8}$.

   $\frac{1}{2}$ is to the right of $\frac{1}{8}$; therefore $\frac{1}{2} > \frac{1}{8}$.

   Prepare a short ten-minute lesson that illustrates order of fractional numbers through the use of a number-line approach.

5. Explain how the density property enables us to use rational numbers as measures of line segments.

6-3

rational numbers as fractions

Any rational number may, by definition, be expressed in the form a/b, where a is an integer and b is a counting number. The expression a/b is called a fraction with **numerator** a and **denominator** b. When considered in terms of "parts of a whole," the denominator names the fractional part (halves, thirds, quarters, etc.) under consideration and the numerator tells the number of those parts. For example, $\frac{3}{4}$ may be thought of as three fourths (i.e., $3 \times \frac{1}{4}$) of a complete unit. In the figure the rectangular region has been divided into fourths. Three of these parts have been shaded to denote $\frac{3}{4}$ of the original rectangular region.

If we subdivide the rectangular region again as in the next figure, we find that

$$\tfrac{3}{4} = \tfrac{6}{8}.$$

Similarly, for the next figure

$$\tfrac{3}{4} = \tfrac{9}{12}.$$

In general, we have the familiar rule

$$\frac{a}{b} \times \frac{k}{k} = \frac{ak}{bk} = \frac{a}{b} \qquad \text{for any integer } k \neq 0.$$

This rule indicates that multiplication by 1 in the form k/k for $k \neq 0$ leaves the number unchanged. We use the rule in the form

$$\frac{a \times k}{b \times k} = \frac{a}{b}$$

to "reduce" fractions by dividing numerator and denominator by k. For example,

$$\frac{12}{30} = \frac{2 \times 6}{5 \times 6} = \frac{2}{5} \times \frac{6}{6} = \frac{2}{5} \times 1 = \frac{2}{5}$$

$$\frac{72}{18} = \frac{4 \times 18}{1 \times 18} = \frac{4}{1} \times \frac{18}{18} = \frac{4}{1} \times 1 = 4$$

$$\frac{16}{80} = \frac{1 \times 16}{5 \times 16} = \frac{1}{5} \times \frac{16}{16} = \frac{1}{5} \times 1 = \frac{1}{5}.$$

A fraction a/b, where b is a counting number, is said to be **reduced to lowest terms** if it is not possible to find an integer $k > 1$ and integers c and d such that

$$\frac{a}{b} = \frac{ck}{dk}.$$

Each of the above fractions was reduced to lowest terms. This rule also allows us to find as many ways of writing a rational number as we like. For example,

$$\tfrac{2}{3} = \tfrac{4}{6} = \tfrac{6}{9} = \tfrac{8}{12} = \tfrac{10}{15} = \cdots$$

This procedure for writing rational numbers in many ways enables us to express any two rational numbers a/b and c/d as fractions with like denominators:

$$\frac{a}{b} = \frac{a \times d}{b \times d} = \frac{ad}{bd}$$

$$\frac{c}{d} = \frac{c \times b}{d \times b} = \frac{b \times c}{b \times d} = \frac{bc}{bd}$$

Example 1: Write $\tfrac{2}{3}$ and $\tfrac{4}{5}$ as fractions with like denominators.

Solution:

$$\frac{2}{3} = \frac{2 \times 5}{3 \times 5} = \frac{10}{15}, \qquad \frac{4}{5} = \frac{4 \times 3}{5 \times 3} = \frac{12}{15}.$$

To add numbers expressed by fractions with like denominators we add the numerators and use the common denominator:

$$\frac{a}{c} + \frac{b}{c} = \frac{a+b}{c}$$

To subtract numbers expressed by fractions with like denominators we subtract the numerators and use the common denominator:

$$\frac{a}{c} - \frac{b}{c} = \frac{a-b}{c}$$

Example 2: Find the sum $\tfrac{3}{8} + \tfrac{2}{8}$.

Solution:

$$\frac{3}{8} + \frac{2}{8} = \frac{3+2}{8} = \frac{5}{8}$$

The problem in Example 2 may be illustrated using a circular region divided into eighths as in the figure. Three of the parts are shaded with horizontal lines to represent $\tfrac{3}{8}$. Two of the parts are shaded with vertical lines to represent $\tfrac{2}{8}$. The sum $\tfrac{3}{8} + \tfrac{2}{8}$ is represented by the five shaded parts; $\tfrac{3}{8} + \tfrac{2}{8} = \tfrac{5}{8}$.

Any two rational numbers may be represented by fractions a/b and c/d, and therefore by fractions with like denominators, that is,

$$\frac{ad}{bd} \quad \text{and} \quad \frac{bc}{bd}.$$

Thus the sum of any two rational numbers may be found using the rule

$$\frac{a}{b} + \frac{c}{d} = \frac{ad}{bd} + \frac{bc}{bd} = \frac{ad + bc}{bd}.$$

Similarly, the difference of any two rational numbers may be found using the rule

$$\frac{a}{b} - \frac{c}{d} = \frac{ad}{bd} - \frac{bc}{bd} = \frac{ad - bc}{bd}.$$

These rules based upon properties of fractions are in accord with the definitions in §6-1. Still another approach is possible using the distributive law and the interpretation of each fraction a/b as $a \times (1/b)$:

$$\begin{aligned}
\frac{a}{b} + \frac{c}{d} &= \frac{ad}{bd} + \frac{bc}{bd} \\
&= ad \times \frac{1}{bd} + bc \times \frac{1}{bd} \\
&= (ad + bc) \times \frac{1}{bd} \\
&= \frac{ad + bc}{bd}
\end{aligned}$$

Example 3: Use the rule for addition of rational numbers to find $\frac{2}{3} + \frac{3}{4}$.

Solution:

$$\frac{2}{3} + \frac{3}{4} = \frac{a}{b} + \frac{c}{d} = \frac{ad + bc}{bd}$$

$$a = 2, \quad b = 3, \quad c = 3, \quad d = 4,$$

$$ad + bc = 2 \times 4 + 3 \times 3, \quad bd = 3 \times 4$$

$$\frac{2}{3} + \frac{3}{4} = \frac{2 \times 4 + 3 \times 3}{3 \times 4} = \frac{8 + 9}{12} = \frac{17}{12}.$$

Integers have the property that

$$a - b = a + (-b).$$

Rational numbers also have this property,

$$\frac{a}{b} - \frac{c}{d} = \frac{a}{b} + \left(-\frac{c}{d}\right).$$

Recall that

$$\frac{a}{b} - \frac{c}{d} = \frac{ad}{bd} - \frac{bc}{bd} = \frac{ad - bc}{bd}.$$

Also

$$\frac{a}{b} + \frac{-c}{d} = \frac{ad}{bc} + \frac{-bc}{bd} = \frac{ad - bc}{bd}.$$

Therefore

$$\frac{a}{b} + \frac{-c}{d} = \frac{a}{b} + \left(-\frac{c}{d}\right) \quad \text{and} \quad \frac{-c}{d} = -\frac{c}{d}.$$

In general, there are three signs associated with any fraction c/d:

1. The sign of the fraction, as in $-\dfrac{c}{d}$.

2. The sign of the numerator, as in $\dfrac{-c}{d}$.

3. The sign of the denominator, as in $\dfrac{c}{-d}$.

From the preceding discussion and multiplication by 1 in the form $-1/-1$,

$$\frac{-c}{d} \times \frac{-1}{-1} = \frac{(-c) \times (-1)}{d \times (-1)} = \frac{c}{-d}$$

and

$$-\frac{c}{d} = \frac{-c}{d} = \frac{c}{-d}.$$

In other words, any two of the three signs associated with a fraction may be changed without changing the number represented by the fraction.

Example 4:   Reduce $\dfrac{6}{-8}$.

Solution: The denominator should be a counting number.

$$\frac{6}{-8} = \frac{-6}{8} = \frac{-3}{4}.$$ This result is often written as $-\dfrac{3}{4}$.

Multiplication of rational numbers causes very little difficulty:

$$\frac{a}{b} \times \frac{c}{d} = \frac{ac}{bd}$$

Division of rational numbers often causes difficulty. Many people "invert and multiply" without really knowing why they do so. We consider this rule in terms of two specific examples. Suppose that we start with two integers such as $6 \div 2$. Then

$$6 \div 2 = \frac{6}{2} = \frac{6 \times 1}{1 \times 2} = \frac{6}{1} \times \frac{1}{2} = 6 \times \frac{1}{2}.$$

Thus $6 \div 2 = 6 \times \frac{1}{2}$ where $\frac{1}{2}$ is the reciprocal of 2. In general, for any integers a and $b \neq 0$,

$$a \div b = a \times \frac{1}{b}.$$

Division of rational numbers may also be expressed as a product. For example, consider

$$\tfrac{2}{3} \div \tfrac{5}{7}.$$

As for integers we may write this quotient in the form

$$\frac{\frac{2}{3}}{\frac{5}{7}}$$

and seek methods for rewriting this new fraction so that its denominator becomes 1. Then we multiply numerator and denominator by the reciprocal of $\frac{5}{7}$, that is, $\frac{7}{5}$. These steps may be displayed as follows:

$$\tfrac{2}{3} \div \tfrac{5}{7} = \frac{\frac{2}{3}}{\frac{5}{7}} \qquad \text{Since these are just two different ways of expressing a quotient.}$$

$$= \frac{\frac{2}{3}}{\frac{5}{7}} \times 1 \qquad \text{By the multiplication property of 1.}$$

$$= \frac{\frac{2}{3}}{\frac{5}{7}} \times \frac{\frac{7}{5}}{\frac{7}{5}} \qquad \text{Since } \tfrac{7}{5} \div \tfrac{7}{5} = 1.$$

$$= \frac{\frac{2}{3} \times \frac{7}{5}}{\frac{5}{7} \times \frac{7}{5}} \qquad \text{By the rule for multiplication of fractional numbers.}$$

$$= \frac{\frac{2}{3} \times \frac{7}{5}}{1} \qquad \text{Since the product of } \tfrac{5}{7} \text{ and } \tfrac{7}{5} \text{ is } \tfrac{35}{35} = 1.$$

$$= \tfrac{2}{3} \times \tfrac{7}{5} \qquad \text{Since any number divided by 1 is equal to that number.}$$

Note that we began with a division problem, and showed that this could be written as an equivalent multiplication problem. In general we have the following rule for division of rational numbers:

$$\frac{a}{b} \div \frac{c}{d} = \frac{a}{b} \times \frac{d}{c}$$

In other words, to divide by a number different from zero we multiply by its reciprocal. This is a formal statement of the common procedure described by "invert and multiply." This procedure may be justified in general as follows:

$$\frac{\frac{a}{b}}{\frac{c}{d}} = \frac{\frac{a}{b} \times \frac{d}{c}}{\frac{c}{d} \times \frac{d}{c}} = \frac{\frac{a}{b} \times \frac{d}{c}}{1} = \frac{a}{b} \times \frac{d}{c}$$

A fraction such as

$$\frac{\frac{2}{3}}{\frac{5}{7}}$$

may also be reduced to lowest terms by multiplying by 1 in a form such that the numerator and the denominator become integers. For example, we could write

$$\frac{\frac{2}{3}}{\frac{5}{7}} = \frac{\frac{2}{3}}{\frac{5}{7}} \times \frac{7}{7} = \frac{\frac{14}{3}}{5} = \frac{\frac{14}{3}}{5} \times \frac{3}{3} = \frac{14}{15}.$$

However, notice that both the numerator and the denominator of the original fraction would become integers if they were multiplied by the least common multiple (L.C.M.) of their own denominators:

L.C.M. $(3, 7) = 21$

$$\frac{\frac{2}{3}}{\frac{5}{7}} \times \frac{21}{21} = \frac{14}{15}$$

In the following exercises and elsewhere the direction "simplify" is used to indicate that all operations are to be performed and the resulting fractions reduced to lowest terms.

EXERCISES

Use the definitions of this section and simplify.

1. $\frac{2}{3} + \frac{1}{2}$ 2. $\frac{5}{6} + \frac{3}{4}$

3. $\frac{6}{18} \times \frac{10}{7}$ 4. $\frac{12}{15} \times \frac{28}{10}$

5. $\frac{2}{3} \div \frac{4}{5}$ 6. $\frac{7}{8} \div \frac{3}{4}$

7. $\frac{5}{12} \div \frac{10}{18}$ 8. $\frac{17}{34} \div \frac{25}{10}$

9. $\frac{5}{6} - \frac{1}{3}$ 10. $\frac{2}{5} - \frac{1}{4}$

11. $\frac{15}{8} - \frac{3}{4}$ 12. $\frac{11}{9} - \frac{7}{5}$

13. $\frac{-3}{4} \times \frac{2}{5}$ 14. $\frac{3}{-2} \times \frac{-3}{5}$

15. $-\left(\frac{2}{3}\right) \times \frac{5}{9}$ 16. $-\left(\frac{7}{8}\right) \times -\left(\frac{3}{4}\right)$

17. $\frac{7}{8} \div \frac{-2}{3}$ 18. $\frac{-9}{16} \div \frac{3}{-2}$

19. $\frac{-3}{4} + \frac{5}{7}$ 20. $\frac{-2}{3} - \frac{-4}{5}$

Complete the indicated operations and simplify.

21. $\frac{2}{3} + \left(\frac{3}{4} + \frac{7}{8}\right)$ 22. $\left(\frac{2}{3} + \frac{3}{4}\right) + \frac{7}{8}$

23. $\frac{7}{8} - \left(\frac{3}{4} - \frac{1}{2}\right)$ 24. $\left(\frac{7}{8} - \frac{3}{4}\right) - \frac{1}{2}$

25. $\frac{1}{2} \times \left(\frac{3}{4} \times \frac{4}{5}\right)$ 26. $\left(\frac{1}{2} \times \frac{3}{4}\right) \times \frac{4}{5}$

27. $\frac{9}{16} \div \left(\frac{1}{2} \div \frac{1}{4}\right)$ 28. $\left(\frac{9}{16} \div \frac{1}{2}\right) \div \frac{1}{4}$

Simplify by multiplying numerator and denominator by the least common denominator of each fraction involved.

29. $\dfrac{\frac{1}{2}}{\frac{3}{4}}$ 30. $\dfrac{\frac{3}{5}}{\frac{2}{3}}$ 31. $\dfrac{\frac{7}{8}}{\frac{7}{8}}$ 32. $\dfrac{\frac{3}{7}}{\frac{5}{6}}$

Simplify:

33. $\dfrac{\frac{1}{2}+\frac{2}{3}}{\frac{3}{4}}$ 34. $\dfrac{\frac{3}{4}}{\frac{1}{2}+\frac{2}{3}}$ 35. $\dfrac{\frac{1}{3}+\frac{3}{4}}{\frac{1}{2}+\frac{2}{3}}$ 36. $\dfrac{\frac{3}{4}-\frac{1}{3}}{\frac{2}{3}-\frac{1}{2}}$

Express as a single quotient.

*37. $\dfrac{a}{b}+\left(\dfrac{c}{d}+\dfrac{e}{f}\right)$ *38. $\left(\dfrac{a}{b}+\dfrac{c}{d}\right)+\dfrac{e}{f}$

*39. $\dfrac{a}{b}\left(\dfrac{c}{d}+\dfrac{e}{f}\right)$ *40. $\left(\dfrac{a}{b}\times\dfrac{c}{d}\right)+\left(\dfrac{a}{b}\times\dfrac{e}{f}\right)$

*41. $\dfrac{\dfrac{a}{b}+\dfrac{c}{d}}{\dfrac{e}{f}}$ *42. $\dfrac{\dfrac{a}{b}}{\dfrac{c}{d}+\dfrac{e}{f}}$

PEDAGOGICAL CONSIDERATIONS

1. Draw a rectangular region 6 centimeters long and 3 centimeters wide for each part of this exercise. Let the entire rectangular region represent 1 unit and shade portions to represent:

   (a) $\frac{1}{2}$ (b) $\frac{1}{3}$ (c) $\frac{2}{3}$ (d) $\frac{1}{6}$

   (e) $\frac{1}{9}$ (f) $\frac{5}{9}$ (g) $\frac{1}{18}$ (h) $\frac{7}{18}$

2. Draw a rectangular region as in Exercise 1 above. Let a part of the region that is 5 centimeters long and 2 centimeters wide represent 1 unit and shade portions to represent:

   (a) 1   (b) $\frac{1}{5}$ (c) $\frac{2}{5}$ (d) $\frac{6}{5}$

   (e) $\frac{1}{2}$ (f) $\frac{3}{2}$ (g) $\frac{1}{10}$ (h) $\frac{13}{10}$

3. Multiplication of two fractional numbers such as $\frac{2}{3}\times\frac{3}{4}$ is sometimes presented to elementary school students by means of diagrams such as the following. First draw a rectangular region and shade $\frac{3}{4}$ of the figure as shown.

Next shade $\frac{2}{3}$ of the figure as in the following diagram.

The final picture shows that the rectangular region has been divided into 12 equal parts, and that six of these parts are shaded in both directions. Thus $\frac{2}{3} \times \frac{3}{4} = \frac{6}{12}$.

Prepare a similar diagram to illustrate the product $\frac{3}{4} \times \frac{5}{7}$.

4. Prepare a collection of visual aids that can be used to present basic concepts of fractional numbers to elementary school children.

5. Find out what you can about the use of Cuisenaire rods in the teaching of computation with fractions. If possible, demonstrate their use in class.

6. The University of Illinois Committee on School Mathematics presents work with fractions through an approach known as "stretchers and shrinkers." Find out what you can about this procedure and explain it in class.

7. Prepare a 30-minute test that can be used to evaluate an elementary student's knowledge of the four fundamental operations with fractions.

8. Here is a flow chart with directions for reducing a fraction.

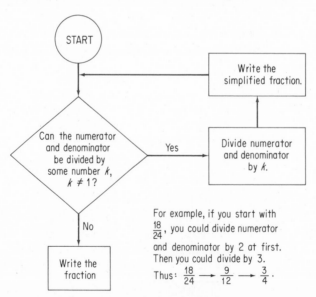

Use the flow chart to reduce the fraction $\frac{102}{186}$.

We have seen that any rational number can be named by a fraction. Any rational number can also be named by a decimal. In particular, fractions that have denominators that are powers of 10 can be named as decimals by our knowledge of place value. For example, consider the following:

$$\tfrac{3}{10} = 0.3, \qquad \tfrac{3}{100} = 0.03, \qquad \tfrac{3}{1000} = 0.003$$

The following chart summarizes the most commonly used decimal values.

Tenths	$\dfrac{1}{10}$	$\dfrac{1}{10}$	0.1
Hundredths	$\dfrac{1}{100}$	$\dfrac{1}{10^2}$	0.01
Thousandths	$\dfrac{1}{1000}$	$\dfrac{1}{10^3}$	0.001
Ten-thousandths	$\dfrac{1}{10,000}$	$\dfrac{1}{10^4}$	0.0001
Hundred-thousandths	$\dfrac{1}{100,000}$	$\dfrac{1}{10^5}$	0.00001
Millionths	$\dfrac{1}{1,000,000}$	$\dfrac{1}{10^6}$	0.000001

We make use of negative exponents in writing decimals as follows:

$$0.1 = \frac{1}{10} = 10^{-1}$$

$$0.01 = \frac{1}{10^2} = 10^{-2}$$

$$0.001 = \frac{1}{10^3} = 10^{-3}$$

Notice that for any integer k we have

$$\frac{1}{10^k} = 10^{-k}.$$

In general we have

$$b^{-k} = \frac{1}{b^k}.$$

For example:

$$2^{-3} = \frac{1}{2^3} = \frac{1}{8}$$

$$5^{-2} = \frac{1}{5^2} = \frac{1}{25}$$

$$3^2 = \frac{1}{3^{-2}}$$

Now we are able to write any number given in decimal notation in expanded notation as in the following example.

Example 1: Write 8027.45 in expanded notation.

Solution:

$$8027.45 = (8 \times 10^3) + (0 \times 10^2) + (2 \times 10^1)$$
$$+ (7 \times 10^0) + (4 \times 10^{-1}) + (5 \times 10^{-2})$$

The decimal in Example 1 is read "eight thousand twenty-seven and forty-five hundredths." We use the word "and" to represent the decimal point.

We now need to discuss procedures for computing with decimals. Consider the addition problem $0.35 + 0.49$. This same problem may be stated and solved using fractions as

$$\tfrac{35}{100} + \tfrac{49}{100} = \tfrac{84}{100}.$$

By a direct translation from these fractions to decimal form we have

$$0.35 + 0.49 = 0.84$$

which we can also obtain by arranging the addends in the form

$$
\begin{array}{r}
0.35 \\
+0.49 \\
\hline
0.84
\end{array}
$$

so that tenths are added to tenths and hundredths are added to hundredths. Notice that we obtain this form by "lining up the decimal points."

The addition problem $0.35 + 0.49$ may also be done through application of the distributive property

$$ac + bc = (a + b)c$$

where

$$0.35 + 0.49 = 35(0.01) + 49(0.01)$$
$$= (35 + 49)(0.01) = 84(0.01) = 0.84.$$

Notice that this use of the distributive property is equivalent to doing the problem in hundredths; that is, $0.35 + 0.49$ is $(35 + 49)$ hundredths.

Example 2: Subtract: $0.73 - 0.28$.

Solution: Three methods of solution are given. In fractions we have

$$\frac{73}{100} - \frac{28}{100} = \frac{73 - 28}{100} = \frac{45}{100} = 0.45.$$

Lining up the decimal points, we have

$$
\begin{array}{r}
0.73 \\
-0.28 \\
\hline
0.45
\end{array}
$$

In hundredths we have

$$
\begin{aligned}
0.73 - 0.28 &= 73(0.01) - 28(0.01) \\
&= 73(0.01) + (-28)(0.01) \\
&= [73 + (-28)](0.01) = 45(0.01) = 0.45.
\end{aligned}
$$

To establish the usual rule for multiplication of decimals, we examine several examples done as fractions and attempt to discover a pattern.

Example 3: Multiply: (a) 0.7×0.8; (b) 0.7×0.08; (c) 0.07×0.08.

Solution: (a) $0.7 \times 0.8 = \frac{7}{10} \times \frac{8}{10} = \frac{56}{100} = 0.56$
Note that a number of *tenths* multiplied by a number of *tenths* gives a number of *hundredths*.

(b) $0.7 \times 0.08 = \frac{7}{10} \times \frac{8}{100} = \frac{56}{1000} = 0.056$

Note that a number of *tenths* multiplied by a number of *hundredths* gives a number of *thousandths*.

(c) $0.07 \times 0.08 = \frac{7}{100} \times \frac{8}{100} = \frac{56}{10000} = 0.0056.$

Note that a number of *hundredths* multiplied by a number of *hundredths* gives a number of *ten-thousandths*.

The above examples illustrate the rule for multiplying two rational numbers that are written as decimals; the number of decimal places in the product is equal to the sum of the numbers of decimal places in the two factors that are multiplied.

Example 4: Place the decimal point in the correct position in the product: $0.23 \times 5.7 = 1311$.

Solution: There are two decimal places in 0.23 and one decimal place in 5.7. The product should then have three decimal places. Thus $0.23 \times 5.7 = 1.311$.

Division with decimals is probably the most difficult arithmetic skill for many people. Once again we can justify the usual algorithm by using fractional equivalents. Consider the division problem $0.46 \div 2$. This problem can be expressed and solved using fractions:

$$\frac{46}{100} \div 2 = \frac{46}{100} \times \frac{1}{2} = \frac{46 \times 1}{100 \times 2} = \frac{23}{100} = 0.23$$

We may think of any division problem in the form

$$\text{divisor}\overline{)\overset{\text{quotient}}{\text{dividend}}}$$

Thus the problem under consideration may be written as

$$2\overline{)\overset{0.23}{0.46}}$$

Notice that the divisor is a whole number; the division may be done as for whole numbers with the decimal point in the quotient placed directly over the decimal point in the dividend. Whenever a decimal is divided by a whole number, we divide as for whole numbers and place the decimal point in the quotient directly over the decimal point in the dividend.

Any problem of division of ordinary decimals may be solved by finding an equivalent problem in which the divisor is a whole number and following the above procedure. Consider the problem

$$1.75 \div 2.5$$

We first write the problem in the form

$$\frac{1.75}{2.5}$$

and multiply the fraction by $\frac{10}{10}$ to obtain a fraction with a whole number as denominator:

$$\frac{1.75}{2.5} = \frac{1.75}{2.5} \times \frac{10}{10} = \frac{17.5}{25}$$

This rewriting of the fraction

$$\frac{1.75}{2.5} \quad \text{in the form} \quad \frac{17.5}{25}$$

corresponds to rewriting the division problem

$$2.5\overline{)1.75} \quad \text{in the form} \quad 25\overline{)17.5}$$

In other words, multiplying both the dividend and the divisor by 10 is equivalent to multiplying the numerator and the denominator of the corresponding fraction by 10. In each case the problem is unchanged.

Most people think of the above process as "shifting" the decimal point one unit to the right in both the dividend and the divisor. In general, since the numerator and the denominator of the corresponding fraction may be multiplied by any power of 10, the decimal point may be shifted as many places as necessary to make the divisor a whole number and shifted the same number of places in the dividend.

Example 5: Divide: $3.726 \div 0.23$.

Solution: We move the decimal point two places to the right in the divisor to make the divisor a whole number. Therefore we move the decimal point two places to the right in the dividend. Then we divide as for whole numbers and place the decimal point in the quotient directly over the new location of the decimal point in the dividend.

$$
\begin{array}{r}
16.2 \\
0.23\overline{\smash{)}3.72\llap{.}6} \\
23 \\
\hline
142 \\
138 \\
\hline
46 \\
46 \\
\hline
\end{array}
\qquad \text{Thus } 3.726 \div 0.23 = 372.6 \div 23 = 16.2.
$$

EXERCISES

Write in decimal notation.

1. 10^{-3}
2. 10^{-5}
3. 4×10^{-2}
4. 5×10^{-1}
5. 7×10^{-3}
6. 8×10^{-4}
*7. 0.2×10^{-3}
*8. 0.01×10^{-2}

Write in expanded notation.

9. 4.23
10. 79.8
11. 423.83
12. 2302.05
13. 0.007
14. 2.0301
15. 0.0001
16. 9000.09
17. 0.101
18. 1.0101

Write in decimal notation.

19. $(1 \times 10^{-1}) + (3 \times 10^{-2}) + (5 \times 10^{-3})$
20. $(5 \times 10^{1}) + (2 \times 10^{0}) + (1 \times 10^{-1}) + (7 \times 10^{-2}) + (3 \times 10^{-3})$
21. $(7 \times 10^{1}) + (0 \times 10^{0}) + (0 \times 10^{-1}) + (8 \times 10^{-2}) + (2 \times 10^{-3})$
22. $(2 \times 10^{-2}) + (5 \times 10^{-3}) + (1 \times 10^{-4})$
23. $(5 \times 10^{3}) + (7 \times 10^{2}) + (0 \times 10^{1}) + (5 \times 10^{0}) + (5 \times 10^{-1})$
24. $(6 \times 10^{0}) + (0 \times 10^{-1}) + (8 \times 10^{-2}) + (5 \times 10^{-3})$

Write as a decimal.

25. Thirty-seven and twenty-nine hundredths

26. Four hundred thirty-eight thousandths

27. Two and nineteen ten-thousandths

28. One hundred six and eighty-six hundredths

Write each decimal in words.

29. 273.87   30. 0.605   31. 9.0018   32. 79.00009

Find each sum or difference by using decimals and by using fractions.

33. $0.7 + 0.9$ 34. $0.12 + 0.15$

35. $0.03 + 0.08$ 36. $0.123 + 0.456$

37. $0.9 - 0.2$ 38. $0.17 - 0.09$

39. $0.18 - 0.11$ 40. $0.423 - 0.178$

Select the best answer.

41. 0.89×0.9 is about (**a**) 0.8; (**b**) 8; (**c**) 0.08; (**d**) 0.9.

42. 79.6×9.8 is about (**a**) 8; (**b**) 80; (**c**) 800; (**d**) 8000.

43. 9.01×6.99 is about (**a**) 0.63; (**b**) 6.3; (**c**) 63; (**d**) 630.

44. 0.09×10.01 is about (**a**) 0.009; (**b**) 0.09; (**c**) 0.9; (**d**) 9.

45. $19.79 \div 3.98$ is about (**a**) 5; (**b**) 0.5; (**c**) 0.05; (**d**) 50.

46. $0.079 \div 0.02$ is about (**a**) 4; (**b**) 40; (**c**) 0.4; (**d**) 0.04.

47. $7.962 \div 1.9$ is about (**a**) 4; (**b**) 40; (**c**) 0.4; (**d**) 0.04.

48. $123.9 \div 0.2$ is about (**a**) 62; (**b**) 620; (**c**) 6.2; (**d**) 0.62.

PEDAGOGICAL CONSIDERATIONS

More and more course guides and curriculum outlines are now being introduced through the use of behavioral objectives. That is, objectives of instruction are stated in terms of desired student behavior. Such objectives are easy to assess in that the statement should indicate the type of performance required by the student to demonstrate attainment of the objective.

There are certain key action words that are hepful to use in constructing behavioral objectives. These include the following:

Identify	Describe
Distinguish	State a principle or rule
Construct	Apply the rule
Name	Demonstrate
Order	Interpret

Objectives as traditionally stated have been vague and difficult to evaluate. For example, the objective "the student should understand the relationship between rational numbers and decimals" does not clearly spell out the expected behavior on the part of the student after studying such a unit of work. As a contrast, consider the objective "the student should be able to demonstrate the expanded notation for a given decimal numeral." Here we know what is expected of the student, and can clearly construct a test item to evaluate his knowledge of this particular item.

1. Write five behavioral objectives for the material of this section as it would be presented to an elementary school class. Use a different action word for each objective.

2. Write one test item that can be used to evaluate a student's attainment of each of the objectives you list.

3. Prepare a 30 minute test that can be used to evaluate an elementary student's knowledge of the four fundamental operations with decimals.

6-5

rational numbers and repeating decimals

There is an interesting relationship between rational numbers and their decimal representations that we shall explore in this section. First we note that every rational number in the form a/b can be written as a decimal. If the denominator b is a multiple of 10, then we may write the decimal directly, as in these cases:

$$\tfrac{3}{10} = 0.3, \qquad \tfrac{13}{100} = 0.13, \qquad \tfrac{123}{1000} = 0.123$$

The denominator may also be a factor of some multiple of 10. If so, we can multiply numerator and denominator by a power of 2 or a power of 5 in order to write an equivalent fraction with denominator that is a power of 10.

Example 1: Write as a decimal: (a) $\tfrac{3}{5}$; (b) $\tfrac{5}{8}$.

Solution: (a) $\tfrac{3}{5} = \tfrac{3}{5} \times \tfrac{2}{2} = \tfrac{6}{10} = 0.6.$
(b) $\tfrac{5}{8} = \tfrac{5}{8} \times \tfrac{125}{125} = \tfrac{625}{1000} = 0.625.$

In Example 1(b) we could also have found the decimal equivalent by division. Thus $\tfrac{5}{8}$ is the indicated division $5 \div 8$.

$$
\begin{array}{r}
0.625 \\
8\overline{)5.000} \\
\underline{48} \\
20 \\
\underline{16} \\
40 \\
\underline{40} \\
\end{array}
$$

$\tfrac{5}{8} = 0.625$

Next let us consider a fraction such as $\frac{1}{3}$. Here there is no whole number that we can multiply numerator and denominator by to obtain a fraction with a power of 10 as denominator. Thus we resort to long division.

$$
\begin{array}{r}
0.3333\ldots \\
3\overline{)1.0000\ldots} \\
9 \\
\overline{10} \\
9 \\
\overline{10} \\
9 \\
\overline{10} \\
9 \\
\overline{1}
\end{array}
$$

We soon note that this division process gives us an endless supply of 3's. In fact, we find that the quotient is an **infinite repeating decimal,** 0.3333 . . . . We may write this as follows:

$$\tfrac{1}{3} = 0.33\overline{3}$$

In an infinite repeating decimal, we use a bar to indicate that a sequence of digits repeats endlessly. The bar is used over the digit or digits that repeat.

Example 2: Write $\frac{3}{11}$ as a repeating decimal.

Solution: By division we find that the quotient repeats the digits 2 and 7 endlessly. We write this quotient as follows:

$$\tfrac{3}{11} = 0.2727\overline{27}$$

Note that it is customary to show several sets of digits before using the bar. However, it would be equally correct to write the quotient as $0.\overline{27}$.

You may, if you wish, think of any terminating decimal as having repeated zeros and thus as being a repeating decimal. For example:

$$\tfrac{1}{4} = 0.25\overline{0}, \qquad \tfrac{3}{8} = 0.375\overline{0}$$

It is easy to show that every rational number in fractional form can be represented as a repeating (or terminating) decimal. Consider, for example, any rational number such as $\frac{12}{7}$. When we divide 12 by 7, the possible remainders are 0, 1, 2, 3, 4, 5, 6. If the remainder is 0, the division is exact; if any remainder occurs a second time, the terms after it will repeat also. Since there are only seven possible remainders when you divide by 7, the remainders must repeat or be exact by the seventh decimal place. Consider the determination of the decimal value of $\frac{12}{7}$ by long division:

217

sec. 6-5
rational
numbers
and
repeating
decimals

$$1.\overline{714285}$$
$$7\,\overline{)12.000000}$$
$$\underline{7}$$
$$⑤0$$
$$\underline{49}$$
$$10$$
$$\underline{7}$$
$$30$$
$$\underline{28}$$
$$20$$
$$\underline{14}$$
$$60$$
$$\underline{56}$$
$$40$$
$$\underline{35}$$
$$⑤$$

The fact that the remainder 5 occurred again implies that the same steps will be used again in the long division process and the digits 714285 will be repeated over and over; that is, $\frac{12}{7} = 1.714285\overline{714285}$. Similarly, any rational number p/q can be expressed as a terminating or repeating decimal, and at most q decimal places will be needed to identify it.

We can also show that any terminating or repeating decimal can be written as a rational number in the form a/b.

If a decimal is terminating, you can write it as a fraction with a power of 10 as the denominator. For example, if $n = 0.75\overline{0}$, then $100n = 75$, and $n = \frac{75}{100}$, which reduces to $\frac{3}{4}$. If a fraction can be expressed as a terminating decimal, its denominator must be a factor of a power of 10.

If a decimal is repeating, it can be written as a rational number. For example, if a decimal n repeats one digit, we can find $10n - n$. Suppose $n = 3.244\overline{4}$; then $10n = 32.444\overline{4}$ and we have:

$$10n = 32.444\overline{4}$$
$$\underline{n = \quad 3.244\overline{4}}$$
$$9n = 29.200\overline{0}$$

$$n = \frac{29.2}{9} = \frac{292}{90} = \frac{146}{45}$$

We can avoid the use of decimals in this way:

$$100n = 324.44\overline{4}$$
$$\underline{10n = \quad 32.44\overline{4}}$$
$$90n = 292$$

$$n = \tfrac{292}{90} = \tfrac{146}{45}$$

If a decimal n repeats two digits, we find $10^2 n - n$; if it repeats three digits, we find $10^3 n - n$; and so forth.

Example 3: Express $0.\overline{36}$ as a quotient of integers.

Solution: Let $n = 0.\overline{36}$; then $100n = 36.\overline{36}$ and we have:

$$100n = 36.\overline{36}$$
$$\underline{n = 0.\overline{36}}$$
$$99n = 36$$
$$n = \tfrac{36}{99} = \tfrac{4}{11}$$

Example 4: Express $0.78346\overline{346}$ as a quotient of integers.

Solution:

$$1000n = 783.46346\overline{346}$$
$$\underline{n = 0.78346\overline{346}}$$
$$999n = 782.68$$
$$n = \frac{782.68}{999} = \frac{78,268}{99,900} = \frac{19,567}{24,975}$$

We may summarize the discussion of this section in this way:

Every rational number can be named by a repeating decimal.
Every repeating decimal is the name of a rational number.

Note that for this summary we classify a terminating decimal, such as 0.25, as a repeating decimal. That is, $0.25 = 0.25\overline{0}$.

EXERCISES

Write each number as a decimal.

1. $\frac{4}{5}$ 2. $\frac{9}{20}$ 3. $\frac{7}{8}$ 4. $\frac{19}{200}$

5. $\frac{27}{15}$ 6. $\frac{1}{6}$ 7. $\frac{3}{7}$ 8. $\frac{1}{13}$

9. $\frac{9}{13}$ 10. $\frac{13}{7}$

Express each number as a quotient of integers.

11. $0.\overline{45}$ 12. $0.\overline{234}$

13. $0.8\overline{1}$ 14. $0.4\overline{14}$

15. $0.5\overline{22}$ 16. $0.\overline{9}$

17. $2.11\overline{1}$ 18. $5.22\overline{2}$

19. $2.14\overline{14}$ 20. $4.25\overline{25}$

21. $0.123\overline{123}$ 22. $65.268\overline{268}$

219

sec. 6-5
rational
numbers
and
repeating
decimals

23. Write the first ten digits to the right of the decimal point in the decimal
(a) $0.\overline{72}$; (b) $0.3\overline{571}$; (c) $0.8\overline{0}$; (d) $0.42\overline{37}$.

24. Write the digit in the fifteenth position to the right of the decimal point
in the decimal (a) $0.52\overline{0}$; (b) $0.7\overline{9}$; (c) $0.\overline{243}$; (d) $0.71\overline{68}$.

25. Arrange in order, from smallest to largest.

$2.37, \quad $2.98, \quad $2.57, \quad $2.63, \quad $2.08

26. Write each decimal to two decimal places and arrange in order, from
smallest to largest.

4.37, \quad 4.42, \quad 4.4, \quad 4.39, \quad 4.51, \quad 4.3

27. Write each decimal to at least three decimal places and arrange in
order, from smallest to largest.

1.45, \quad $1.\overline{4}$, \quad $1.\overline{5}$, \quad $1.4\overline{5}$, \quad $1.\overline{45}$

28. Write each decimal to at least six decimal places and arrange in order,
from smallest to largest.

0.234, \quad $0.\overline{234}$, \quad $0.\overline{23}$, \quad $0.\overline{24}$, \quad 0.24

*Name a rational number that lies between each of these pairs of rational
numbers.*

*29. 0.234 and 0.235

*30. $0.\overline{234}$ and $0.\overline{235}$

*31. $0.23\overline{0}$ and $0.24\overline{0}$

*32. 0.234 and $0.\overline{234}$

*33. Consider the decimal 0.252252225 . . . . This is an infinite decimal that
has a pattern but that does not repeat a fixed set of digits, and thus
does not represent a rational number. (a) How many 2's will there be
immediately preceding the fifteenth 5? (b) How many 2's will there
be altogether preceding the fifteenth 5?

*34. Name an infinite decimal for a number between 0.1 and 0.2 that is not
a rational number.

PEDAGOGICAL CONSIDERATIONS

1. Write the decimal representation for each of the rational numbers
$\frac{1}{7}, \frac{2}{7}, \frac{3}{7}, \frac{4}{7}, \frac{5}{7}$, and $\frac{6}{7}$. See if you can find a pattern that describes the
manner in which the digits in each representation are related.

2. Repeat Exercise 1 above for the multiples of $\frac{1}{13}$ from $\frac{1}{13}$ to $\frac{12}{13}$.

3. Show that $0.99\overline{9} = 1$. Then try to prepare an explanation of this fact
to satisfy a seventh or eighth grader. Consider these possibilities:

$$\frac{1}{3} = 0.33\overline{3}$$
$$+\frac{2}{3} = 0.66\overline{6}$$
$$\overline{1 = 0.99\overline{9}}$$

$$\frac{1}{3} = 0.33\overline{3}$$
$$3 \times \frac{1}{3} = 3 \times 0.33\overline{3}$$
$$1 = 0.99\overline{9}$$

4. Find out how decimals are used in scientific notation. Then prepare a lesson plan showing how you would introduce this topic to a sixth grade class.

5. If the denominator of a rational number can be expressed as the product of powers of 2 and 5 only, then it can be named by a terminating decimal. **(a)** Show that this is so for several examples, such as $\frac{3}{50}$ and $\frac{19}{40}$. **(b)** Explain why

$$\frac{n}{2^q 5^p}$$

for any whole numbers n, p, and q can be represented by a terminating decimal.

6-6

rational numbers as per cents

Any rational number can be expressed as a decimal and therefore as a number of hundredths:

$$\frac{3}{4} = 0.75 = \frac{75}{100}$$

$$\frac{3}{2} = 1.5 = 1.50 = \frac{150}{100}$$

$$\frac{1}{3} = 0.333\overline{3} = \frac{33.3\overline{3}}{100} = \frac{33\frac{1}{3}}{100}$$

Any fraction may be considered as a **ratio** of its numerator to its denominator. For example, $\frac{1}{2}$ is the ratio of 1 to 2; $\frac{5}{8}$ is the ratio of 5 to 8. Then any number of hundredths is the ratio of that number to 100 and is called a **per cent** (from the Latin *per centum* for per hundred). The symbol for per cent is %. Thus:

$$\frac{3}{4} = \frac{75}{100} = 75\%$$

$$\frac{3}{2} = \frac{150}{100} = 150\%$$

$$\frac{1}{3} = \frac{33\frac{1}{3}}{100} = 33\frac{1}{3}\%$$

To express a rational number as a per cent, write the number as a decimal or as a number of hundredths, and then as a per cent. To express a per cent as a fraction, write the per cent as a number of hundredths and reduce the fraction, if possible:

$$50\% = \frac{50}{100} = \frac{1}{2}$$

$$100\% = \frac{100}{100} = 1$$

$$0.3\% = \frac{0.3}{100} = \frac{3}{1000}$$

A statement such as

$$50\% = \tfrac{1}{2}$$

may be expressed as

$$\tfrac{50}{100} = \tfrac{1}{2}$$

and read as "50 is to 100 as 1 is to 2." Such an equality of ratios (fractions) is often called a **proportion.** By the definition of the equality of fractions we have the **proportion property**

$$\frac{a}{b} = \frac{c}{d} \quad \text{if and only if} \quad ad = bc.$$

The problem of expressing a fraction as a per cent may be considered as *solving* (finding a missing part of) a proportion.

Example 1:   Write $\tfrac{5}{8}$ as a per cent.

Solution:

$$\frac{5}{8} = \frac{n}{100}$$
$$5 \times 100 = 8 \times n$$
$$8n = 500 \qquad \text{(Symmetric, =)}$$
$$n = 62\tfrac{1}{2}$$
$$\tfrac{5}{8} = 62\tfrac{1}{2}\%$$

Another way to express $\tfrac{5}{8}$ as a per cent is to recognize the fraction as the indicated quotient $5 \div 8$. Then divide, and recall that per cent means hundredths.

$$
\begin{array}{r}
0.625 \\
\hline
8\,\overline{)5.000} \\
4\,8 \\
\hline
20 \\
16 \\
\hline
40 \\
40 \\
\hline
\end{array}
$$

Thus $\tfrac{5}{8} = 0.625 = 62.5\%$, or $62\tfrac{1}{2}\%$.

We are now ready to solve problems that involve per cents. Most such problems can be solved by just remembering that a per cent is the ratio of some number to 100. In the following three examples we will solve three different "types" of problems by setting up proportions and using the proportion property.

Example 2: Find 25% of 80.

Solution: We wish to find a number n that compares to 80 in the same way that 25% compares to 100%. The following line segment is used to represent 80, which we consider to be 100% or all of the number.

```
0        n                    80
├────────┼────────────────────┤
0%      25%                  100%
```

Thus:

$$\frac{n}{80} = \frac{25}{100}$$

(n compares to 80 in the same way that 25% compares to 100%)

$$100 \times n = 80 \times 25$$
$$100n = 2000$$
$$n = 20 \quad (2000 \div 100 = 20)$$

Therefore 25% of 80 is 20.

Example 3: 20 is what per cent of 80?

Solution: Consider this diagram, with the known parts shown.

```
0      20                     80
├──────┼──────────────────────┤
0%    n%                     100%
```

Thus:

$$\frac{20}{80} = \frac{n}{100}$$

(20 compares with 80 in the same way that $n\%$ compares with 100%)

$$20 \times 100 = 80 \times n$$
$$80n = 2000$$
$$n = 25 \quad (2000 \div 80 = 25)$$

Therefore 20 is 25% of 80.

Example 4: 20 is 25% of what number?

Solution: Again, consider the diagram comparing known parts.

```
0      20                      n
├──────┼──────────────────────┤
0%    25%                    100%
```

Thus:

$$\frac{20}{n} = \frac{25}{100}$$

(20 compares with n in the same way that 25% compares with 100%)

$$20 \times 100 = n \times 25$$
$$25n = 2000$$
$$n = 80 \quad (2000 \div 25 = 80)$$

Therefore 20 is 25% of 80.

In each of the examples shown, other means of solution are possible. Because the numbers selected in these examples caused no problem, the other methods may be easier. For example, consider again the problem of finding 25% of 80. We can also obtain the answer by either of the following methods:

Using fractions: $25\% = \frac{25}{100} = \frac{1}{4}$
$\frac{1}{4}$ of $80 = \frac{1}{4} \times 80 = 20$

Using decimals: $25\% = 0.25$
25% of $80 = 0.25 \times 80 = 20$

Although these approaches may seem somewhat easier here, the procedure described in Examples 2, 3, and 4 are applicable to all problems, regardless of their complexity. Consider, for example, one additional problem solved by the proportion method and where the answer is not readily seen in advance.

Example 5: There are 22 students that arrive by school bus. These students represent 55% of the students in the class. How many students are there in the class?

Solution: Here we consider the ratio of students bussed to the total number in the class as $22/n$. This is equal to the ratio of 55% to the total, which is always 100%. Thus we have the following:

$$\frac{22}{n} = \frac{55}{100}$$
$$22 \times 100 = 55 \times n$$
$$55n = 2200$$
$$n = 40 \qquad (2200 \div 55 = 40)$$

There are 40 students in the class.

EXERCISES

Write each per cent as a decimal.

1. 37% 2. 87% 3. 5% 4. 175% 5. 0.95%
6. 1% 7. 150% 8. 225% 9. 100% 10. 0.5%

Write each per cent as a fraction in simplest form.

11. 50% 12. 20% 13. 85% 14. 55% 15. 99%
16. 120% 17. 150% 18. 225% 19. 5% 20. 0.5%

Write each decimal as a per cent.

21. 0.45   **22.** 0.03   **23.** 0.65   **24.** 1.45   **25.** 0.7

26. 0.09   **27.** 2.35   **28.** 0.005   **29.** 0.001   **30.** 1.00

Write each fraction as a per cent.

31. $\frac{3}{100}$ **32.** $\frac{76}{100}$ **33.** $\frac{92}{100}$ **34.** $\frac{125}{100}$ **35.** $\frac{150}{100}$

36. $\frac{12}{20}$ **37.** $\frac{18}{50}$ **38.** $\frac{9}{10}$ **39.** $\frac{13}{10}$ **40.** $\frac{32}{20}$

Write a proportion and solve it for each of the following problems.

41. Find 40% of 80.   **42.** Find 80% of 120.

43. Find 35% of 60.   **44.** Find 120% of 85.

45. 40 is what per cent of 160?   **46.** 15 is what per cent of 120?

47. 120 is what per cent of 160?   **48.** 160 is what per cent of 120?

49. 15 is 20% of what number?   **50.** 60 is 75% of what number?

51. 60 is 125% of what number?   **52.** 80 is 10% of what number?

53. Bill spent 40% of his allowance for a pen. If his allowance is $2.00, how much did he spend for the pen?

54. Doris' savings account pays 5% interest per year. Last year she left her account untouched and received $2.80 interest. How much money did she have in the account?

55. Karen spent $1.50 for a book, accounting for 30% of the money that she had with her. How much money did she have with her?

56. Wendy bought a coat on sale for $35.55. The regular price of the coat was $45.00.
    (a) What per cent discount did she receive? That is, the amount she saved on this sale is what per cent of the regular price of the coat?
    (b) The sale price is what per cent of the regular price of the coat?

57. The school population in a certain elementary school rose from 500 to 600 in a recent year. What was the per cent of increase? That is, the increase was what per cent of the original enrollment?

58. Jan went on a diet and reduced her weight from 150 pounds to 135 pounds? What was the per cent of decrease in her weight? That is, her loss of weight was what per cent of her original weight?

PEDAGOGICAL CONSIDERATIONS

1. Many traditional junior high school texts solve per cent problems by considering three different "cases." Find out what these three types of problems are, and then identify the examples of this section according to this classification.

2. Per cent problems are often solved by using the formula $p = r \times b$, where p is the percentage, r is the rate, and b is the base. For example, consider the statement 25% of 80 is 20. Here 25 is the rate, 80 is the base, and 20 is the percentage. Now use this formula to solve each of the examples given in this section.

3. Prepare a bulletin board display showing various uses of per cent found in local newspapers and magazines.

6-7
properties of the set of rational numbers

We may summarize the properties of the set of rational numbers as follows:

Closure, +: The sum of any two rational numbers is a rational number.

Closure, ✕: The product of any two rational numbers is a rational number.

Commutative, +: $a + b = b + a$ for all rational numbers a and b.

Commutative, ✕: $a \times b = b \times a$ for all rational numbers a and b.

Associative, +: $a + (b + c) = (a + b) + c$ for all rational numbers, a, b, and c.

Associative, ✕: $a \times (b \times c) = (a \times b) \times c$ for all rational numbers a, b, and c.

Identity, +: The set of rational numbers contains an element 0 such that $a + 0 = 0 + a = a$ for every rational number a.

Identity, ✕: The set of rational numbers contains an element 1 such that $a \times 1 = 1 \times a = a$ for every rational number a.

Inverse, +: For every rational number a, there is another rational number, $-a$, such that $a + (-a) = 0$.

Inverse, ✕: For every rational number a, $a \neq 0$, there is another rational number $\frac{1}{a}$ such that $a \times \frac{1}{a} = 1$.

Distributive: For all rational numbers a, b, and c we have $a \times (b + c) = (a \times b) + (a \times c)$.

By studying these properties we note that the set of rational numbers forms a group under addition. However this set does not form a group with respect to multiplication since zero does not have a multiplicative inverse in the set.

We describe most of the properties of the set of rational numbers by saying that the set of rational numbers forms a field. In general, any set of numbers forms a **field** if:

The set is closed under addition and multiplication.

The associative law holds for addition and multiplication.

The commutative law holds for addition and multiplication.

The identity elements for addition (0) and multiplication (1) are members of the set.

Each element of the set has its inverse under addition in the set, and each element except 0 has its inverse under multiplication in the set.

The elements of the set satisfy the distributive law for multiplication with respect to addition.

In addition to the field properties, we have found that the set of rational numbers has the density property. Between any two rational numbers there is always another rational number. Despite this fact, the rational numbers do not "fill" the number line. Every rational number is the coordinate of some point on the number line, but not every point on the number line can be named by a rational number. In the next chapter we shall extend our number system so as to establish a one-to-one correspondence between the points of the number line and a new set of numbers, thereby completing the number line.

EXERCISES

Copy the following table. Use "✓" to show that the set of elements named at the top of the columns has the property listed at the side. Use "✕" if the set does not have the property.

	Property	Counting numbers	Whole numbers	Integers	Positive rationals	Rational numbers
1.	Closure, +					
2.	Associative, +					
3.	Identity, +					
4.	Inverse, +					
5.	Commutative, +					
6.	Commutative group, +					
7.	Closure, ×					
8.	Associative, ×					
9.	Identity, ×					
10.	Inverse, ×					
11.	Commutative, ×					
12.	Commutative group, ×					
13.	Distributive					

PEDAGOGICAL CONSIDERATIONS

Computers have come to play a significant role in our mathematics curriculum at the college level. In recent years many secondary schools and some elementary schools have started to make use of computers and calculators in their instructional programs. One of the most widely used languages for the computer is called BASIC. A few of the symbols used in this language are as follows:

Addition: $+$
Subtraction: $-$
Multiplication: $*$
Division: $/$

A vertical arrow is used to indicate that a number is being raised to a power. For example:

$10 \uparrow 2$ means 10^2 $5 \uparrow 3$ means 5^3

In the language of BASIC we may write:

$(a + b) \uparrow 2$ to mean $(a + b)^2$
$(a \uparrow 2) + (b \uparrow 2)$ to mean $a^2 + b^2$

The elements of this language can be introduced to above average elementary school students and used as a vehicle for development of computational skills, as well as concepts on mathematical operations.

1. Compute:

(a) $(2 * 3) \uparrow 2$ **(b)** $(12/4) \uparrow 3$ **(c)** $(12 \uparrow 2)/(3 * 8)$

2. For further information on BASIC read "Introduction to an Algorithmic Language," a publication of the National Council of Teachers of Mathematics, 1201 Sixteenth Street, N.W., Washington, D.C.

CHAPTER TEST

1. Classify each statement as true or false:
   (a) Every rational number is positive or negative.
   (b) Between every two rational numbers there is always another rational number.
   (c) There is a one-to-one correspondence between the set of rational numbers and the points on the number line.
   (d) The number 0 does not have a multiplicative inverse.
   (e) Every rational number has an additive inverse.

227

2. Name the property illustrated by each statement for rational numbers a, b, and c: (a) $a + (b + c) = (b + c) + a$; (b) $a(b + c) = ab + ac$.

3. Copy the following table. Use "✓" to show that the number at the top of the column is a member of the set named at the side. Use "✕" if the number is not a member of the set.

	(a)	(b)	(c)	(d)
Set	$-\frac{2}{3}$	0	5	-4
Counting numbers				
Whole numbers				
Integers				
Rational numbers				

4. Express each number as a quotient of integers: (a) $0.\overline{63}$, (b) $0.6\overline{12}$.

5. Express as a repeating decimal:
   (a) $\frac{1}{2}$ (b) $\frac{2}{3}$ (c) $\frac{1}{9}$ (d) $\frac{7}{12}$ (e) $\frac{1}{13}$

6. Find all possible replacements for n that will make each of the sentences below true, where n must be a member of the set named at the top of each column. When there is no such replacement, write "none."

	(a) Counting numbers	(b) Integers	(c) Rational numbers
Sentence			
$n - 3 = 0$			
$n + 5 = 5$			
$2n = 9$			

7. Express as a single quotient.

   (a) $\dfrac{a}{b} + \dfrac{c}{d}$ (b) $\dfrac{a}{b} \div \dfrac{c}{d}$

8. Write in decimal notation:
   (a) $(2 \times 10^{-1}) + (3 \times 10^{-2}) + (4 \times 10^{-3})$
   (b) $(5 \times 10^{1}) + (3 \times 10^{0}) + (6 \times 10^{-1})$

9. Write each per cent as a decimal and as a fraction in simplest form:
   (a) 45%   (b) 100%   (c) 8%   (d) 120%   (e) 0.1%

10. Write a proportion and solve each problem.
    (a) What is 30% of 80?   (b) 12 is 15% of what number?

The Set of Real Numbers

In the preceding chapter we studied the set of rational numbers and we found that this set suffices for most of our daily arithmetic needs. However there are a number of mathematical problems that we are still unable to solve using only rational numbers. For example, there is no rational number n for which either of the following equations are true:

$$n^2 - 3 = 0, \quad (n^2 = 3)$$
$$n^2 + 3 = 0, \quad (n^2 = -3)$$

That is, there is no rational number a/b such that $(a/b)^2$ is equal to 3 or to -3.

In this chapter we shall extend our number system to make the solution of such equations possible and to obtain coordinates for all points of a number line. Such extensions will allow us to consider most of the fundamental operations and concepts of elementary algebra, which we shall do in the chapters that follow.

7-1
introducing the real numbers

In mathematics we are concerned with the four fundamental operations $(+, -, \times, \div)$. The rational numbers have been introduced so that these operations are always possible with the exception of division by zero. However, using just the set of rational numbers, we are still unable to find a

229

replacement for n so as to make this sentence true:

$$n^2 = 2$$

That is, we are unable to find a rational number a/b such that

$$\frac{a}{b} \times \frac{a}{b} = 2.$$

We call the positive number whose square is 2 the **square root** of 2, and express this number symbolically as $\sqrt{2}$. We wish to show that it is possible to locate a point on the number line whose coordinate is $\sqrt{2}$, and also that $\sqrt{2}$ does not name a rational number. To do so we will need to make use of the **Pythagorean theorem.** This theorem deals with the relationship among the lengths of the sides of a right triangle. Informally the theorem states that the sum of the squares of the two legs of a right triangle is equal to the square of the hypotenuse. In the figure the hypotenuse has a measure of c units and the two legs have measures of a and b units each.

$$a^2 + b^2 = c^2$$

To locate $\sqrt{2}$ on the number line, we construct a right triangle with sides measuring one unit each. By the Pythagorean theorem we find the length of the hypotenuse, x, to be $\sqrt{2}$. Next we draw an arc of a circle, as in the figure, to locate a point with coordinate $\sqrt{2}$ on the number line.

$$x^2 = 1^2 + 1^2$$
$$x^2 = 1 + 1$$
$$x^2 = 2$$
$$x = \sqrt{2}$$

The **radical symbol** $\sqrt{}$ is used to denote the nonnegative square root of a number. Thus $\sqrt{9} = 3$, $-\sqrt{9} = -3$, and $\pm\sqrt{9} = 3$ or -3. This restriction of the radical symbol to indicate only the positive square root of a positive number is part of the present effort to let mathematical expressions have only one meaning.

The set of all numbers that are coordinates of points on the number line is the set of **real numbers.** Some real numbers such as

$$3, \tfrac{2}{3}, 0, \text{ and } -5$$

are rational numbers; all other real numbers are **irrational numbers.** For

231

sec. 7-1
*introducing
the
real
numbers*

example,

$$\sqrt{2}, \quad 3\sqrt{2}, \quad \tfrac{1}{2}\sqrt{2}, \quad \sqrt{3}, \quad \sqrt{5}, \text{ and } \pi$$

are irrational numbers. The set of real numbers is the union of the set of rational numbers and the set of irrational numbers as indicated by this array:

The one-to-one correspondence between the elements of the set of real numbers and the set of points on the number line is a major feature that distinguishes the real numbers from the set of rational numbers. Each real number is the coordinate of a point on the number line; each point on the number line is the graph of a real number. Accordingly, the number line is often referred to as the **real number line** and is said to be **complete.**

Next we wish to prove that $\sqrt{2}$ does not name a rational number. To do so we make use of an indirect proof first used by Euclid. The proof is not difficult, but is fairly subtle and will require some careful thought on the reader's part. We assume the opposite of what we are trying to prove. That is, we assume that $\sqrt{2}$ is a rational number and show that this leads to a contradiction. Thus suppose that $\sqrt{2} = a/b$, a rational number, where a/b has been reduced to lowest terms. Then, in particular, a and b are not both even integers.

If $\sqrt{2} = \dfrac{a}{b}$, then $2 = \dfrac{a^2}{b^2}$ by squaring both sides of the equation. (Remember that $\sqrt{2} \times \sqrt{2} = 2$.)

Then $2b^2 = a^2$, and $2b^2$ is even since it has a factor 2.

Now a^2 is even since it is equal to $2b^2$. However, if the square of an integer is even, then the integer is even. (See §5-6, Exercise 23.) Thus a is even.

Since a is even, we may write $a = 2k$, where k is some integer. Thus

$$2b^2 = (2k)^2 \text{ by this substitution.}$$
$$2b^2 = 4k^2 \text{ by squaring } 2k.$$
$$b^2 = 2k^2 \text{ by division of each member}$$
of the equation by 2.

Now b^2 is even since it is equal to a number with a factor of 2. Again, if the square of an integer is even, then the integer is even; thus b is even. This is contrary to the assumption that a and b were not both even. Therefore our assumption that there exists a rational number whose square is 2 has led to a contradiction. In other words, the real number whose square is 2 can not be a rational number, and is therefore an irrational number.

The proof just developed is often called an *existence proof* in that it demonstrated the existence of an irrational number. We can use the fact that $\sqrt{2}$ is an irrational number to prove the existence of many other irrational numbers, as in this example.

Example 1: Prove that $1 + \sqrt{2}$ is an irrational number.

Solution: We use an indirect proof and assume that $1 + \sqrt{2}$ is rational. Thus we assume it can be written in the form a/b, as the quotient of two integers, with $b \neq 0$. Assume

$$1 + \sqrt{2} = \frac{a}{b}.$$

Then

$$\sqrt{2} = \frac{a}{b} - 1.$$

But $(a/b) - 1$ names a rational number because the set of rational numbers is closed with respect to subtraction. Therefore $\sqrt{2}$ must be a rational number. But this contradicts the known fact that $\sqrt{2}$ is an irrational number. Therefore our original assumption was incorrect, and $1 + \sqrt{2}$ must be an irrational number.

Note that we can locate (graph) a point on the number line whose coordinate is $1 + \sqrt{2}$. We repeat the construction used earlier, but shifted one unit to the right.

Example 2: Graph $1 + \sqrt{2}$ on the number line.

Solution:

EXERCISES

Classify each statement as true or false.

1. Every rational number is a real number.
2. Every real number is a rational number.
3. Every integer is a real number.
4. Every real number is either a rational number or an irrational number.
5. Every point on the number line has a real number as its coordinate.

233

sec. 7-1
introducing
the
real
numbers

Copy the following table. Use "✓" to show that the number at the top is a member of the set listed at the side. Use "✗" if it is not a member of the set.

6. 7. 8. 9. 10.

Set	3	$-\frac{2}{3}$	$\sqrt{9}$	0	$\sqrt{5}$
Counting numbers					
Whole numbers					
Integers					
Rational numbers					
Real numbers					

Find a replacement for n that will make each of the sentences below true when n must be a member of the set named at the top of the column. If there is no such replacement, write "none."

Sentence	Counting numbers	Whole numbers	Integers	Rational numbers	Real numbers
11. $n + 3 = 1$					
12. $3n = 8$					
13. $n \times n = 2$					
14. $n \times n = -1$					

Copy the following table. Use "✓" to show that the set of elements named at the top of the column has the property listed at the side. Use "✗" if the set does not have the property.

15.   16.   17.   18.

Property	Positive integers	Positive odd integers	Negative integers	Real numbers
Closure, +				
Closure, +				
Inverse, +				
Inverse, ×				
Dense				

Prove that each of the numbers is an irrational number.

19. $\sqrt{2} + 2$ 20. $\sqrt{2} - \frac{1}{2}$

Construct a number line and graph each number.

21. $3 + \sqrt{2}$ 22. $\frac{1}{2} + \sqrt{2}$

PEDAGOGICAL CONSIDERATIONS

1. The construction for $\sqrt{2}$ can be extended to find segments whose measures are $\sqrt{3}$, $\sqrt{4}$, $\sqrt{5}$, etc. Merely continue to construct right triangles, using the hypotenuse of the preceding triangle as one leg and a segment of one unit as the other leg. The following figure shows the construction for $\sqrt{3}$.

Continue this method of construction to find lengths of $\sqrt{5}$ and $\sqrt{7}$.

2. An alternative proof to show that $\sqrt{2}$ is an irrational number is one that involves a method of exhaustion. We begin by assuming $\sqrt{2} = a/b$, where a and b are integers, $b \neq 0$, and again where a and b have no factors in common, that is, the fraction a/b has been reduced to lowest terms. Then there are only four possible combinations for a and b:

  (i) a and b are both even.
  (ii) a and b are both odd.
  (iii) a is odd and b is even.
  (iv) a is even and b is odd.

Show that each possibility leads to a contradiction. Therefore, since all cases are exhausted, the original assumption that $\sqrt{2}$ is a rational number must be incorrect.

7-2

graphs of sets of real numbers

Now that we have introduced the set of real numbers we are able to extend our ideas of graphs on the number line for different sets of numbers. For example, let us consider the set of numbers that are less than or equal to 3. Notice how the graph of this set changes as we use different sets of numbers as the basis for discussion.

Counting numbers: $\{1, 2, 3\}$

Whole numbers: $\{0, 1, 2, 3\}$

Integers: $\{\ldots, -3, -2, -1, 0, 1, 2, 3\}$

The heavily shaded arrowhead at the left is used to indicate that the set extends indefinitely to the left, including all negative integers in this case.

Real numbers:

The heavily shaded arrow drawn to the left indicates that the set includes 3 and all real numbers that are less than 3.

The following examples show other graphs on the number line.

Example 1:   Graph the set of real numbers from 2 through 5.

Solution: The graph is a line segment with endpoints indicated by the heavy dots at the points with coordinates 2 and 5.

Note in Example 1 that the language "from 2 through 5" is used to indicate that the endpoints are to be included in the graph. If they are to be excluded, the set would be described as in Example 2.

Example 2:   Graph the set of real numbers between 2 and 5.

Solution: This graph may be obtained from the graph in Example 1 by excluding the points with coordinates 2 and 5. Use hollow dots to show that the points with coordinates 2 and 5 are not included in the graph.

235

Example 3: Graph the set of real numbers greater than or equal to 2.

Solution: A heavy dot is placed at the point with coordinate 2 to indicate that this point is included in the graph of the set. Then a heavily shaded arrow is drawn to indicate that all points with coordinates greater than 2 are to be included.

Example 4: Graph the set of real numbers less than 1.

Solution:

All of the preceding graphs were on a number line. We can also use real numbers to locate points on a plane. To do so we represent the points of a plane by ordered pairs of real numbers in the following way.

First we draw two perpendicular number lines, called the *x*-axis and the *y*-axis, as in the figure. These **coordinate axes** divide the entire plane into four parts, called **quadrants,** that are numbered as shown. The point where the two axes intersect is called the **origin** and is said to have coordinates (0, 0).

We locate points on this plane by means of ordered pairs of numbers (x, y), where we agree that the order is first x, the signed distance along the *x*-axis, and then y, the signed distance along the *y*-axis. Each element of the ordered pair tells us how many units to move from the origin, and in which direction.

The points A: $(3, 2)$, B: $(-3, 2)$, C: $(-3, -2)$, and D: $(3, -2)$ are located in the next four graphs. Notice that one point is located in each of the four quadrants.

A: (3, 2): Start at the origin. Move three units to the right and two units up. The coordinates of point *A* are (3, 2).

B: (−3, 2): Start at the origin. Move three units to the left and two units up. The coordinates of point *B* are (−3, 2).

C: (−3, −2): Start at the origin. Move three units to the left and two units down. The coordinates of point *C* are (−3, −2).

D: (3, −2): Start at the origin. Move three units to the right and two units down. The coordinates of point D are (3, −2).

EXERCISES

Graph each of the following sets of numbers on a number line.

1. The set of counting numbers less than 5.
2. The set of counting numbers less than or equal to 3.
3. The set of counting numbers from 2 through 5.
4. The set of counting numbers between 2 and 5.
5. The set of whole numbers less than 4.
6. The set of whole numbers less than or equal to 3.
7. The set of whole numbers between 0 and 4.
8. The set of whole numbers from 0 through 4.
9. The set of integers between −2 and 3.
10. The set of integers from −3 through 2.
11. The set of integers from −7 through −1.
12. The set of integers less than −2.

Graph the real numbers.

13. Between −5 and 3.
14. Less than 2.
15. Greater than −1.
16. Less than or equal to 0.
17. Greater than or equal to −3.
18. From −3 through 5.

Draw a pair of coordinate axes and locate each of the following points:

19. P: (2, −1)
20. Q: (−3, −4)
21. R: (4, −3)
22. S: (5, 2)

Use the figure below and give the coordinates of each point shown.

23. A **24.** B **25.** C **26.** D **27.** E **28.** F

PEDAGOGICAL CONSIDERATIONS

1. Think of a city street (call it School Street) and use city blocks as units of distance even though the blocks may not be of equal lengths. Suppose that School Street extends east and west with your school at one of the major street corners. Describe how you might use coordinates on School Street as a number line to direct a stranger how to drive from your school to any street that crosses School Street. Assume that only one street crosses School Street at each intersection and that each street actually crosses.

2. As in Exercise 1 describe how you might direct the stranger to any desired point on School Street.

3. Think of a city with Main Street and School Street intersecting at right angles at the center of the city. Suppose that the streets east of Main Street are named in order First Street, Second Street, Third Street, and so forth. Also suppose that the streets north of School Street are named in order First Avenue, Second Avenue, Third Avenue, and so forth. Draw a map of the northeast part of the city. Describe how you might use coordinates to direct a stranger from the intersection of School Street and Main Street to any desired street intersection in the northeast part of the city. (*Note:* This procedure may be extended for all parts of the city using a street plan such as the one in Salt Lake City, Utah.)

4. As in Exercise 3 describe how you might direct the stranger to any desired point on any street in the northeast part of the city.

In Chapter 6 we studied the relationship between the set of rational numbers and the set of repeating decimals. Thus we know that every rational number can be named by a decimal that repeats. Recall that we agreed that a terminating decimal may be thought of as a decimal that repeats the digit zero. Thus, for example, we have the following decimal representations of several rational numbers:

$$\tfrac{3}{4} = 0.75\overline{0}, \qquad \tfrac{1}{3} = 0.3\overline{3}, \qquad \tfrac{1}{7} = 0.142857\overline{142857}$$

We also found a way to express every repeating decimal as the quotient of two integers, and thus as the name of a rational number. For example, consider the repeating decimal $0.72\overline{72}$.

$$
\begin{aligned}
100n &= 72.72\overline{72} \\
-n &= 0.72\overline{72} \\
\hline
99n &= 72 \\
n &= \tfrac{72}{99} = \tfrac{8}{11} \qquad \text{Thus } 0.72\overline{72} = \tfrac{8}{11}.
\end{aligned}
$$

Now we wish to consider the decimal representation for irrational numbers. To do so, we return again to the number $\sqrt{2}$ and try to find its decimal equivalent through arithmetic procedures. First we know that $\sqrt{2}$ must fall somewhere between 1 and 2 since $1^2 = 1$, $2^2 = 4$, and $(\sqrt{2})^2 = 2$. Thus we may write this first inequality:

$$1 < \sqrt{2} < 2$$

Next, by arithmetic, we may verify that $(1.4)^2 = 1.96$ and $(1.5)^2 = 2.25$. This gives rise to the next inequality:

$$1.4 < \sqrt{2} < 1.5$$

Continuing in this way, we can verify each of the following by multiplication:

$$
\left.
\begin{aligned}
(1.41)^2 &= 1.9881 \\
(1.42)^2 &= 2.0164
\end{aligned}
\right\} \qquad 1.41 < \sqrt{2} < 1.42
$$

$$
\left.
\begin{aligned}
(1.414)^2 &= 1.999396 \\
(1.415)^2 &= 2.002225
\end{aligned}
\right\} \qquad 1.414 < \sqrt{2} < 1.415
$$

$$
\left.
\begin{aligned}
(1.4142)^2 &= 1.99961644 \\
(1.4143)^2 &= 2.00034449
\end{aligned}
\right\} \qquad 1.4142 < \sqrt{2} < 1.4143
$$

$$
\left.
\begin{aligned}
(1.41421)^2 &= 1.9999899241 \\
(1.41422)^2 &= 2.0000182084
\end{aligned}
\right\} \qquad 1.41421 < \sqrt{2} < 1.41422
$$

We are now caught up in a never-ending process. The decimal representation for $\sqrt{2}$ does not terminate, and it does not repeat. Given enough time and patience we may express $\sqrt{2}$ correct to as many decimal places as we wish, but at each step we merely have a rational number approximation to this irrational number. For example, correct to seven decimal places we have the following:

$$\sqrt{2} = 1.4142135\ldots$$

241

sec. 7-3
decimal
representation
of real
numbers

We use the three dots to show that the decimal is an infinite one. The absence of a bar over any of the digits shows that it is a nonrepeating decimal.

We first defined an irrational number as a real number that is *not* a rational number. We may now define an irrational number directly by saying that an irrational number is a real number that is named by a nonrepeating, nonterminating decimal.

In summary, every real number can be represented by a decimal. If the real number is a rational number, then it may be represented by a terminating or by a repeating decimal. If the real number is an irrational number, then it may be represented by a nonterminating, nonrepeating decimal.

It is often convenient to represent irrational numbers by a sequence of digits that has a pattern but does not repeat any particular sequence of digits. For example, each of the following names an irrational number:

$$0.20220222022220222220\ldots$$
$$0.305300530005300005\ldots$$
$$0.404004000400004000004\ldots$$

Example 1: How many zeros are there altogether between the decimal point and the one hundredth 5 in this sequence?

$$0.05005000500005\ldots$$

Solution: There is one zero preceding the first 5, two zeros preceding the second 5, etc. The total number of zeros is the sum

$$1 + 2 + 3 + \cdots + 100.$$

By the method used in §1 of the introductory chapter, the sum is 5050.

We may use decimal representations to show the density of the set of real numbers. That is, the set of real numbers is said to be dense; between any two real numbers, there is always another real number. We have already shown that this is true for two rational numbers (see §6-2). In Example 2 we find an irrational number between two given rational numbers.

Example 2: Find an irrational number between 0.47 and 0.48.

Solution: It is easier to see the solution if the given numbers are written as decimals that repeat the digit 0.

$$0.47 = 0.470000\overline{0}$$
$$0.48 = 0.480000\overline{0}$$

We need to write a decimal for an irrational number that is greater than 0.47 and less than 0.48. This decimal must be a nonrepeating, nonterminating decimal. There are many possible answers. Here are two:

$$0.47247224722247222247\ldots$$
$$0.47505005000500005\ldots$$

The pattern shown allows you to write as many digits of each decimal as you wish. Each pattern insures that the decimal will never terminate nor repeat any particular sequence of digits.

EXERCISES

Classify each decimal as the name of a rational number or an irrational number.

1. 0.745

2. $0.745\overline{45}$

3. 0.7454554555 . . .

4. $0.745\overline{745}$

5. 0.7454554555

6. 0.7454454445 . . .

Tell whether each number can be represented by (a) a terminating decimal; (b) a repeating decimal; (c) a nonterminating, nonrepeating decimal. Give the decimal representation of each number that can be expressed as a terminating or a repeating decimal.

7. $\frac{3}{4}$

8. $\frac{5}{12}$

9. $\sqrt{81}$

10. $\sqrt{3}$

11. $\sqrt{10}$

12. $\frac{13}{16}$

Write the first fifteen digits to the right of the decimal point in each decimal.

13. $0.\overline{75}$

14. $0.2\overline{143}$

15. 0.353553555 . . .

16. 0.203200320003 . . .

Write the digit in the fifteenth position to the right of the decimal point for each decimal.

17. $0.1\overline{2}$

18. $5.3\overline{21}$

19. 0.737737773 . . .

20. 0.928922892228 . . .

Perform the necessary multiplications to verify each inequality.

21. $1.7 < \sqrt{3} < 1.8$

22. $1.73 < \sqrt{3} < 1.74$

23. $2.2 < \sqrt{5} < 2.3$

24. $2.23 < \sqrt{5} < 2.24$

Use the method shown in this section to find each square root, correct to the nearest tenth.

25. $\sqrt{7}$

26. $\sqrt{10}$

27. $\sqrt{11}$

28. $\sqrt{17}$

List the numbers of each set in order, from smallest to largest.

29. 0.35, 0.353553555 . . . , 0.35355, $0.\overline{35}$, $0.3\overline{5}$.

243

sec. 7-4
properties
of the
real
numbers

30. 1.414, 1.4141, 1.$\overline{41}$, 1.414114111 . . . , 1.4.

31. 0.078, 0.07$\overline{8}$, 0.0787887888 . . . , 0.07, 0.$\overline{07}$.

32. Which of the following names a rational number between 0.37 and 0.38? **(a)** 0.375; **(b)** 0.$\overline{37}$; **(c)** 0.373773777 . . . ; **(d)** 0.37$\overline{8}$.

33. Which of the following names an irrational number between 0.234 and 0.235?   **(a)** 0.2345;   **(b)** 0.$\overline{234}$;   **(c)** 0.234040040004 . . .;   **(d)** 0.23454554555 . . . .

34. Name two rational numbers between 0.523 and 0.524.

35. Name two irrational numbers between 0.38 and 0.39.

36. Name two irrational numbers between 0.$\overline{78}$ and 0.$\overline{79}$.

37. Name two rational numbers between 0.$\overline{356}$ and 0.$\overline{357}$.

PEDAGOGICAL CONSIDERATIONS

1. Examine several introductory algebra texts and try to find two different procedures (algorithms) for taking square roots of numbers.

2. There are many irrational numbers that play important roles in mathematics. One of these is π (pi), the ratio of the circumference of a circle to the length of its radius. Write a report on the history of π. Compare the decimal representation for π with the decimal for $\frac{22}{7}$, the commonly used rational approximation for π.

3. Prepare a report on the Pythagoreans, a group of ancient Greek mathematicians who are credited with the discovery of irrational numbers.

7-4

properties
of
the
real
numbers

We began our study of the number system in Chapter 5 with the set of counting numbers, and have now progressed through the set of real numbers. At each step we discussed the various properties of the set of numbers under study, and then made further extensions in order to be able to solve certain types of problems. As a summary of these developmental stages, use the chart on page 244 to note the steps that we have gone through. In each case we wish to find a replacement for *n* so as to make the sentence true, using the set of numbers listed. In each case there is no possible solution, thus justifying as extension of the number system.

Equation	Set of numbers	Solution	Extensions
$n + 2 = 2$	Counting numbers	None	For whole numbers, we have $n = 0$.
$n + 3 = 2$	Whole numbers	None	For integers, we have $n = -1$.
$2n = 3$	Integers	None	For rational numbers, we have $n = 3/2$
$n^2 = 2$	Rational numbers	None	For real numbers, we have $n = \pm \sqrt{2}$
$n^2 = -2$	Real numbers	None	Another extension is needed! (See § 7-5)

Each extension of the number system has been made so that as many as possible of the properties of the previous set of numbers are retained as properties of the new set of numbers. Furthermore, in each case one or more new properties are acquired. For example, the extension of the set of counting numbers to the set of whole numbers was made to obtain 0, the identity for addition. The extension of the set of whole numbers to the set of integers was made to obtain the additive inverses of the counting numbers, that is, to make subtraction always possible. The extension of the set of integers to the set of rational numbers was made to obtain density and multiplicative inverses for each number different from zero, that is, to be able to divide by any number different from zero. The extension of the set of rational numbers to the set of real numbers was made to obtain completeness, the one-to-one correspondence of the set of real numbers with the set of points on a number line.

In each of these extensions it has been possible to retain all of the basic properties of the previous sets of numbers. In particular, the set of rational numbers forms a field (Chapter 6) and the set of real numbers forms a field. To review and summarize, the set of real numbers R with operations $+$ and $\times$ defined on R forms a field because the following 11 properties hold for all elements a, b, and c of R.

Closure, $+$: $a + b$ is a unique element of R.

Closure, $\times$: $a \times b$ is a unique element of R.

Associative, $+$: $(a + b) + c = a + (b + c)$.

Associative, $\times$: $(a \times b) \times c = a \times (b \times c)$.

Commutative, $+$: $a + b = b + a$.

Commutative, $\times$: $a \times b = b \times a$.

Identity, $+$: There is an element 0 in R such that $0 + a = a + 0 = a$.

Identity, $\times$: There is an element 1 in R such that $1 \times a = a \times 1 = a$.

Inverse, $+$: For each element a there is an element $(-a)$ in R such that $a + (-a) = 0$.

245

*sec. 7-4
properties
of the
real
numbers*

Inverse ($\neq$ 0), $\times$: For each element a different from zero, the additive identity, there is an element $\dfrac{1}{a}$ in R such that

$$a \times \frac{1}{a} = 1.$$

Distributive: $a \times (b + c) = (a \times b) + (a \times c)$.

In addition to these field properties, we have seen that the set of real numbers has the density property. Between any two real numbers there is always another real number. Furthermore the set of real numbers has the completeness property. That is, there is a one-to-one correspondence between the set of real numbers and the set of points on the number line. As we noted earlier, it is the completeness property that is unique for the set of real numbers, and that distinguishes the real numbers from the rational numbers.

Computations with radicals are restricted in this book to statements such as $(\sqrt{2})^2 = 2$. When values are needed for real numbers such as $\sqrt{2}$, $3\sqrt{2}$, or π, either the decimal representations are "rounded off" or other approximations such as $\frac{22}{7}$ for π are used.

We have seen that the set of real numbers may be classified in several different ways. Any real number is

1. Positive, negative, or zero.

2. A rational number or an irrational number.

3. Expressible as a terminating, a repeating, or a nonterminating, nonrepeating decimal.

The relationship of the set of real numbers to some of its subsets that we have studied is shown in the following array:

EXERCISES

Classify each statement as true or false.

1. Every real number is either an irrational number or an integer.

2. Every real number is either positive or negative.

3. Every integer is a rational number.

4. Every negative integer is a rational number.

5. The set of real numbers contains the additive inverse of each of its elements.

6. The set of rational numbers contains the multiplicative inverse of each of its elements.

7. Every irrational number is a real number.

8. The set of rational numbers is both dense and complete.

9. Every irrational number is either positive or negative.

10. The set of real numbers is a subset of the set of rational numbers.

Complete the chart to summarize the various properties of the number systems that we have studied thus far. In each case use "✓" to show that the set of elements named at the top of the column has the property listed at the side. Use "✕" if the set does not have the stated property. Note that the first 11 properties listed are the field properties.

	Counting numbers	Whole numbers	Integers	Rational numbers	Real numbers
11. Closure, +					
12. Closure, ✕					
13. Commutative, +					
14. Commutative, ✕					
15. Associative, +					
16. Associative, ✕					
17. Identity, + (0)					
18. Identity, ✕ (1)					
19. Inverse, +					
20. Inverse, ✕					
21. Distributive					
22. Field					
23. Density					
24. Completeness					

State whether each sentence is true or false. If true, give an example to illustrate the property. If false, give a counterexample to verify your conclusion.

25. The set of rational numbers is closed under multiplication.

26. The set of irrational numbers is closed under addition.

27. The set of irrational numbers is closed under multiplication.

28. The set of rational numbers is closed under division.

29. The set of integers forms a field.

30. The set of real numbers forms a field.

A topic that often causes difficulty in elementary school mathematics is that of "rounding off" numbers. A number line picture is helpful here to explain this process for students. Thus in the figure below, we wish to round 22 and 27 to the nearest 10; that is, we wish to find the nearest multiple of 10. Since 22 is closer to 20 than it is to 30, we round this number down to 20. On the other hand, 27 is closer to 30 than it is to 20 so that we round this up to 30.

The natural question that arises is the case of a number that is exactly halfway between two multiples of 10. The usual practice in such situations is to automatically round up. For example, we would round 25 to 30, and 35 to 40. In later work in mathematics, especially in statistics, a bias in data is introduced if this is always done. Can you see why this is so? The usual practice followed there is to round a number that has a 5 in the right-hand digit's place so that the preceding numeral is even. In other words, rounded to the nearest 10, we would round 25 down to 20, but would round 35 up to 40. This insures that half of the time we round down and half of the time we round up. Use this procedure to round in the exercises that follow.

1. Round to the nearest 100: **(a)** 348; **(b)** 257; **(c)** 175; **(d)** 265.

2. Round to the nearest tenth: **(a)** 0.37; **(b)** 0.23; **(c)** 0.55; **(d)** 0.85.

3. Round to the nearest hundredth: **(a)** 0.376; **(b)** 0.284; **(c)** 0.025; **(d)** 0.015.

7-5

the set of complex numbers

The set of real numbers suffices for almost all of the work of elementary mathematics. However, we shall make one more extension of our number system in order to be able to solve equations such as $n^2 + 1 = 0$. For the set of real numbers, there is no replacement for n that will make this sentence true. We therefore invent a number i such that $i^2 + 1 = 0$; that is, $i^2 = -1$ and $i = \sqrt{-1}$. For the sentence $n^2 + 1 = 0$, $n^2 = -1$ and $n = \pm\sqrt{-1}$; that is, $n = \pm i$.

As further illustrations of the notation for the square roots of negative numbers, consider:

$$\sqrt{-4} = \sqrt{4} \cdot \sqrt{-1} = 2i, \qquad \sqrt{-9} = \sqrt{9} \cdot \sqrt{-1} = 3i,$$
$$\sqrt{-5} = \sqrt{5} \cdot \sqrt{-1} = \sqrt{5}\,i$$

A number that may be expressed in the form bi, where b is a real number different from zero and $i = \sqrt{-1}$, is called a **pure imaginary number**. The sum of a real number different from zero and a pure imaginary number, such as $2 + 5i$, is called a **mixed imaginary number**. The union of the set of real numbers and the set of imaginary numbers is called the set of **complex numbers**. In general, a complex number $a + bi$, where a and b are real numbers and $i = \sqrt{-1}$, is

a *real number* if $b = 0$;
an *imaginary number* if $b \neq 0$;
a *pure imaginary number* if $a = 0$ and $b \neq 0$; and
a *mixed imaginary number* if $a \neq 0$ and $b \neq 0$.

Example 1: Write $2 + \sqrt{-9}$ in the form $a + bi$.

Solution:

$$\sqrt{-9} = \sqrt{9} \cdot \sqrt{-1} = 3i$$
$$2 + \sqrt{-9} = 2 + 3i$$

Example 2: Write 5 in the form $a + bi$.

Solution:

$$5 + 0i$$

Example 3: Write $\sqrt{-3}$ in the form $a + bi$.

Solution:

$$\sqrt{-3} = \sqrt{3} \cdot \sqrt{-1} = \sqrt{3}\, i = 0 + \sqrt{3}\, i$$

The importance of recognizing the set of possible replacements being used is shown by the solutions of the equations in the following array. The word "none" implies that for the set of numbers named at the left of the row there is no replacement for n that will make the sentence true.

Replacements	$n + 1 = 1$	$n + 1 = 0$	$2n = 1$	$n^2 - 1 = 0$	$n^2 + 9 = 0$
Counting numbers	none	none	none	1	none
Whole numbers	0	none	none	1	none
Integers	0	-1	none	$1, -1$	none
Rational numbers	0	-1	$\frac{1}{2}$	$1, -1$	none
Real numbers	0	-1	$\frac{1}{2}$	$1, -1$	none
Complex numbers	0	-1	$\frac{1}{2}$	$1, -1$	$3i, -3i$

EXERCISES

Express each number in the form a + bi.

1. $2 + \sqrt{-9}$ 2. $3 + \sqrt{-4}$
3. $\sqrt{9}$ 4. 0
5. $5 + \sqrt{-3}$

Classify each statement as true or false.

6. Every rational number is a complex number.
7. Every irrational number is a complex number.
8. Every complex number is a rational number.
9. Every real number can be written in the form $a + bi$.
10. Every integer can be written in the form $a + bi$.

Copy the following table. Use "✓" to show that the number at the top of the column is a member of the set named at the side. Use "✗" if the number is not a member of the set.

| | 11. | 12. | 13. | 14. | 15. |
Set	-7	$\frac{3}{5}$	0	$\sqrt{5}$	$\sqrt{-9}$
Counting numbers					
Whole numbers					
Integers					
Rational numbers					
Irrational numbers					
Real numbers					
Imaginary numbers					
Complex numbers					

Find a replacement for n that will make each sentence true, if n must be a member of the set named at the top of the column. When there is no such replacement, write "none."

	Sentence	Counting numbers	Integers	Rational numbers	Real numbers	Complex numbers
16.	$n + 2 = 0$					
17.	$n + 5 = 2$					
18.	$3n - 1 = 0$					
19.	$n^2 = 4$					
20.	$n^2 + 4 = 0$					

249

PEDAGOGICAL CONSIDERATIONS

Elementary school students should distinguish between operations that are impossible and operations that they are not yet prepared to do. For example,

$\frac{5}{0}$ is impossible.

$\frac{0}{0}$ is impossible.

$5 - 8$ cannot be expressed as a whole number but can be expressed as an integer.

$5 \div 3$ cannot be expressed as an integer but can be expressed as a rational number.

$\sqrt{2}$ cannot be expressed as a rational number but can be expressed as a real number.

$\sqrt{-4}$ cannot be expressed as a real number but can be expressed as a complex number.

Complex numbers are important to elementary school teachers to help them distinguish between the "impossible" and the "possible with a suitable set of numbers."

Many educators feel that students should not be told and especially should not be drilled upon statements that will have to be "unlearned" later. Thus it is important that elementary school teachers avoid statements such as

> You can't subtract 8 from 5.
> It is not possible to subtract 8 from 5.
> You can't divide 5 by 3.

Such statements can certainly be defended as correct under a certain interpretation but, more important, they can be misunderstood and thereby create mental blocks to future progress of the student. Teachers should make statements such as

> $5 - 8$ is not one of our whole numbers.
> You do not have a number to represent $5 - 8$.
> $5 \div 3$ is not an integer.

Semantic problems of this sort arise in many ways and at most grade levels.

Write out appropriate explanations for each problem (some are possible in the usual manner), including reasons, other representations, and so forth that may be needed to help the student understand the problem. The set of numbers assumed to be known to the child is indicated in parentheses when it is not mentioned in the statement.

1. $\frac{12}{2}$ (whole numbers)

2. $\frac{12}{0}$ (whole numbers)

3. $12 - 15$ (whole numbers)

4. $12 - 15$ (integers)

5. $12 \div 5$ (integers)

6. $12 \div 5$ (rational numbers)

7. $5 = (\sqrt{5})(\sqrt{5})$ (real numbers)

8. The only positive integral factors of 5 are 5 and 1.

9. Any positive rational number is a divisor of 5.

10. Any rational number different from 0 is a factor of 5.

11. 5 does not have a rational number as a square root.

12. 5 does have a real number as a square root.

13. -5 does not have a real number as a square root.

14. -5 does have a complex number as a square root.

7-6

transfinite cardinal numbers

Whenever two sets are such that their members may be placed in a one-to-one correspondence with each other, we say that the two sets are *equivalent sets* (§1-3). Thus the set of letters in the word "Thursday" is equivalent to the set of the first eight counting numbers.

```
T   h   u   r   s   d   a   y
↕   ↕   ↕   ↕   ↕   ↕   ↕   ↕
1   2   3   4   5   6   7   8
```

Also, the set of counting numbers is equivalent to the set of even counting numbers (§1-4).

```
1   2   3   4   5   6  · · ·  n  · · ·
↕   ↕   ↕   ↕   ↕   ↕         ↕
2   4   6   8  10  12  · · · 2n · · ·
```

Each counting number n can be matched with an even counting number $2n$.

At the turn of the twentieth century, a German mathematician named Georg Cantor developed an entire theory of infinite sets of numbers. He assigned a cardinal number to the set of counting numbers, namely, $\aleph_0$. This is read "aleph-null" and is actually a *transfinite* (beyond the finite) *cardinal number*. It is correct then to say that there are $\aleph_0$ counting numbers, just as you might say that there are seven days of the week or ten fingers on your hands. Furthermore, any set that can be matched in a one-to-one correspondence with the set of counting numbers is also of size $\aleph_0$ (**countably infinite**) and is said to be a **denumerably infinite set**.

The discussion of transfinite numbers gives rise to some very interesting paradoxes. One of the most famous of these is the story of the infinite house. This is a house that contains an infinite number of rooms, numbered 1, 2, 3, 4, 5, and so on. Each room is occupied by a single tenant. That is, there is a one-to-one correspondence between rooms and occupants. There are

$\aleph_0$ rooms, and $\aleph_0$ occupants. One day a stranger arrived at the house and asked to be admitted. The caretaker was an amateur mathematician and was able to accommodate this visitor in the following manner. He asked the occupant of room 1 to move to room 2, the occupant of room 2 to move to room 3, the occupant of room 3 to move to room 4, and in general the occupant of room n to move to room $n + 1$. Now everyone had a room and room number 1 was available for the visitor! In other words, we have demonstrated this interesting fact:

$$\aleph_0 + 1 = \aleph_0$$

Several days later an infinite number of visitors arrived at the house, all demanding individual rooms! Again the caretaker was able to accommodate them. He merely asked each tenant to move to a room with a number that was double his current room number. That is, the occupant of room 1 moved to room 2, the occupant of room 2 moved to room 4, the occupant of room 3 moved to room 6, and in general the occupant of room n moved to room $2n$. After this move was made, the new arrivals were placed in rooms 1, 3, 5, and so forth. Here we have an example of another interesting fact of transfinite arithmetic:

$$\aleph_0 + \aleph_0 = \aleph_0$$

As another example of this last fact note that the set of even counting numbers is of size $\aleph_0$, and the set of odd counting numbers is also of size $\aleph_0$. But the even counting numbers together with the odd counting numbers form the set of all counting numbers, which we have agreed is of size $\aleph_0$. Therefore, again, we see that $\aleph_0 + \aleph_0 = \aleph_0$.

Cantor used an interesting scheme to show that the set of rational numbers can be placed in a one-to-one correspondence with the set of counting numbers. That is, the set of rational numbers is equivalent to the set of counting numbers and is therefore denumerably infinite and thus of size $\aleph_0$. Cantor's scheme was to arrange the positive rational numbers as follows, and then to consider their listing in a diagonal manner as indicated by the arrows.

$$
\begin{array}{ccccccc}
\frac{1}{1} & \frac{1}{2} & \frac{1}{3} & \frac{1}{4} & \frac{1}{5} & \cdots \\
\frac{2}{1} & \frac{2}{2} & \frac{2}{3} & \frac{2}{4} & \frac{2}{5} & \cdots \\
\frac{3}{1} & \frac{3}{2} & \frac{3}{3} & \frac{3}{4} & \frac{3}{5} & \cdots \\
\frac{4}{1} & \frac{4}{2} & \frac{4}{3} & \frac{4}{4} & \frac{4}{5} & \cdots \\
\frac{5}{1} & \frac{5}{2} & \frac{5}{3} & \frac{5}{4} & \frac{5}{5} & \cdots \\
\end{array}
$$

The set of rational numbers with many numbers repeated, as in the array, can now be placed in a one-to-one correspondence with the set of counting numbers:

$$\frac{1}{1} \ , \ \frac{2}{1} \ , \ \frac{1}{2} \ , \ \frac{1}{3} \ , \ \frac{2}{2} \ , \ \frac{3}{1} \ , \ \frac{4}{1} \ , \ \frac{3}{2} \ldots$$
$$\updownarrow \quad \updownarrow \quad \updownarrow \quad \updownarrow \quad \updownarrow \quad \updownarrow \quad \updownarrow \quad \updownarrow$$
$$1 \ , \ 2 \ , \ 3 \ , \ 4 \ , \ 5 \ , \ 6 \ , \ 7 \ , \ 8 \ldots$$

This matching shows that there are no more rational numbers than counting numbers. Since the set of counting numbers is a subset of the set of rational numbers, there cannot be more counting numbers than rational numbers. Therefore, regardless of how much this defies the imagination as well as our intuition, it appears that there are just as many rational numbers as there are counting numbers!

Cantor thought that the set of real numbers might also be of size $\aleph_0$. However by a very ingenious proof he showed that such is not the case and that there are more real numbers than there are rational numbers. If the reader is not yet too weary, he can proceed with the following proof that the set of real numbers is not countable. Otherwise, this may be a good time to skip over to the exercises.

Cantor asserted, as we know from §7-3, that every real number can be expressed as a nonterminating decimal. He then assumed that a one-to-one correspondence existed between the set of counting numbers and the set of real numbers written in decimal form. Under this assumption the correspondence may be indicated in an array such as the following, where the digits of each decimal are represented by letters with subscripts.

$$1 \longleftrightarrow 0. \quad a_1 \quad a_2 \quad a_3 \quad a_4 \quad a_5 \ldots$$

$$2 \longleftrightarrow 0. \quad b_1 \quad b_2 \quad b_3 \quad b_4 \quad b_5 \ldots$$

$$3 \longleftrightarrow 0. \quad c_1 \quad c_2 \quad c_3 \quad c_4 \quad c_5 \ldots$$

$$4 \longleftrightarrow 0. \quad d_1 \quad d_2 \quad d_3 \quad d_4 \quad d_5 \ldots$$

$$5 \longleftrightarrow 0. \quad e_1 \quad e_2 \quad e_3 \quad e_4 \quad e_5 \ldots$$

Next Cantor proceeded to show that this assumption of a one-to-one correspondence must be incorrect since a real number can be created that could not possibly be in the array listed. He did this by considering the diagonal of the array on page 254.

$$0. \quad a_1 \quad a_2 \quad a_3 \quad a_4 \quad a_5 \ldots$$
$$0. \quad b_1 \quad b_2 \quad b_3 \quad b_4 \quad b_5 \ldots$$
$$0. \quad c_1 \quad c_2 \quad c_3 \quad c_4 \quad c_5 \ldots$$
$$0. \quad d_1 \quad d_2 \quad d_3 \quad d_4 \quad d_5 \ldots$$
$$0. \quad e_1 \quad e_2 \quad e_3 \quad e_4 \quad e_5 \ldots$$

He then formed a new decimal

$$N = 0.x_1 x_2 x_3 x_4 x_5 \ldots$$

where $x_1 \neq a_1$, $x_2 \neq b_2$, $x_3 \neq c_3$, $x_4 \neq d_4$, $x_5 \neq e_5$, etc. Thus N cannot be a member of the set of decimals listed since it differs from the first decimal in at least the first place; it differs from the second decimal in at least the second place; etc. That is, the assumption that the set of real numbers is countable (can be placed in a one-to-one correspondence with the set of counting numbers) is proved to be incorrect and the set of real numbers is not countable.

Cantor assigned the letter C to represent the size of the set of real numbers. Thus C is a transfinite cardinal number which is larger than $\aleph_0$. The fact that there are at least two transfinite cardinal numbers indicates that it is not appropriate to use the symbol ∞ as a number without being very explicit as to its meaning and use in a particular situation. Actually it is now known that there are infinitely many different transfinite cardinal numbers.

EXERCISES

1. Show that the set of odd counting numbers has the transfinite cardinal number $\aleph_0$.

2. Show that there is a one-to-one correspondence between the set of counting numbers and (a) the set of whole numbers; (b) the set of integers greater than or equal to -1.

3. Use the results of Exercise 2(a) to show that $\aleph_0 + 1 = \aleph_0$.

4. Use the results of Exercise 2(b) to show that $\aleph_0 + 2 = \aleph_0$.

Use a one-to-one correspondence with the set of counting numbers to show each of these facts.

5. $\aleph_0 + 3 = \aleph_0$

6. $\aleph_0 + 5 = \aleph_0$

7. $\aleph_0 - 1 = \aleph_0$

8. $\aleph_0 - 3 = \aleph_0$

9. $\aleph_0 + 10 = \aleph_0$

10. $\aleph_0 - 10 = \aleph_0$

PEDAGOGICAL CONSIDERATIONS

The concept of infinity is a very difficult one for young children to acquire. Initially, almost any large number serves as their concept of the infinite. Thus it is wise to provide experience with very large numbers to contrast the idea of very large with that of the infinite. As one attempt along these lines, a certain class engaged in the process of trying to collect one million bottle-caps. Among the many interesting questions that can be asked concerning this project is that of the size container needed to hold this number of caps. For example, would the classroom be large enough to hold one million bottle-caps?

One of the largest numbers ever named is a **googol**. This has been defined as 1 followed by one hundred zeros:

1000000000000000000000000000000000000000000000000000
000000000000000000000000000000000000000000000000

This number is larger than what is considered to be the total number of protons or electrons in the universe! A googol can be expressed, using exponents, as

$$10^{100}.$$

Even larger than a googol is a **googolplex**, defined as 1 followed by a googol of zeros. One famous mathematician claimed that there would not even be room between the earth and the moon to write all the zeros in a googolplex!

1. Describe at least three different experiences so as to develop the concept of very large but finite sets with elementary school students. One example that generally creates much interest is to ask them to guess how long it will take to spend $1,000,000 at the rate of one dollar per minute. Thereafter they can perform the necessary computations to arrive at the correct answer.

2. Describe at least three different experiences to develop the concept of an infinite set with elementary school students.

3. Students in the senior high school and beyond incorrectly use the symbol ∞ to represent infinity as the name of a specific number. Write a brief explanation of the distinction between ∞ and $\aleph_0$.

CHAPTER TEST

1. Classify each statement as true or false:
   (a) Every real number is a complex number.
   (b) Every complex number is a real number.

(c) There is a one-to-one correspondence between the set of rational numbers and the set of points on the number line.

(d) The set of irrational numbers is a subset of the set of real numbers.

(e) Every irrational number is the coordinate of some point on the number line.

2. Construct a number line and locate a point whose coordinate is $3 + \sqrt{2}$.

3. Prove that $\sqrt{2} - 1$ is irrational.

4. Graph each set on a number line.

   (a) The set of integers between -2 and 3.

   (b) The set of real numbers greater than or equal to -1.

5. Use the given figure and give the coordinates of each of these points:

   (a) A (b) B (c) C (d) D (e) E

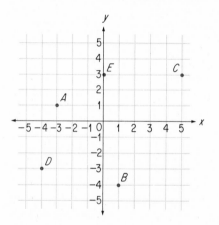

6. Name two irrational numbers between $0.\overline{46}$ and $0.\overline{47}$.

7. Copy the following table. Use "✓" to show that the number at the top of the column is a member of the set named at the side. Use "✗" if the number is not a member of the set.

Set	(a) $-\frac{4}{5}$	(b) 0	(c) $\sqrt{9}$	(d) $\sqrt{7}$	(e) $\sqrt{-5}$
Counting numbers					
Whole numbers					
Integers					
Rational numbers					
Real numbers					
Complex numbers					

8. Find all possible replacements for n that will make each of the sentences below true, where n must be a member of the set named at the top of each column. When there is no such replacement, write "none."

      (a)   **(b)**   **(c)**   **(d)**   **(e)**

Sentence	Counting numbers	Integers	Rational numbers	Real numbers	Complex numbers
$n - 1 = 0$					
$n + 3 = 3$					
$3n = 8$					
$n^2 = 3$					
$n^2 + 9 = 0$					

9. Copy the following table. Use "✓" to show that the set of elements named at the top of each column has the property listed at the side. Use "✗" if the set does not have the property.

      (a)   **(b)**   **(c)**   **(d)**   **(e)**

Property	Counting numbers	Whole numbers	Integers	Rational numbers	Real numbers
Identity, +					
Inverse, +					
Identity, ×					
Inverse, ×					
Density					
Completeness					

10. Express in the form $a + bi$:

(a) $3 + \sqrt{-16}$ (b) $\sqrt{7}$

(c) $\sqrt{-4}$ (d) $2 + \sqrt{-3}$

ALGEBRA, GEOMETRY, STATISTICS, PROBABILITY

8

Sentences in One Variable

A symbol that may be replaced by any member of a set of elements is a variable. Variables such as □, *n*, *x*, and *t* are often used in arithmetic. However, the study of sentences involving variables is a part of algebra. Thus our present study of sentences in one variable represents a shift of emphasis from arithmetic to algebra. This is part of the shift from teaching arithmetic in elementary school to teaching mathematics in elementary school.

Algebra is an extension of the arithmetic of integers, rational numbers, and real numbers that we considered in Chapters 5, 6, and 7. Arithmetic sentences involve only numbers and operations on numbers. Each such sentence can be identified as true or false. Algebraic sentences may involve variables and be true for some replacements of the variable and false for others.

8-1
sentences and statements

Sentences that can be identified as true or identified as false are called **statements.** Each of the following is a statement:

(a) $7 + 3 = 3 + 7$ (a true statement of equality)
(b) $7 - 3 = 3 - 7$ (a false statement of equality)

Statements of equality, whether true or false, are called **equations.** We may also write **statements of inequality** using any one of these symbols:

$$\neq: \quad \text{is not equal to}$$
$$>: \quad \text{is greater than}$$
$$<: \quad \text{is less than}$$

Each of the following is a statement of inequality:

(a) $7 + 2 > 7 - 2$ (a true statement of inequality)
(b) $5 + 3 \neq 3 + 5$ (a false statement of inequality)
(c) $8 - 2 < 8$ (a true statement of inequality)

We formally define the **order** of any two real numbers a and b, where a is less than b, as follows:

$a < b$ (and $b > a$) if and only if there is a positive number c such that $a + c = b$.

For example, $2 < 5$ (or $5 > 2$) and there is a positive number **3** such that $2 + 3 = 5$. This concept of order is consistent with the ordering of integers in §5–4 and the ordering of rational numbers in §7–1. Indeed, the **trichotomy law,** cited in §5–4 for counting numbers, holds for all real numbers: If a and b are any two real numbers, then exactly one of the relations

$$a < b, \qquad a = b, \qquad a > b$$

must hold.

Many sentences cannot be classified as being true or as being false and are known as **open sentences.** Here are some examples of open sentences:

(a) $x + 2 = 5$ (c) $x + 1 > 4$
(b) $x - 2 \neq 3$ (d) $x - 1 < 3$

In the preceding examples the symbol x represents a number and is called a *variable*. A variable is a placeholder for its replacements. Any open sentence is neither true nor false until a replacement is made for the variable. After such a replacement is made, we have a statement, that is, a sentence that can be classified as either true or false.

Example 1: Find a replacement for x that will make the sentence $x + 2 = 5$ a true statement of equality.

Solution: When $x = 3$, we have the true statement $3 + 2 = 5$. When $x \neq 3$, we have a false statement of equality. We call 3 the *solution* of the given open sentence.

Given any open sentence, we need to determine the set of replacements that can be used for the variable. This set of permissible replacements is the **replacement set** of the variable. For example, consider this open sentence:

Christmas is in the month of _____.

The set of possible replacements of the variable (indicated by the blank) in this open sentence is the set of the 12 months. The statement is true when the replacement is December; the statement is false for other replacements.

Next let us consider the sentence $x - 1 < 3$ for various replacement sets. Note that for different replacement sets there may be different sets of replacements that make the sentence true. In each case the set of numbers that make the sentence true is called the **solution set** or the **truth set** of the given sentence for that replacement set. Recall that the elements of a set are generally listed within a pair of braces as shown in the array.

Sentence	Replacement set	Solution set
$x - 1 < 3$	Counting numbers	$\{1, 2, 3\}$
$x - 1 < 3$	Whole numbers	$\{0, 1, 2, 3\}$
$x - 1 < 3$	Integers	$\{\ldots, -2, -1, 0, 1, 2, 3\}$
$x - 1 < 3$	Real numbers	All real numbers less than 4

Note that the sentence $x - 1 > 3$ has the solution set $\{5, 6, 7, \ldots\}$ whether the replacement set is the set of counting numbers, whole numbers, or integers.

Example 2: Find the solution set for $x + 1 > 4$ when x is a whole number.

Solution: When $x = 3$, $3 + 1 = 4$. The solution set S consists of all whole numbers greater than 3; that is,

$$S = \{4, 5, 6, \ldots\}.$$

Example 3: Find the solution set for $n + 2 = 2 + n$ when n is a real number.

Solution: This sentence is a true statement of equality for *all* possible replacements of the variable, since it is an application of the commutative property of addition. The solution set is the set of real numbers.

An equation that is true for all possible replacements of the variable (as in Example 3) is called an **identity**. A statement of inequality, such as $n + 2 > n$, may also be classified as an

---

Example 4: For what values of x is the statement $x + 2 = x$ true for an integer x?

Solution: Regardless of the integer selected as the replacement for x, the sentence $x + 2 = x$ is always false. The solution set is the empty set, $\varnothing$.

Statements of inequality may also have the empty set as their solution set. For example, there is no replacement for x that will make the sentence $x + 2 < x$ true; the solution set is the empty set.

EXERCISES

1. Give three statements of equality for which the solution set is the empty set.

2. Give three sentences that are identities.

3. Classify each of these statements as (i) a true statement of equality; (ii) a false statement of equality; (iii) a true statement of inequality; or (iv) a false statement of inequality.

(a) $7 - 3 \neq 3 - 7$ (b) $3 - 7 < 7 - 3$

(c) $\frac{23}{25} > \frac{24}{25}$ (d) $-\frac{23}{25} > -\frac{24}{25}$

(e) $19 \times 21 = 20^2 - 1$ (f) $13^2 = 5^2 + 12^2$

(g) $\frac{432}{796} > \frac{432}{795}$ (h) $-\frac{432}{796} < -\frac{432}{795}$

For which of these sentences is the solution set the empty set? Which sentences are identities?

4. $x + 3 > x$ *5.* $x + 2 \neq x$

6. $x + 2 \neq 2 + x$ *7.* $x > x + 3$

8. $x < x + 1$ *9.* $x + 1 < x + 2$

10. $x + 2 < x + 1$ *11.* $x > x - 1$

Find the solution set for each sentence when the replacement set is the set of whole numbers.

12. $x + 2 = 5$ *13.* $x - 2 = 3$

14. $x + 1 < 4$ *15.* $x - 1 < 4$

16. $x - 2 < 3$ *17.* $x - 2 > 3$

Find the solution set for each sentence when the replacement set is the set of integers.

18. $x + 1 < 4$ *19.* $x - 1 < 4$

20. $x - 2 < 3$ *21.* $x - 2 > 1$

Describe the solution set for each sentence when the replacement set is the set of real numbers.

22. $x + 1 < 4$ *23.* $x + 1 > 4$

24. $x - 2 < 3$ *25.* $x - 2 > 3$

26. $x + 2 \neq 5$ *27.* $x + 1 \neq 1$

PEDAGOGICAL CONSIDERATIONS

1. It is now common for primary grade students to work with sentences such as the following:

$$7 + \boxed{} = 12$$

Review an elementary mathematics book from grades 1, 2, or 3 and collect examples that illustrate this early introduction of the concept of a variable.

2. The University of Illinois Committee on School Mathematics (U.I.C.S.M.) has prepared experimental curricula for grades 9 through 12 with a heavy emphasis on the discovery approach. In addition, they have completed informal experimentation with children from ages 4 through 11. Their approach features a heavy emphasis on frames as a means of developing number concepts. Read the U.I.C.S.M. staff report, *Arithmetic With Frames*, which is Chapter 3 of the twenty-seventh yearbook of the National Council of Teachers of Mathematics.

3. Prepare a brief report on the contributions made by several ancient civilizations to the development of algebra.

4. Refer to a history of mathematics book and try to find a reference to the "rule of false position" method for solving equations. Then demonstrate the use of this procedure for finding the solution of an equation.

8-2
graphs of sentences

The solution set of any open sentence in one variable may be graphed on a number line. That is, we may draw a graph to represent the set of points that correspond to the solution set of a sentence. We often refer to this graph simply as the *graph* of the equation or inequality.

The following examples illustrate various types of graphs. In each example we are to find and graph the solution set of the given open sentence for the given replacement set. The graph of the sentence is the graph of its solution set.

Example 1: Graph the sentence $x + 3 = 5$, x an integer.

Solution: The solution set consists of a single element, $\{2\}$. We graph this solution set on a number line by drawing a solid dot at 2. The graph consists of this single *point*.

The equation in Example 1 can be thought of as a **set-selector;** it selects from the replacement set of x just those numbers that make the sentence true when used as replacements for x. The selected set is the solution set of the equation. The solution set $\{2\}$ is "the set of all x such that $x + 3 = 5$." We may designate this solution set in **set-builder notation** as

$$\{x \mid x + 3 = 5\}.$$

The replacement set of the variable may be indicated in the set-builder notation as follows:

$$\{x \mid x + 3 = 5, x \text{ an integer}\}$$

We may also write

$$\{x \mid x + 3 = 5\} = \{2\}.$$

That is, the set of all x such that $x + 3 = 5$ is the set consisting of the element 2.

Example 2:   Graph the sentence $x + 3 > 5$, x a real number.

Solution: The solution set consists of all real numbers greater than 2. The graph of the solution set is drawn by first placing a hollow dot at 2 on the number line to indicate that this point is not a member of the solution set. We then draw a heavily shaded arrow to show that all numbers greater than 2 satisfy the given inequality.

For any real number b the figure formed by the graph of an inequality of the form $x > b$ (see Example 2) is called a **half-line.** (The figure formed by the graph of an inequality of the form $x < b$ is also a half-line.)

Example 3:   Graph the sentence $x + 3 \geq 5$, x a real number.

Solution: This sentence is read "$x + 3$ is greater than or equal to 5." Therefore, it is true when $x = 2$, and it is true when $x > 2$.

For any real number b the figure formed by the graph of an inequality of the form $x \geq b$ (see Example 3) is called a **ray.** (The figure formed by the graph of an inequality of the form $x \leq b$ is also a ray. The sentence $x \leq b$ is read "x is less than or equal to b.")

Example 4:   Graph the sentence $x + 3 \not< 5$, x a real number.

Solution: This sentence is read "$x + 3$ is not less than 5." This is equivalent to saying that $x + 3$ is greater than or equal to 5; $x + 3 \geq 5$. Thus the solution set is the same as for Example 3.

Example 5:   Graph the sentence $-1 \leq x \leq 3$, x a real number.

Solution: This sentence is true when x is replaced by any real number *from* -1 *through* 3.

For any real numbers a and b where $a < b$ the graph of the sentence $a \leq x \leq b$ (see Example 5) is called a **line segment**.

Example 6:   Graph the sentence $-1 < x < 3$, x a real number.

Solution: The solution set consists of all the real numbers *between* -1 and 3.

The points with coordinates -1 and 3 are the **endpoints** of the line segment graphed in Example 5. Notice that the graph in Example 6 can be obtained from the graph in Example 5 by removing the endpoints of the line segment. When we wish names for each of these graphs, a line segment with its endpoints is called a **closed line segment**; a line segment without either endpoint is an **open line segment**; a line segment with one endpoint but not both is neither closed nor open. For example, the following graph of $-1 \leq x < 3$ is neither closed nor open.

Example 7:   Graph the sentence $x + 2 = 2 + x$, x a real number.

Solution: This sentence is true for all replacements of x; it is an identity. The solution set is the set of all real numbers. Thus the graph is the entire number line.

267

Example 8: Graph the sentence $x + 3 \leq 5$, x a whole number.

Solution: The solution set is $\{0, 1, 2\}$ and is graphed as a set of three points.

Example 9: Graph the sentence $x + 3 > 5$, x a whole number.

Solution: The solution set is $\{3, 4, 5, \ldots\}$, an infinite set of whole numbers. The sentence may be graphed as follows, where the heavily shaded arrowhead is used to indicate that the graph continues in the manner shown. The solution set consists of the set of all whole numbers greater than or equal to 3.

We summarize our discussion of types of graphs by considering again the solution set for the sentence $x - 1 < 3$ (see §8–1) for different replacement sets.

Replacement set	Solution set	Graph of solution set
Counting numbers	$\{1, 2, 3\}$	
Whole numbers	$\{0, 1, 2, 3\}$	
Integers	$\{\ldots, -2, -1, 0, 1, 2, 3\}$	
Real numbers	All real numbers less than 4	

EXERCISES

Identify the graph of the solution set of each sentence as a point, a half-line, a ray, a line segment, or a line.

1. $x + 3 = 7$
2. $x - 2 > 7$
3. $x + 2 \geq 5$
4. $x < x + 3$
5. $-2 \leq x \leq 5$
6. $x + 3 \leq 5$
7. $x + 1 > x$
8. $x - 2 = 7$
9. $x - 1 < 5$
10. $-3 \leq x \leq 0$

Graph each set for whole numbers x.

11. $\{x \mid x + 2 \leq 5\}$
12. $\{x \mid x - 1 < 5\}$

13. $\{x \mid 2 \leq x \leq 7\}$

14. $\{x \mid 0 < x \leq 3\}$

15. $\{x \mid x + 2 < 3\}$

16. $\{x \mid x + 1 < 5\}$

Graph each sentence for real numbers x.

17. $x + 1 > 2$

18. $x - 2 \leq 3$

19. $-2 \leq x \leq 5$

20. $-3 < x < 3$

21. $x + 2 \geq 5$

22. $x - 1 < 4$

23. $x + 1 > 4$

24. $x + 3 \not< 4$

25. $x + 2 > x$

26. $x + 2 = x$

27. $x + 1 \not> 3$

28. $x + 2 \neq 3$

29. $x - 2 < 1$

30. $x - 1 < 2$

Graph each set for real numbers.

*31. $\{x \mid 3x - 2 \geq -5\}$

*32. $\{x \mid 2x + 3 \leq 1\}$

*33. $\{x \mid x^2 = 25\}$

*34. $\{x \mid x^2 + 3 = 19\}$

*35. $\{x \mid x^2 < 9\}$

*36. $\{x \mid x^2 \geq 16\}$

8-3

**compound
sentences**

In Chapter 2 (§2-1) we considered compound statements. Let us now turn our attention to compound sentences in algebra. For example, consider this compound sentence:

$$x + 1 > 2 \quad \text{and} \quad x - 2 < 1$$

The sentence $x + 1 > 2$ is true for all x greater than 1; the sentence $x - 2 < 1$ is true for all x less than 3. Recall that a compound sentence of the form $p \wedge q$ (p and q) is true only when both parts are true. Thus the given compound sentence is true for the set of elements in the *intersection* of the two sets. Graphically we can show this as follows:

The graph of the compound sentence consists of an open interval and can be described as

$$1 < x < 3.$$

The solution set of this compound sentence can be written in set-builder notation as

$$\{x \mid x > 1\} \cap \{x \mid x < 3\}.$$

This is read as "the intersection of the set of all x such that x is greater than 1 and the set of all x such that x is less than 3." Note that no replacement set for x was specified here. When such is the case, we shall assume the replacement set to be the set of real numbers.

Example 1: Find the solution set

$$x \geq -2 \quad \text{and} \quad x + 1 \leq 4, \qquad x \text{ an integer.}$$

Solution: Here we want the set of integers that are greater than or equal to -2 but at the same time are less than or equal to 3. (If $x + 1 \leq 4$, then $x \leq 3$.) The solution set is $\{-2, -1, 0, 1, 2, 3\}$.

Example 2: Find the solution set for the replacement set of real numbers

$$x + 3 < 5 \quad \text{and} \quad 3 < 1.$$

Solution: You will note that the second part of this sentence ($3 < 1$) is false. If part of a sentence of the form $p \wedge q$ is false, then, as you may recall, the entire sentence is false. Thus the solution set is the empty set.

Next we consider a compound sentence involving the connective "or":

$$x + 1 < 2 \quad \text{or} \quad x - 2 > 1$$

The sentence $x + 1 < 2$ is true for all x less than 1; the sentence $x - 2 > 1$ is true for all x greater than 3. Recall that a sentence of the form $p \vee q$ (p or q) is true unless both parts are false. Thus the given compound sentence is true for the set of elements in the *union* of the two sets. Graphically we have the following:

The graph of the compound sentence consists of the union of two half-lines. The solution set can be written in set-builder notation as

$$\{x \mid x < 1\} \cup \{x \mid x > 3\}.$$

This is read as "the union of the set of all x such that x is less than 1 and the set of all x such that x is greater than 3."

Example 3: Find the solution set for the replacement set of real numbers

$$5 > 1 \quad \text{or} \quad x + 2 < 5.$$

270

Solution: Since the first part of this sentence is true, the whole sentence is true. The solution set is the entire set of real numbers.

Example 4: Graph the solution set

$$x \leq -1 \quad \text{or} \quad x \geq 2.$$

Solution: The graph is the union of two rays.

Example 5: Graph the sentence

$$x = -2 \quad \text{or} \quad x \geq 1.$$

Solution: Since no replacement set is specified, we assume that the replacement set is the set of real numbers.

Example 6: Find the solution set

$$x + 2 \neq 2 + x \quad \text{or} \quad 7 < 2.$$

Solution: Both parts of the sentence are false; the solution set is the empty set.

Example 7: Find the solution set

$$\{x \mid (x - 2)(x + 1) = 0\}.$$

Solution: The product of two numbers is zero if and only if at least one of the numbers is zero. That is, for all real numbers a and b, if $a \times b = 0$ then $a = 0$, or $b = 0$, or $a = 0$ and $b = 0$. Therefore, if $(x - 2)(x + 1) = 0$, then $x - 2 = 0$ or $x + 1 = 0$ and $x = 2$ or $x = -1$. The solution set is $\{2, -1\}$.

EXERCISES

List the elements in the solution set for integers x.

1. $x \geq 1$ and $x + 1 \leq 5$
2. $x \leq -1$ and $x > -5$
3. $x > 1$ and $x < 3$
4. $x \geq 2$ or $x \leq 1$

Graph each sentence for real numbers x.

5. $x \geq 1$ and $x \leq 5$
6. $x \leq 1$ or $x \geq 5$

7. $x + 1 < 3$ and $x \geq 0$ 8. $-5 > 0$ and $x + 2 = 7$

9. $x + 1 > x$ and $5 > 1$ 10. $x - 1 > 3$ or $x < 0$

11. $x + 2 \geq 5$ and $x - 1 \leq 5$ 12. $7 > 3$ or $x + 2 \leq 5$

13. $0 < -2$ or $x + 1 < 3$ 14. $x + 1 \geq 2$ or $x = -1$

15. $x - 2 = 2 - x$ or $3 > 5$ 16. $x + 1 = x$ or $-2 > 2$

Graph each set for real numbers x.

17. $\{x \mid x > 1\} \cap \{x \mid x < 3\}$ 18. $\{x \mid x \leq -1\} \cup \{x \mid x \geq 2\}$

19. $\{x \mid x + 2 < 3\} \cup \{x \mid x \geq 5\}$ 20. $\{x \mid x - 1 < 4\} \cap \{x \mid x \geq -2\}$

21. $\{x \mid x(x + 2) = 0\}$ 22. $\{x \mid (x - 1)(x + 3) = 0\}$

23. $\{x \mid (x + 1)(x + 5) = 0\}$ 24. $\{x \mid (x - 3)(x - 2) = 0\}$

*25. $\{x \mid x(x + 2) \geq 0\}$ *26. $\{x \mid (x - 1)(x + 3) \leq 0\}$

*27. $\{x \mid (x + 1)(x + 5) < 0\}$ *28. $\{x \mid (x - 3)(x - 2) > 0\}$

PEDAGOGICAL CONSIDERATIONS

Compound sentences are common at all grade levels. Compound number sentences occur at many grade levels. Find the solution set of each sentence where $n \in \{1, 2, 3, 4, 5, \ldots, 25\}$.

1. n is a multiple of 5 and an odd number.

2. n is less than 17 but greater than 11.

3. $2n$ is an even number and n is a multiple of 3.

4. $2n$ is an odd number or $3n$ is an even number.

5. $5n$ is an even number and n is less than 20.

6. n is less than 5 or greater than 20.

7. $2n$ is less than 5 or greater than 20.

8. n is divisible by 3 and by 2.

9. n is divisible by 5 or by 7.

10. n is divisible by 3 or by 11.

8-4

sentences involving absolute value

The absolute value of a number x was defined on the number line in §5-8 as the distance from the origin of the point with coordinate x. Thus $|3| = 3$, $|-3| = 3$, and in general for any real number x:

273

sec. 8-4
sentences
involving
absolute
values

$$|x| = x, \qquad \text{if } x \text{ is positive}$$
$$|x| = 0, \qquad \text{if } x = 0$$
$$|x| = -x, \qquad \text{if } x \text{ is negative}$$

Note that if x is negative, then $-x$ is the opposite of x and therefore is a positive number. Thus if $x = -3$, $|-3| = -(-3) = 3$. Accordingly, the absolute value of any real number different from zero is a positive number; $|0| = 0$.

Example 1:   Find the solution set: $|x| = 3$.

Solution: There are two numbers whose absolute value is 3, namely, 3 and -3. (On a number line, the points with coordinates 3 and -3 are each three units away from the origin.) The solution set is $\{-3, 3\}$.

Note that the solution set of Example 1 can be described by the compound sentence

$$x = -3 \quad \text{or} \quad x = 3.$$

The instruction to find the solution set of a sentence may be indicated briefly as "**solve.**" Similarly, the instruction to graph a sentence may be abbreviated as "**graph.**"

Example 2:   Graph $|x| \geq 3$.

Solution: We are to find all points on the number line that are three units or more than three units from the origin.

Note that the solution set of Example 2 can be described by the compound sentence

$$x \leq -3 \quad \text{or} \quad x \geq 3.$$

In set-builder notation we may write this as

$$\{x \mid x \leq -3\} \cup \{x \mid x \geq 3\}.$$

As in Example 2, we have for any nonnegative number k:

$$\{x \mid |x| \geq k\} = \{x \mid x \leq -k\} \cup \{x \mid x \geq k\}$$

Example 3: Graph $|x| \leq 3$.

Solution: We are to find all points on the number line that are three units or less than three units from the origin; these are the points that are "at most three units" from the origin.

The solution set of Example 3 can be described in each of these ways:

$$x \geq -3 \quad \text{and} \quad x \leq 3$$
$$\{x \mid x \geq -3\} \cap \{x \mid x \leq 3\}$$
$$\{x \mid -3 \leq x \leq 3\}$$

As in Example 3, we have for any nonnegative number k:

$$\{x \mid |x| \leq k\} = \{x \mid x \geq -k\} \cap \{x \mid x \leq k\}$$

Example 4: Solve $|x - 1| = 3$.

Solution: As in Example 1, if the absolute value of a number is 3, then that number is 3 or -3. If $x - 1 = 3$, then $x = 4$; if $x - 1 = -3$, then $x = -2$. Thus the solution set is $\{-2, 4\}$.

The points with coordinates -2 and 4 of the solution set for Example 4 are each at a distance of 3 units from the point with coordinate 1. Thus the given sentence

$$|x - 1| = 3$$

may be interpreted in terms of distance on a number line.

The distance between the point with coordinate x and the point with coordinate 1 is 3. This interpretation is illustrated in the following figure.

The absolute value notation may be used to indicate the distance between any two points on the number line. Thus $|x - y|$ represents the distance between the point with coordinate x and the point with coordinate y.

275

sec. 8-4
sentences
involving
absolute
values

Notice that the distance between the points with coordinates 2 and 5 may be denoted either as $|5 - 2|$ or as $|2 - 5|$, since $|5 - 2| = |3| = 3$ and $|2 - 5| = |-3| = 3$. In general, $|x - y| = |y - x|$ and the distance between any two distinct points is a positive number.

If we are told that $|x + 1| = 3$, we may consider this sentence in the form $|x - (-1)| = 3$; that is, the distance between the point with coordinate x and the point with coordinate -1 is 3. Thus, as shown in the figure, $x = -4$ or $x = 2$.

The problem $|x + 1| = 3$ can also be solved by using the fact that if the absolute value of a number is 3, then that number must be 3 or -3. If $x + 1 = 3$, then $x = 2$; if $x + 1 = -3$, then $x = -4$. Thus if $|x + 1| = 3$, then $x = -4$ or $x = 2$.

Example 5: Graph $\{x \mid |x - 1| \geq 3\}$.

Solution: We may rewrite the given statement as

$$\{x \mid x - 1 \geq 3\} \cup \{x \mid x - 1 \leq -3\}.$$

If $x - 1 \geq 3$, then $x \geq 4$; if $x - 1 \leq -3$, then $x \leq -2$. Thus the solution set is $\{x \mid x \geq 4 \text{ or } x \leq -2\}$. The graph of the solution set is:

Note that in Example 5 the graph of $|x - 1| \geq 3$ is the union of two rays and consists of the points of the number line that are at a distance of three units or more than three units from the point with coordinate 1.

Example 6: Graph $|x - 1| \leq 3$.

Solution: The graph of $|x - 1| \leq 3$ consists of the points of the number line that are at a distance of at most three units from the point with coordinate 1. Thus the graph is a line segment.

EXERCISES

List the set of integers for which each sentence is true.

1. $|x| = 2$ 2. $|x| \leq 3$

3. $|x| < 5$ 4. $|x| = -3$

5. $|x - 1| = 3$ 6. $|x + 1| = 5$

7. $|x + 2| = 1$ 8. $|x - 2| = 3$

Evaluate:

9. $|-7|$ 10. $|-2| + |-3|$

11. $|(-2) + (-3)|$ 12. $(|-5|)^2$

13. $|(-5)^2|$ 14. $|-5| \times |-3|$

Identify the graph of each sentence on a real number line as two points, a line segment, or the union of two rays.

15. $|x| \leq 7$ 16. $|x - 2| = 5$

17. $|x| \geq 1$ 18. $|x + 3| \leq 1$

Graph each sentence on a real number line.

19. $|x| \leq 2$ 20. $|x| \geq 1$

21. $|x| > 3$ 22. $|x| < 5$

23. $|x - 1| \geq 2$ 24. $|x + 1| \leq 3$

25. $|x| = -2$ 26. $|x| \geq 0$

27. $|x + 2| \geq 3$ 28. $|x - 2| \leq 1$

29. $|x - 3| < 2$ 30. $|x + 1| > 2$

*31. $2 \leq |x| \leq 3$ *32. $3 \leq |x - 1| \leq 5$

*33. $|x| + 2 = |x + 2|$ *34. $|x| = -x$

*35. $|x| + |x - 3| = 3$ *36. $|x^2 - 10| \leq 6$

PEDAGOGICAL CONSIDERATIONS

1. Assume that integers have been studied and graphed on a number line. Prepare a brief introduction of absolute value for use with elementary school students with this background. Include in your report suggestions for the use of visual aids to help get this concept across.

2. In a first course in algebra the following definition is usually made:

$$\sqrt{x^2} = |x|$$

For example, as a result of this definition we may write

$$\sqrt{(3)^2} = |3| = 3 \quad \text{and} \quad \sqrt{(-3)^2} = |-3| = 3.$$

Many students carelessly and incorrectly write $\sqrt{x^2} = x$. Show how this incorrect statement can contradict the use of the radical symbol to denote the nonnegative square root of a number. (See §7–1.)

8-5
sentences of the first degree

So far we have been solving sentences by informal methods. For example, if we are given that

$$x - 3 = 7$$

we recognize that x may be obtained as $(x - 3) + 3$; that is,

$$(x - 3) + 3 = x + (-3 + 3) = x + 0 = x.$$

Therefore, we may solve the given equation as follows:

$$x - 3 = 7$$
$$(x - 3) + 3 = 7 + 3$$
$$x = 10$$

We have added 3 to both members of the equation to obtain an **equivalent sentence,** that is, a sentence with the same solution set as the given sentence. Formally, we have used a property of equations. Here are some of the common properties of statements of equality involving real numbers a, b, c, and d:

$$\textbf{Identity, } =: \quad a = a.$$
$$\textbf{Addition, } =: \quad \text{If } a = b \text{ and } c = d, \text{ then } a + c = b + d.$$
$$\textbf{Multiplication, } =: \quad \text{If } a = b \text{ and } c = d, \text{ then } ac = bd.$$

The uses of these properties together with the field properties of Chapter 7 are shown in the examples that follow.

Example 1: Solve and give a reason for each step in the solution of $x + 3 = 7$.

Solution:

Statements	Reasons
$x + 3 = 7$	Given.
$(x + 3) + (-3) = 7 + (-3)$	Addition, =.
$x + [3 + (-3)] = 7 + (-3)$	Associative, +.
$x + 0 = 4$	Addition.
$x = 4$	Zero, +. (Addition property of zero.)

277

Example 2: Solve $2x - 3 = 7$, and explain each step.

Solution:

Statements	Reasons
$2x - 3 = 7$	Given.
$2x + (-3) = 7$	Definition, $-$.
$[2x + (-3)] + 3 = 7 + 3$	Addition, $=$.
$2x + [(-3) + 3] = 7 + 3$	Associative, $+$.
$2x + 0 = 10$	Addition.
$2x = 10$	Zero, $+$.
$\frac{1}{2}(2x) = \frac{1}{2} \times 10$	Multiplication, $=$.
$(\frac{1}{2} \times 2) \cdot x = \frac{1}{2} \times 10$	Associative, $\times$.
$1x = 5$	Multiplication.
$x = 5$	One, $\times$. (Multiplication property of one.)

Notice that subtraction is not needed when solving equations since

$$a - b = a + (-b).$$

Similarly, division is not needed since

$$a \div b = a \times \frac{1}{b}.$$

The methods used in Examples 1 and 2 may be used to solve any sentence that is an equation of the **first degree in one variable** x, that is, any sentence that can be expressed in the form

$$ax + b = 0, \qquad a \neq 0.$$

The expression $ax + b$ is called an **expression of the first degree** in x.

Sentences involving inequalities may also be solved by making use of certain basic properties. First we list some of the common properties of order relations among real numbers a, b, c, and d:

Addition, $<$: If $a < b$ and $c = d$, then $a + c < b + d$.
Addition, $>$: If $a > b$ and $c = d$, then $a + c > b + d$.
Multiplication, $<$: If $a < b$, then $ac < bc$ if $c > 0$.
  If $a < b$, then $ac > bc$ if $c < 0$.
Multiplication, $>$: If $a > b$, then $ac > bc$ if $c > 0$.
  If $a > b$, then $ac < bc$ if $c < 0$.

In general, if the same number is added to both members of a statement of inequality, the "sense" of the inequality is preserved. Thus in the following examples the first sentence listed is true, and the second one must also be true.

$$2 < 5, \qquad 2 + 3 < 5 + 3$$
$$8 > 3, \qquad 8 + 5 > 3 + 5$$
$$-3 < 5, \qquad -3 + 1 < 5 + 1$$

278

279

sec. 8-5
sentences
of the
first
degree

If both members of a true statement of inequality are multiplied by the same positive number, the resulting sentence is also true. If both members of an inequality are multiplied by a negative number, it is necessary to reverse the sense of the inequality to obtain an equivalent sentence. Thus consider these sets of equivalent sentences:

$$3 < 7, \qquad 2 \times 3 < 2 \times 7$$
$$5 > -1, \qquad 3 \times 5 > 3 \times -1$$
$$2 < 8, \qquad -3 \times 2 > -3 \times 8$$

The uses of these properties of order relations are shown in the examples that follow.

Example 3: Solve $\frac{1}{2}x - 2 > 5$ and explain each step.

Solution:

Statements	Reasons
$\frac{1}{2}x - 2 > 5$	Given.
$\frac{1}{2}x + (-2) > 5$	Definition, $-$.
$[\frac{1}{2}x + (-2)] + 2 > 5 + 2$	Addition, $>$.
$\frac{1}{2}x + [(-2) + 2] > 5 + 2$	Associative, $+$.
$\frac{1}{2}x + 0 > 7$	Addition.
$\frac{1}{2}x > 7$	Zero, $+$.
$2(\frac{1}{2}x) > 2 \times 7$	Multiplication, $>$.
$(2 \times \frac{1}{2})x > 2 \times 7$	Associative, $\times$.
$1x > 14$	Multiplication.
$x > 14$	One, $\times$.

Example 4: Solve $-2x + 3 > 7$ and explain each step.

Solution:

Statements	Reasons
$-2x + 3 < 7$	Given.
$(-2x + 3) + (-3) < 7 + (-3)$	Addition, $<$.
$-2x + [3 + (-3)] < 7 + (-3)$	Associative, $+$.
$-2x + 0 < 4$	Addition.
$-2x < 4$	Zero, $+$.
$-\frac{1}{2}(-2x) > (-\frac{1}{2})4$	Multiplication, $<$.
$(-\frac{1}{2} \times -2)x > (-\frac{1}{2})4$	Associative, $\times$.
$1x > -2$	Multiplication.
$x > -2$	One, $\times$.

Note the change in the sense of the inequality from $<$ to $>$ when both members were multiplied by the negative number, $-\frac{1}{2}$. Also note that both members were multiplied by $-\frac{1}{2}$ inasmuch as the product $(-\frac{1}{2} \times -2)x = 1x$, thus allowing us to solve for x as required.

EXERCISES

State a reason for each step in the following solutions.

1. $x + 2 = 7$
   (a) $(x + 2) + (-2) = 7 + (-2)$
   (b) $x + [2 + (-2)] = 7 + (-2)$
   (c) $x + 0 = 5$
   (d) $x = 5$

2. $x + 2 < 7$
   (a) $(x + 2) + (-2) < 7 + (-2)$
   (b) $x + [2 + (-2)] < 7 + (-2)$
   (c) $x + 0 < 5$
   (d) $x < 5$

3. $2x + 3 = 7$
   (a) $(2x + 3) + (-3) = 7 + (-3)$
   (b) $2x + [3 + (-3)] = 7 + (-3)$
   (c) $2x + 0 = 4$
   (d) $2x = 4$
   (e) $\frac{1}{2}(2x) = \frac{1}{2}(4)$
   (f) $(\frac{1}{2} \times 2)x = \frac{1}{2}(4)$
   (g) $1x = 2$
   (h) $x = 2$

4. $-\frac{2}{3}x + 2 < 8$
   (a) $(-\frac{2}{3}x + 2) + (-2) < 8 + (-2)$
   (b) $-\frac{2}{3}x + [2 + (-2)] < 8 + (-2)$
   (c) $-\frac{2}{3}x + 0 < 6$
   (d) $-\frac{2}{3}x < 6$
   (e) $(-\frac{3}{2})(-\frac{2}{3}x) > (-\frac{3}{2}) \times 6$
   (f) $[(-\frac{3}{2})(-\frac{2}{3})] \times x > (-\frac{3}{2}) \times 6$
   (g) $1x > -9$
   (h) $x > -9$

Solve for x.

5. $3x + 2 = 11$

6. $2x - 3 = 8$

7. $-3x + 1 = 7$

8. $-2x + 5 = 9$

9. $\frac{1}{2}x - 2 = 5$

10. $\frac{2}{3}x - 1 = 5$

11. $2x + 3 < 7$

12. $3x - 2 < 7$

13. $3x - 1 > 5$

14. $2x + 3 > 8$

15. $-2x + 3 < 7$

16. $-3x + 1 < 5$

17. $-\frac{1}{2}x + 2 < 5$

18. $-\frac{1}{3}x + 1 > 2$

19. $-\frac{3}{4}x - 1 > 2$

20. $-\frac{2}{3}x - 3 < 9$

1. Many slow learners are helped to write and solve equations through the use of flow charts. For example, note the equations that the following flow charts describe.

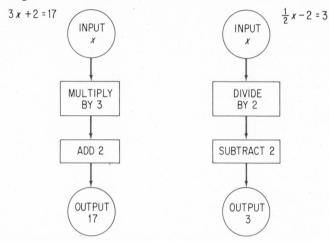

Draw flow charts to describe each of the following equations.

(a) $2x + 3 = 7$ **(b)** $3x - 5 = 10$

(c) $\frac{1}{2}x - 1 = 7$ **(d)** $\frac{1}{4}x + 2 = 6$

2. Students can be aided in their attempts to solve equations by drawing reverse flow charts. Thus they work backwards, using opposite operations as in these examples:

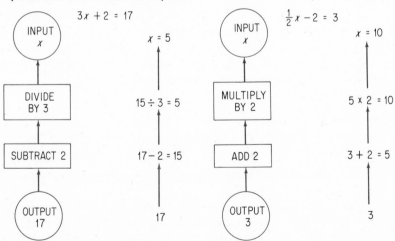

Draw reverse flow charts to solve each of the equations in Exercise 1 above.

8-6

products of expressions of the first degree

The expressions

$$x, \qquad x+2, \qquad 2x+1, \qquad 5-x$$

are expressions of the first degree in x. The expressions

$$x^2, \qquad x(x+1), \qquad (x-1)(x+1), \qquad x^3$$

are products of expressions of the first degree but are not themselves of the first degree. Special procedures are needed to solve or graph sentences involving products of expressions of the first degree. We consider only a few special types of such sentences.

The real number system has the property that a product of real numbers is zero if and only if at least one of the factors is zero. For example, if $a \cdot b = 0$, then $a = 0$, or $b = 0$, or both $a = 0$ and $b = 0$. We may use this property to solve certain types of sentences.

Example 1: Solve $2x(x-2)(x+3) = 0$.

Solution: This sentence is a statement that a certain product of real numbers is zero. The factors of the product are

$$\begin{array}{ll} 2 & \text{which is never zero} \\ x & \text{which is zero if } x = 0 \\ x - 2 & \text{which is zero if } x = 2 \\ x + 3 & \text{which is zero if } x = -3 \end{array}$$

The solution set of the given sentence is $\{0, 2, -3\}$.

In Example 1 the factors could also have been taken as $2x$, $x - 2$, and $x + 3$. As in the solution of Example 1, any sentence in which a product of factors of the first degree is zero may be solved by solving the equations obtained by setting each of the factors equal to zero. The union of these solutions is the solution set of the original sentence.

Products of the form

$$(x - a)(x + a)$$

occur so frequently that we learn to recognize them in other forms.

$$\begin{aligned}
(x - a)(x + a) &= (x - a) \times x + (x - a) \times a & &\text{Distributive.} \\
&= (x^2 - ax) + (xa - a^2) & &\text{Distributive.} \\
&= (x^2 - ax) + (ax - a^2) & &\text{Commutative, } \times. \\
&= [x^2 + (-ax)] + [ax + (-a^2)] & &\text{Definition, } -. \\
&= x^2 + \{(-ax) + [ax + (-a^2)]\} & &\text{Associative, } +. \\
&= x^2 + \{[(-ax) + ax] + (-a^2)\} & &\text{Associative, } +. \\
&= x^2 + \{0 + (-a^2)\} & &\text{Addition.} \\
&= x^2 + (-a^2) & &\text{Zero, } +. \\
&= x^2 - a^2 & &\text{Definition, } -.
\end{aligned}$$

sec. 8-6
products
of
expressions
of the
first
degree

Usually we find this product as follows:

$$
\begin{array}{r}
x + a \\
x - a \\
\hline
-ax - a^2 \\
x^2 + ax \\
\hline
x^2 \qquad - a^2
\end{array}
$$

In either case,

$$(x - a)(x + a) = x^2 - a^2.$$

Example 2: Solve $x^2 - 9 = 0$.

Solution: $x^2 - 9 = (x - 3)(x + 3) = 0$.

If $x - 3 = 0$, then $x = 3$.
If $x + 3 = 0$, then $x = -3$.

The solution set is $\{3, -3\}$.

The method used in Example 2 may be used for any sentence that can be expressed in the form $x^2 - a^2 = 0$.

Example 3: Solve $5x^2 + 3 = 23$.

Solution:

$5x^2 + 3 = 23$	Given.
$[5x^2 + 3] + (-23) = 23 + (-23)$	Addition, $=$.
$5x^2 + [3 + (-23)] = 0$	Addition and associative, $+$.
$5x^2 + (-20) = 0$	Addition.
$5x^2 - 20 = 0$	Definition, $-$.
$\frac{1}{5}(5x^2 - 20) = \frac{1}{5} \times 0$	Multiplication, $=$.
$x^2 - 4 = 0$	Multiplication.
$(x - 2)(x + 2) = 0$	$x^2 - a^2 = (x - a)(x + a)$.
$x = 2, -2$	A product is zero only if at least one of its factors is zero.

These methods may be extended to obtain a formula for solving any sentence that can be written in the form

$$ax^2 + bx + c = 0, \qquad a \neq 0$$

that is, any **quadratic equation.** The derivation of the formula is considered in the exercises. The formula, called the **quadratic formula**, is usually expressed as

$$x = \frac{-b \pm \sqrt{b^2 - 4ac}}{2a}.$$

In other words, the solution set is

$$\left\{ \frac{-b + \sqrt{b^2 - 4ac}}{2a}, \frac{-b - \sqrt{b^2 - 4ac}}{2a} \right\}.$$

Example 4: Use the quadratic formula to solve $2x^2 - 5x + 3 = 0$.

Solution: We compare $ax^2 + bx + c$ and $2x^2 - 5x + 3$ to obtain $a = 2$, $b = -5$, and $c = 3$. Then

$$x = \frac{-(-5) \pm \sqrt{(-5)^2 - 4(2)(3)}}{2(2)} = \frac{5 \pm \sqrt{25 - 24}}{4}$$

$$= \frac{5 \pm \sqrt{1}}{4} = \frac{5 \pm 1}{4} = \frac{6}{4} \quad \text{or} \quad \frac{4}{4}.$$

The solution set is $\{1, \frac{3}{2}\}$.

If any product of two expressions of the first degree is equal to zero, the sentence can be solved by the quadratic formula. There exist very complicated formulas for products of three expressions of the first degree and products of four expressions of the first degree. There do not exist any general formulas for solving sentences obtained from products of five or more expressions of the first degree equal to zero.

EXERCISES

Solve.

1. $(x - 3)(x - 5) = 0$
2. $(x + 2)(x - 3) = 0$
3. $(x + 1)(x + 5) = 0$
4. $(x - 6)(x + 4) = 0$
5. $(x + 2)(x - 1)(x + 3) = 0$
6. $(x + 1)(x - 2)(x + 4) = 0$
7. $2x(x + 5)(x - 2) = 0$
8. $3x(x - 1)(x - 2)(x - 5) = 0$
9. $(2x - 1)(3x - 6)(x + 5) = 0$
10. $(3x - 1)(2x + 5)(x - 7) = 0$
11. $x^2 - 25 = 0$
12. $x^2 - 49 = 0$
13. $2x^2 - 72 = 0$
14. $3x^2 = 12$
15. $2x^2 + 5 = 23$
16. $3x^2 - 1 = 47$

Write in the form $ax^2 + bx + c$.

17. $(x + 1)(x + 2)$
18. $(x + 5)(x + 5)$
19. $(x - 3)(x - 3)$
20. $\left(x + \dfrac{b}{2a}\right)\left(x + \dfrac{b}{2a}\right)$

285

sec. 8-6

products
of
expressions
of the
first
degree

The equation $ax^2 + bx + c = 0$, $a \neq 0$ is given. State a reason for each numbered step in the following derivation of the quadratic formula.

$$ax^2 + bx + c = 0 \qquad \text{Given.}$$

21. $x^2 + \dfrac{b}{a}x + \dfrac{c}{a} = 0 \times \dfrac{1}{a}$

22. $x^2 + \dfrac{b}{a}x + \dfrac{c}{a} = 0$

23. $\left(x^2 + \dfrac{b}{a}x + \dfrac{c}{a}\right) + \left(-\dfrac{c}{a}\right) = 0 + \left(-\dfrac{c}{a}\right)$

24. $\left(x^2 + \dfrac{b}{a}x\right) + \left[\dfrac{c}{a} + \left(-\dfrac{c}{a}\right)\right] = 0 + \left(-\dfrac{c}{a}\right)$

25. $\left(x^2 + \dfrac{b}{a}x\right) + 0 = 0 + \left(-\dfrac{c}{a}\right)$

26. $x^2 + \dfrac{b}{a}x = -\dfrac{c}{a}$

27. $x^2 + \dfrac{b}{a}x + \dfrac{b^2}{4a^2} = -\dfrac{c}{a} + \dfrac{b^2}{4a^2}$

28.
$$= \dfrac{b^2}{4a^2} + \left(-\dfrac{c}{a}\right)$$

$$= \dfrac{b^2}{4a^2} + \left(\dfrac{-c}{a}\right) \qquad \text{Equivalent fractions, } -\dfrac{c}{a} = \dfrac{-c}{a}.$$

29.
$$= \dfrac{b^2}{4a^2} + \dfrac{-4ac}{4a^2}$$

30.
$$= \dfrac{b^2 - 4ac}{4a^2}$$

31. $\left(x + \dfrac{b}{2a}\right)^2 + \left(-\dfrac{b^2 - 4ac}{4a^2}\right) = \dfrac{b^2 - 4ac}{4a^2} + \left(-\dfrac{b^2 - 4ac}{4a^2}\right)$

32. $\hspace{6cm} = 0$

33. $\left(x + \dfrac{b}{2a}\right)^2 - \dfrac{b^2 - 4ac}{4a^2} = 0$

$\left(x + \dfrac{b}{2a}\right)^2 - \left(\dfrac{\sqrt{b^2 - 4ac}}{2a}\right)^2 = 0 \qquad$ Properties of square root.

$x + \dfrac{b}{2a} = \dfrac{\sqrt{b^2 - 4ac}}{2a} \quad \text{or} \quad \dfrac{-\sqrt{b^2 - 4ac}}{2a} \qquad$ As in Example 2.

$x = \dfrac{-b}{2a} + \dfrac{\sqrt{b^2 - 4ac}}{2a} \quad \text{or} \quad \dfrac{-b}{2a} - \dfrac{\sqrt{b^2 - 4ac}}{2a} \qquad$ Addition, =; associative, +; addition; zero, +.

That is, $x = \dfrac{-b \pm \sqrt{b^2 - 4ac}}{2a}$.

Use the quadratic formula and solve.

34. $x^2 + x - 12 = 0$

35. $x^2 - 2x - 15 = 0$

36. $2x^2 + 3x - 5 = 0$

37. $3x^2 + x - 13 = 0$

PEDAGOGICAL CONSIDERATIONS

Many mathematical "tricks" can be explained and developed through the use of algebraic techniques. Consider, for example, this set of directions:

> Think of a number.
> Multiply your number by 2.
> Add 9.
> Subtract 3.
> Divide by 2.
> Subtract the number you started with.
> Your final answer will always be 3.

1. Use n to represent the original number, and write a mathematical phrase to represent each step in the set of directions given above. For example, the first three steps would give n, $2n$, and $2n + 9$. From this show why the result is always 3, regardless of the number one starts with.

2. Make up several similar "tricks" of your own.

Many interesting discovery lessons can be built around a study of the **Fibonacci sequence**:

$$1, 1, 2, 3, 5, 8, 13, 21, 34, \ldots$$

Note that after the first two terms, each new term is obtained as the sum of the preceding two terms. Thus in the sequence shown, the next term would be found as $21 + 34$, that is, 55.

3. Considering how the sequence is formed, can you predict something about the distribution of odd and even numbers? For example, approximately what per cent of the first 1000 terms will be even?

4. A sequence can be formed using the rule given above, and starting with any two numbers. Consider, for example, these sequences:

$$3, 4, 7, 11, 18, 29, 47, 76, 123, 199, \ldots$$
$$1, 4, 5, 9, 14, 23, 37, 60, 97, 157, \ldots$$
$$5, 6, 11, 17, 28, 45, 73, 118, 191, 309, \ldots$$

In each case find the sum of the first 10 terms. Then compare this sum with the product of 11 times the seventh term. What do you discover? Test your discovery with one of your own Fibonacci sequences.

*5. Prove your discovery of Exercise 4 by finding the sum of the first 10 terms of this general Fibonacci sequence:

$$a, b, a + b, a + 2b, 2a + 3b, 3a + 5b, \ldots$$

Show that this sum is equal to 11 times the seventh term.

CHAPTER TEST

1. Which sentences have the empty set as their solution set? Which sentences are identities?
   (a) $x + 1 = 1 + x$ (b) $x + 3 \neq x$
   (c) $x > x + 1$ (d) $x - 1 < x$

2. Describe the graph of the solution set for each sentence in terms of a specific geometric figure:
   (a) $x - 2 < 5$ (b) $x + 2 = 2 + x$
   (c) $2 \leq x \leq 5$ (d) $x + 1 \geq 3$

3. List the elements in the solution set of each sentence for whole numbers x:
   (a) $x + 1 \leq 5$ (b) $1 < x \leq 5$
   (c) $x - 1 > 2$ (d) $|x| \leq 3$

4. Classify as true or false:
   (a) $3 > 1$ and $-2 < -3$ (b) $3 < 7$ and $|-2| = 2$
   (c) $-1 > 0$ or $0 > -2$ (d) $2 < 0$ or $|5| = -5$

Graph the solution set for each sentence in Exercises 5 through 7 for real numbers x.

5. (a) $x + 2 > 3$ (b) $x - 1 \leq 3$
6. (a) $x \leq 2$ or $x + 1 > 5$ (b) $x - 1 \leq 2$ and $x > 0$
7. (a) $|x| \leq 5$ (b) $|x + 1| \geq 2$
8. Graph the solution set for (a) A, (b) B, (c) $A \cup B$, (d) $A \cap B$, where sets A and B are defined for real numbers x as

$$A = \{x \mid x + 2 \leq 7\}, \qquad B = \{x \mid x - 1 > 2\}.$$

Solve:

9. (a) $3x + 1 = 7$ (b) $-\frac{3}{4}x + 1 > 7$
10. (a) $(x + 7)(x - 5) = 0$ (b) $5x^2 = 45$

9

Sentences
in
Two
Variables

Sentences in two variables arise in many ways. For example, the sentence $d = 4h$ may be used to express the distance d in miles that John walks in h hours at 4 miles per hour. The sentence $p = 4s$ may be used to express the perimeter p of a square in terms of the side s. The sentence $I = 0.04P$ may be used to express the simple interest I on P dollars for one year at 4 per cent per year. The sentence $C = \frac{5}{9}(F - 32)$ may be used to express the temperature C in degrees Celsius (or Centigrade) in terms of the temperature F in degrees Fahrenheit.

In the seventeenth century the French mathematician René Descartes (1596–1650) made the initial discoveries of the branch of mathematics known as analytic geometry (Chapter 12). Basically he provided a geometric interpretation for algebraic sentences in two variables. In the present chapter we shall have the opportunity to study such sentences and their graphs in light of Descartes' geometry.

9-1

solution sets

We discussed open sentences in one real variable in Chapter 8. Now let us consider open sentences in two real variables, such as

$$x + y = 8.$$

If a replacement is made for only one of the variables, the sentence $x + y = 8$ remains an open sentence. For example, if x is replaced by 5, the sentence becomes $5 + y = 8$, which is neither true nor false, since no value has been specified for the real number y. Thus, a *pair* of replacements is needed for an open sentence in two variables before we can determine whether it is true or false for these replacements.

If x is replaced by 5 and y is replaced by 3, the sentence $x + y = 8$ is true. Usually we think of the variables as an *ordered pair* (x, y) and speak of the replacements as an ordered pair of numbers $(5, 3)$. By convention the variable x is assumed to be the first of the two variables, and the first number in the ordered pair of numbers is the replacement for x; the variable y is then taken as the second variable, and the second number in the ordered pair is taken as the replacement for y. Thus for the ordered pair $(7, 1)$ we have $x = 7$ and $y = 1$; for the ordered pair $(1, 7)$ we have $x = 1$ and $y = 7$. The sentence $x + y = 8$ is true or false for specified ordered pairs of numbers. It is true for $(5, 3)$, since $5 + 3 = 8$; it is false for $(2, 7)$, since $2 + 7 \neq 8$. The solution set for the sentence $x + y = 8$ is the set of ordered pairs of numbers for which the sentence is true. Indeed, the solution set for any sentence in two variables is a set of ordered pairs.

Just as with sentences in one variable, the solution sets of sentences in two variables depend upon the replacement sets of the variables used. For example, if the replacement set for x and y is the set of counting numbers, then the solution set for $x + y = 8$ is

$$\{(1, 7), (2, 6), (3, 5), (4, 4), (5, 3), (6, 2), (7, 1)\}.$$

Example 1: Find the solution set for the sentence $x + y < 3$ if the replacement set for x and y is the set of whole numbers.

Solution: When $x = 0$, y must be less than 3. Thus the ordered pairs of numbers $(0, 0)$, $(0, 1)$, and $(0, 2)$ make the sentence true. Similarly, when $x = 1$, y must be less than 2; when $x = 2$, y must be less than 1. Can $x = 3$? Is $x > 3$ possible? Thus the solution set is

$$\{(0, 0), (0, 1), (0, 2), (1, 0), (1, 1), (2, 0)\}.$$

Example 2: Find the solution set for the sentence $x + y \leq 3$ when x and y are counting numbers.

Solution: $\{(1, 1), (1, 2), (2, 1)\}.$

EXERCISES

Find the solution set for each sentence for the replacement set $\{1, 2, 3, 4\}$.

1. $x + y = 3$　　　　　　　　　　2. $2x + y \leq 4$

3. $x + y \leq 4$

5. $x + y < 2$

4. $y = x + 1$

6. $y < x$

Find the solution set for each sentence, using the set of counting numbers as the replacement set for x and y.

7. $x + y = 5$

9. $y \leq 3 - x$

8. $x + y < 4$

10. $y = 4 - x$

For the replacement set $\{-4, -3, -2, -1, 0, 1, 2, 3, 4\}$ find the solution set for each sentence.

11. $y = |x|$

13. $y = -|x|$

15. $y = x^2$

12. $y = |-x|$

14. $y = |x| - 3$

16. $y = x^2 - 5$

PEDAGOGICAL CONSIDERATIONS

1. The use of two variables in a mathematical expression can be introduced on an intuitive basis through the use of flow charts. Consider, for example, the following:

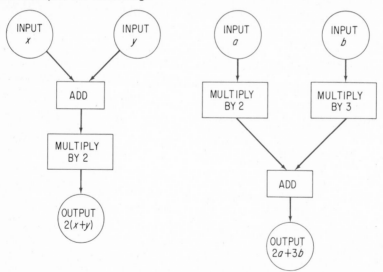

Copy and complete each table of values, using the flow charts above.

x	3	5	-2	3	-1
y	4	1	7	-2	-4
$2(x+y)$					

a	2	3	-5	-9	-7
b	3	2	4	-1	-5
$2a+3b$					

2. Draw a flow chart to show each of the following expressions:

(a) $3x + 5y$ (b) $3(x + y)$ (c) $\frac{1}{2}x + 3y$ (d) $3(2x + 3y)$

9-2

ordered pairs   A set of ordered pairs may be obtained from any given set of numbers. For example, consider the set of numbers

$$U = \{1, 2, 3\}.$$

We may form a set of all ordered pairs of numbers whose members are both elements of U. A **tree diagram** is helpful in showing the possible pairs of numbers involved.

First element	Second element	Ordered pairs
1	1	(1, 1)
	2	(1, 2)
	3	(1, 3)
2	1	(2, 1)
	2	(2, 2)
	3	(2, 3)
3	1	(3, 1)
	2	(3, 2)
	3	(3, 3)

Each element of U may serve as the first member of an ordered pair and may be matched with each of the elements of U as the second member. This set of ordered pairs is called the **Cartesian product** of U and U, is written as $U \times U$, and is read "U cross U." The set of all ordered pairs whose co-ordinates belong to the given set U is $U \times U$:

$$U \times U = \{(1, 1), (1, 2), (1, 3), (2, 1), (2, 2), (2, 3), (3, 1), (3, 2), (3, 3)\}$$

We shall refer to the set U as the **universal set**.

Example 1:   Let $U = \{1, 2\}$ and list the elements in $U \times U$.

Solution: $\{(1, 1), (1, 2), (2, 1), (2, 2)\}$.

Example 2:   List the elements in the solution set for $x + y = 4$ if $U = \{1, 2, 3, 4\}$.

Solution: We are permitted to replace x and y only by elements of U. Thus for $x = 1$, $y = 3$; for $x = 2$, $y = 2$; for $x = 3$, $y = 1$. The solution set is $\{(1, 3), (2, 2), (3, 1)\}$. Why can we not let $x = 4$?

The Cartesian product $A \times B$ is the set of all ordered pairs with first elements from A and second elements from B.

Example 3: Let $A = \{1, 2\}$ and $B = \{3, 4, 5\}$. Then list the elements of $A \times B$.

Solution: $\{(1, 3), (1, 4), (1, 5), (2, 3), (2, 4), (2, 5)\}$.

The graph of the Cartesian product $U \times U$ for $U = \{1, 2, 3\}$ consists of a set of nine points as in the figure.

This figure can now be used to graph the solution sets of sentences for this given universe. For example, the solution set of $x + y = 3$ for $U = \{1, 2, 3\}$ is $\{(1, 2), (2, 1)\}$. The graph of this solution set is drawn by placing heavily shaded, or brown, dots at the points that correspond to the ordered pairs $(1, 2)$ and $(2, 1)$.

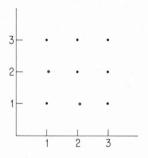

The universal set may include negative numbers. For example, if $U = \{-3, -2, -1, 0, 1, 2, 3\}$, then $U \times U$ consists of a set of 49 ordered pairs of numbers. The graph of $\{(x, y) \mid y = x\}$ for this universal set may be shown as follows:

292

Example 4: Graph $\{(x, y) \mid y \geq x + 2\}$ for the universal set

$$U = \{-3, -2, -1, 0, 1, 2, 3\}.$$

Solution: It is helpful to consider first the corresponding statement of equality, $y = x + 2$.

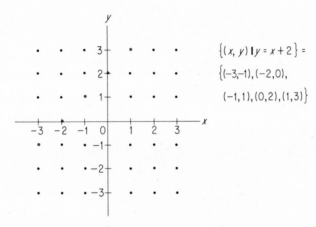

$$\{(x, y) \mid y = x + 2\} =$$
$$\{(-3,-1), (-2,0),$$
$$(-1,1), (0,2), (1,3)\}$$

The graph of $y \geq x + 2$ consists of the points that are solutions of $y = x + 2$, and also all the points in $U \times U$ that are above the points of the graph of $y = x + 2$.

1. Let $U = \{1\}$ and list the elements in $U \times U$.

2. Let $A = \{1, 2\}$ and $B = \{3, 4\}$. List the elements in $A \times B$.

Graph each set for $U = \{-3, -2, -1, 0, 1, 2, 3\}$.

3. $\{(x, y) \mid y = x + 1\}$ 4. $\{(x, y) \mid y = x - 1\}$

5. $\{(x, y) \mid y \geq x\}$ 6. $\{(x, y) \mid y < x\}$

7. $\{(x, y) \mid y \leq x - 2\}$ 8. $\{(x, y) \mid x + y \geq 4\}$

9. $\{(x, y) \mid y = x - 3\}$ 10. $\{(x, y) \mid y = x + 3\}$

11. $\{(x, y) \mid y \geq x - 3\}$ 12. $\{(x, y) \mid y \leq x + 3\}$

List the elements in each solution set for $U = \{1, 2, 3, 4\}$.

13. $x + y = 5$ 14. $y \geq x$

15. $y > x + 1$ 16. $x + y \leq 5$

*17. An *ordered triplet* of numbers is a set of three numbers in order, such as $(1, 3, 2)$. List all the possible ordered triplets that can be formed from the set $U = \{1, 2\}$.

PEDAGOGICAL CONSIDERATIONS

Ordered pairs may be used to show patterns. Some of these patterns may be expressed by algebraic sentences in two variables. Elementary school students usually enjoy the game of "trees" (originally developed by the University of Illinois Committee on School Mathematics). The teacher (or a selected student) thinks of a sentence in two variables. Initially two or three ordered pairs of integers that satisfy the sentence are stated and graphed. If a pair (x, y) satisfies the sentence, then "x trees y." If a pair does not satisfy the sentence, then "x does not tree y." Other pairs are suggested by members of the class as potential pairs for satisfying the sentence that is not known to the class. If for a suggested pair "x trees y," then the pair is graphed. Each student is encouraged to guess privately the sentence (or an equivalent) that is under discussion. If someone thinks that they know the sentence, he is asked a question such as

Does -5 tree 3?

Or, he may be given instructions such as

Let x be 7 and give a value for y.

After many people think that they know the sentence, the sentence is discussed openly.

Try this game of "trees" with some of your classmates using sentences such as:

1. $x + y = 6$

2. $y = x + 1$

3. $y = -x$

4. $y = 2 - x$

5. $y > x$

6. $y \leq x + 2$

7. $y = |x|$

8. $y > |x|$

9. $y \leq 1 + |x|$

10. $y < 2 + |x|$

9-3

graphs on a plane

In §9-2 we graphed the solution sets of sentences using finite coordinate systems. Now we shall extend these concepts and consider U as the set of real numbers. Then $U \times U$ becomes an infinite collection of ordered pairs of numbers and its graph consists of an entire plane. In honor of René Descartes, such a plane is frequently referred to as the **Cartesian plane.** The numbers of each ordered pair of real numbers are the **Cartesian coordinates** of a point of the plane. Each point of a Cartesian plane can be represented by (has as its *coordinates*) an ordered pair of real numbers, and each ordered pair of real numbers can be used to identify (locate) a unique point of the plane. We follow the custom of speaking of "the point with coordinate (x, y)" as "the point (x, y)."

Let us now discuss the graphs of sentences for the universe of real numbers. Consider, for example, the sentence $y = x + 2$. We may list in a table of values several ordered pairs of numbers that are solutions of the sentence. Thus for $x = -1$, we have $y = -1 + 2 = 1$ and for $x = 2$, we have $y = 2 + 2 = 4$. Confirm the other entries given in this table:

$y = x + 2$:

x	-3	-2	-1	0	1	2	3
y	-1	0	1	2	3	4	5

Each ordered pair of numbers (x, y) from the table can then be graphed. These points may be connected in the order of the x-coordinates and the graph of $y = x + 2$ obtained as in the figure at the top of page 296.

295

The graph of $y = x + 2$ is a straight line and extends indefinitely as indicated by the arrowheads. Often these arrowheads are omitted. In general, for the universe of real numbers, any equation that can be expressed in the form

$$ax + by + c = 0,$$

where a and b are not both zero, is said to be a **linear equation** and has a straight line as its graph. We shall consider U to be the set of real numbers from now on unless otherwise specified.

Inasmuch as a straight line is determined by two points, we can graph a linear equation after locating two of its points. When $a \neq 0$ and $b \neq 0$ the most convenient points to locate are the point $(0, y)$ at which the graph crosses the y-axis and the point $(x, 0)$ at which the graph crosses the x-axis. The graph of $y = x + 2$ crosses the y-axis when $x = 0$, that is, at the point A: $(0, 2)$, where 2 is the y-**intercept** of the graph. The graph of $y = x + 2$ crosses the x-axis when $y = 0$, that is, at the point B: $(-2, 0)$, where -2 is the x-**intercept** of the graph.

Example 1: Graph $\{(x, y) \mid x - 2y = 4\}$.

Solution: The equation may be expressed as $x - 2y - 4 = 0$ and thus has a line as its graph. For $x = 0$, $-2y = 4$ and $y = -2$. For $y = 0$, $x = 4$. The x-intercept 4 and the y-intercept -2 determine the line as shown in the graph at the top of page 297.

Usually it is desirable to graph a third point as a check of your work. For example, when $x = 2$ we have $2 - 2y = 4$, $-2y = 2$, and $y = -1$. Thus $(2, -1)$ may be graphed and the point should be on the line.

Statements of inequality can also be graphed for the universe of real numbers. Consider the sentence $y \leq x + 2$. Here the graph consists of all the points on the line $y = x + 2$, as well as the points in the **half-plane** below the line as indicated by the shaded portion of the graph. Thus the graph of $y \leq x + 2$ is the union of a line and a half-plane.

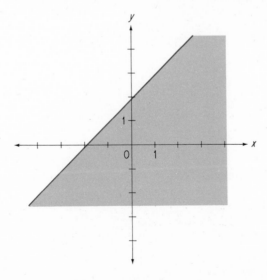

Example 2: Graph $y > x - 1$.

Solution: It is helpful to draw first the corresponding statement of equality, $y = x - 1$, as a dotted line. This line divides the plane into

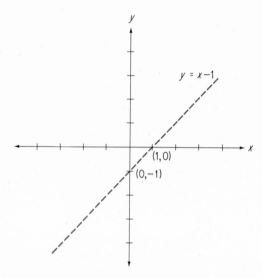

two half-planes. To determine which half-plane to shade, we note that the problem requests all the values of y *greater than* $x - 1$. Thus we shade the half-plane above the line to represent the solution set.

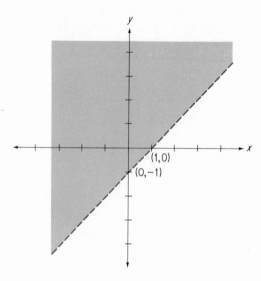

The line in the graph for Example 2 is dotted, since the points of the line *are not* points of the graph. The line in the graph for Example 3 is solid, since the points of the line *are* points of the graph.

Example 3: Graph $x - 2y - 4 \geq 0$.

Solution: This sentence is equivalent to $x - 2y \geq 4$. From Example 1 we have the graph of the line $x - 2y = 4$. To determine which half-plane to shade we may proceed in one of two ways. We can place the given sentence in **y-form** as follows:

$$x - 2y \geq 4$$
$$-2y \geq -x + 4 \qquad \text{Why?}$$
$$y \leq \frac{1}{2}x - 2 \qquad \text{Why?}$$

This tells us that the half-plane *below* the line is included in the graph of the solution set. An alternate plan for determining which half-

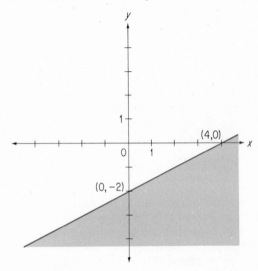

plane is included in the solution set is to select a particular point of one half-plane and see whether or not that point belongs in the graph. The origin, $(0, 0)$, is a convenient point to consider. When $x = 0$ and $y = 0$, we have the false statement $0 - 0 \geq 4$. Thus the points of the half-plane that contains $(0, 0)$ are not in the solution set and again we conclude that the graph consists of the line and the half-plane below the line.

EXERCISES

For the graph of each sentence, find (a) the x-intercept; (b) the y-intercept.

1. $x + y = 5$

2. $x - y = 2$

3. $2x - 3y = 6$

4. $3x - 2y = 6$

5. $y = 2x + 3$

6. $y = 3x + 2$

7. $3x - y + 6 = 0$ 8. $x + 2y - 5 = 0$

9. $5x + 4y = 9$ *10. $ax + by + c = 0$

Write each sentence in y-form, that is, solve for y.

11. $2x + y = 7$ 12. $4x - 2y = 8$ 13. $3x + y \geq 6$

14. $x + 2y \leq 4$ 15. $2x - y \leq 4$ 16. $3x - 2y \geq 6$

Graph each set for the universe of real numbers.

17. $\{(x, y) \mid y = x + 1\}$ 18. $\{(x, y) \mid y = x - 2\}$

19. $\{(x, y) \mid x + y = 3\}$ 20. $\{(x, y) \mid x - y = 3\}$

21. $\{(x, y) \mid 2x + 3y - 6 = 0\}$ 22. $\{(x, y) \mid 3x - 2y - 6 = 0\}$

23. $\{(x, y) \mid x = 2\}$ 24. $\{(x, y) \mid y = -1\}$

25. $\{(x, y) \mid y \geq x\}$ 26. $\{(x, y) \mid y \leq x - 2\}$

27. $\{(x, y) \mid 2x + y > 4\}$ 28. $\{(x, y) \mid x - 2y < 2\}$

29. $\{(x, y) \mid x - y - 1 \leq 0\}$ 30. $\{(x, y) \mid 3x - 4y - 12 \geq 0\}$

PEDAGOGICAL CONSIDERATIONS

1. Prepare a report on the life of René Descartes, stressing his contributions in the field of analytic geometry.

2. When junior high school students are first introduced to the concept of rectangular coordinates, they enjoy an exercise that involves plotting points that form figures. Thus a set of ordered pairs of numbers are given that are to be located, then connected in order. The outcome is some figure, such as an animal, a Christmas tree, etc. Prepare a set of such ordered pairs that gives rise to some figure.

3. An interesting exercise for upper elementary grade students is to have them draw an object in the first quadrant of a Cartesian plane. Then they must attempt to draw the reflection of this object in each of the other quadrants, using the x-axis and y-axis as axes of symmetry. For example, here is the reflection of the letter F in each quadrant:

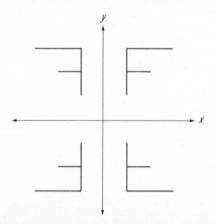

Prepare a similar drawing for the letters P and R. Are there any letters that will look the same in all four quadrants?

4. Try to write the word SEVEN as it would appear if held up to a mirror. Then use a mirror to check your result. Such activities help develop concepts of reflection and symmetry with elementary school students.

9-4
graphs involving absolute value

Not all sentences with two variables have graphs that are straight lines—only those sentences that can be written in the form $ax + by + c = 0$. In §8-4 we considered graphs of sentences in one variable that involved absolute value. Now let us explore sentences with two variables that involve absolute value, such as $y = |x|$. As an aid to graphing this sentence for the universe of real numbers we first consider the graph for $U = \{-3, -2, -1, 0, 1, 2, 3\}$. The values to be graphed are summarized in the table of values below.

Recall the definition of $|x|$:

$$|x| = x \quad \text{for} \quad x \geq 0$$
$$|x| = -x \quad \text{for} \quad x < 0$$

x	-3	-2	-1	0	1	2	3
y	3	2	1	0	1	2	3

These ordered pairs (x, y) such that $y = |x|$ can be graphed on $U \times U$ as follows:

If U is the set of real numbers, our graph consists of an infinite collection of points. Using the preceding graph as a guide, we are now able to graph $y = |x|$ for the set of real numbers as at the top of page 302.

The graph of $y \geq |x|$ is completed by shading in the portion of the plane above the graph, as shown.

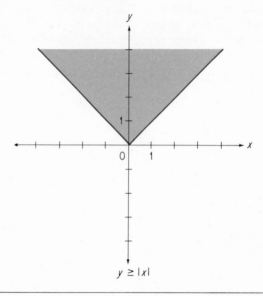

$y \geq |x|$

Example 1:   Graph $\{(x, y) \mid y = |x + 1|\}$.

Solution: Recall that we assume the universal set to be the set of real numbers unless otherwise specified.

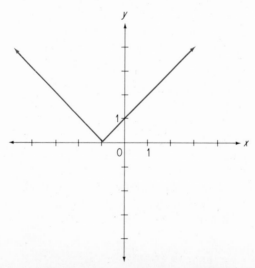

Notice that if $x = -1$, then $y = 0$; if $x = 0$ or -2, then $y = 1$; if $x = 1$ or -3, then $y = 2$; and so forth.

Example 2: Graph $\{(x, y) \mid y \geq |x + 1|\}$.

Solution: From Example 1 we know the graph of $y = |x + 1|$. The graph of $y \geq |x + 1|$ is completed by shading in the portion of the plane above the graph as shown.

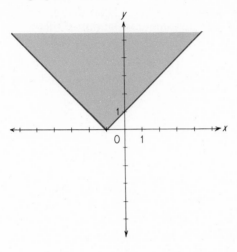

EXERCISES

Graph the solution set for $U = \{-4, -3, -2, -1, 0, 1, 2, 3, 4\}$.

1. $y = |x| + 1$

2. $y = |x| - 1$

3. $y = |x + 1|$

4. $y = |x - 1|$

5. $y = -|x|$

6. $y = -|x + 1|$

7. $y \geq |x| + 2$

8. $y \leq |x| - 2$

Graph each set for the universe of real numbers.

9. $\{(x, y) \mid y = |x| + 1\}$

10. $\{(x, y) \mid y = |x| - 1\}$

11. $\{(x, y) \mid y = |x + 1|\}$

12. $\{(x, y) \mid y = |x - 1|\}$

13. $\{(x, y) \mid y > |x + 2|\}$

14. $\{(x, y) \mid y < |x - 2|\}$

15. $\{(x, y) \mid |x| \leq 2\}$

16. $\{(x, y) \mid |y| \geq 1\}$

*17. $\{(x, y) \mid |x + y| = 1\}$

*18. $\{(x, y) \mid |x| + |y| = 1\}$

*19. $\{(x, y) \mid |x| + |y| \leq 1\}$

*20. $\{(x, y) \mid |x| - |y| = 1\}$

303

1. The overhead projector is an effective device for showing graphs on a plane, especially through the use of translations and rotations. For example, prepare a set of coordinate axes and grid lines. On a separate sheet of acetate draw the graph of $y = |x|$. Then demonstrate each of the following graphs by appropriate translations of this basic curve:

(a) $y = |x| + 1$
Shift curve one unit up.

(b) $y = |x| - 1$
Shift curve one unit down.

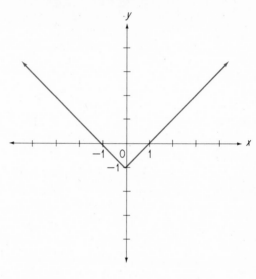

(c) $y = |x + 1|$
Shift curve one unit to the left.

(d) $y = |x - 1|$
Shift curve one unit to the right.

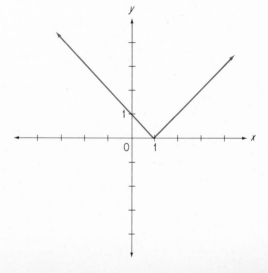

2. A 180° rotation of $y = |x|$ about the x-axis will give the graph of $y = -|x|$. Use this idea to show each of the following by appropriate rotations and translations:

(a) $y = -|x|$

(b) $y = -|x| + 1$

(c) $y = -|x| - 1$

(d) $y = -|x + 1|$

9-5

linear systems

In Chapter 8 we considered compound sentences that involved one variable. Now we shall turn our attention to compound sentences involving two variables. Consider, for example, the compound sentence

$$x + y - 3 = 0 \quad \text{and} \quad x - y - 1 = 0.$$

This is frequently written in the following form:

$$\begin{cases} x + y - 3 = 0 \\ x - y - 1 = 0 \end{cases}$$

and is referred to as a **system of linear equations** or as a set of **simultaneous linear equations**. To solve a set of two simultaneous equations we find the set of ordered pairs that are solutions of both equations. Our approach in this section will be a graphical one rather than an algebraic one; that is, we draw the graph of each sentence and identify the point of intersection, if any, of the two lines. For the given system of equations we have this graph:

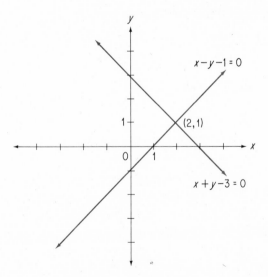

The point located at $(2, 1)$ is on both lines and $(2, 1)$ is the solution of the given system of equations. We may express this fact as

$$\{(x, y) \mid x + y - 3 = 0\} \cap \{(x, y) \mid x - y - 1 = 0\} = \{(2, 1)\}.$$

305

Consider next the compound sentence

$$x + y = 3 = 0 \quad \text{or} \quad x - y - 1 = 0.$$

This word "or" indicates that we are to find the set of ordered pairs of numbers that are solutions of either one or of both of the given equations; that is,

$$\{(x, y) \mid x + y - 3 = 0\} \cup \{(x, y) \mid x - y - 1 = 0\}.$$

The graph of the solution set consists of all the points that are on at least one of the two lines; that is, the graph is the union of the points of the two lines drawn in the preceding figure.

Example 1: Graph $x - y - 1 = 0$ or $x - y + 2 = 0$.

Solution: The graph consists of the union of the points of the two lines.

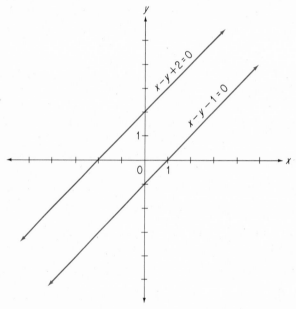

Notice that the two lines in the solution for Example 1 are parallel. The solution set of the sentence "$x - y - 1 = 0$ *and* $x - y + 2 = 0$" would be the empty set.

Example 2: Graph $\{(x, y) \mid (x - y - 1)(x - y + 2) = 0\}$.

Solution: First we note that the product $a \cdot b = 0$ implies that $a = 0$ or $b = 0$. (Recall that the use of the word "or" also includes the case that $a = 0$ *and* $b = 0$.) Thus the given sentence can be written in the form $x - y - 1 = 0$ or $x - y + 2 = 0$. But this is precisely the problem in Example 1. Thus the graph of the solution set consists of the two lines drawn in the figure given there.

Systems of inequalities can also be solved graphically as the union or intersection of half-planes. Consider the system:

$$\begin{cases} x + y - 3 > 0 \\ x - y - 1 > 0 \end{cases}$$

The corresponding statements of equality have been graphed earlier in this section. The graphs of the inequalities are shown in these figures.

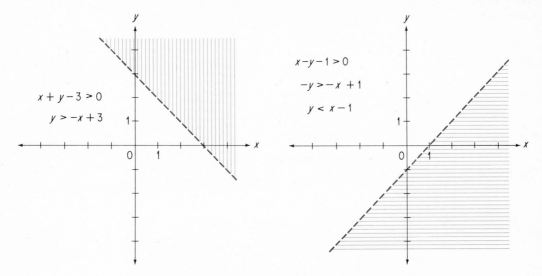

Note that the corresponding lines are dotted since the points of the lines are *not* included in the graphs. Also one graph has vertical shading and the other graph has horizontal shading. The graph of the system con-

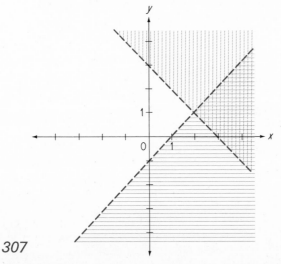

sists of the intersection of these two graphs; that is, the points of the region that is shaded both vertically and horizontally when the graphs of the two inequalities are drawn on the same coordinate plane.

We can also determine from this figure the graph of the sentence $x + y - 3 > 0$ or $x - y - 1 > 0$, namely the union of the points of the two shaded regions. The unshaded region (including the dotted lines) represents the graph of the solution set of the system.

$$\begin{cases} x + y - 3 \leq 0 \\ x - y - 1 \leq 0 \end{cases}$$

Example 3: Graph

$$\{(x, y) \mid x - y + 2 \leq 0\} \cap \{(x, y) \mid 2x + y - 4 \geq 0\}.$$

Solution: The graph of $x - y + 2 \leq 0$ has horizontal shading in the figure; the graph of $2x + y - 4 \geq 0$ has vertical shading. The graph of the solution set consists of the points in the region with both hori-

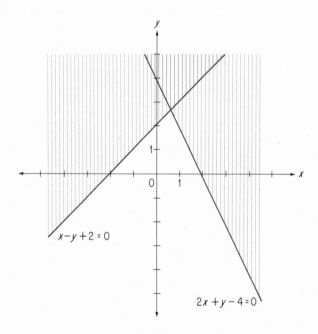

zontal and vertical shading and includes the points of the two rays that serve as the boundary of this region.

Example 4: Graph

$$\{(x, y) \mid x - y + 2 \leq 0\} \cup \{(x, y) \mid 2x + y - 4 \geq 0\}.$$

Solution: Consider the graph for Example 3. The solution set in Example 4 consists of all the points in the regions with horizontal shading or vertical shading, or both, together with the points of the boundary of this region. That is, all points are included *except* those in the unshaded region.

EXERCISES

Graph the solution set for each system.

1. $\begin{cases} x + y - 2 = 0 \\ x - y - 4 = 0 \end{cases}$

2. $\begin{cases} x + y - 5 = 0 \\ x - y - 1 = 0 \end{cases}$

3. $\begin{cases} x + y = 1 \\ x + 2y = 4 \end{cases}$

4. $\begin{cases} 2x - y = 6 \\ x + 2y = -2 \end{cases}$

5. $\{(x, y) \mid x + y - 5 = 0\} \cup \{(x, y) \mid x + y + 2 = 0\}$

6. $\{(x, y) \mid x + 2y - 1 = 0\} \cup \{(x, y) \mid 2x + 4y - 2 = 0\}$

7. $\{(x, y) \mid 2x - y - 4 = 0\} \cap \{(x, y) \mid 4x - 2y + 4 = 0\}$

8. $\{(x, y) \mid (x - y + 3)(2x - 3y + 6) = 0\}$

9. $\begin{cases} x - y + 2 > 0 \\ x + y - 2 > 0 \end{cases}$

10. $\begin{cases} 2x + y - 4 < 0 \\ x - 2y + 4 < 0 \end{cases}$

11. $\begin{cases} 3x - 2y - 6 \geq 0 \\ 2x + 3y + 6 \leq 0 \end{cases}$

12. $\begin{cases} x + 2y - 4 \leq 0 \\ 2x - y + 2 \geq 0 \end{cases}$

13. $\{(x, y) \mid x - y - 3 \leq 0\} \cap \{(x, y) \mid x + y - 3 \leq 0\}$

14. $\{(x, y) \mid x + 2y - 6 \geq 0\} \cap \{(x, y) \mid 2x - y + 2 \leq 0\}$

15. $\{(x, y) \mid 3x - y + 3 \leq 0\} \cup \{(x, y) \mid 3x + 4y - 12 \geq 0\}$

16. $\{(x, y) \mid x + 4y - 4 \geq 0\} \cup \{(x, y) \mid 4x - 2y + 8 \geq 0\}$

*17. $\{(x, y) \mid (2x - 3y + 6)(x + 2y - 4) \leq 0\}$

*18. $\{(x, y) \mid y \geq |x + 1|\} \cap \{(x, y) \mid y \leq 3\}$

*19. $\{(x, y) \mid y \geq |x| - 1\} \cap \{(x, y) \mid y \leq -|x| + 1\}$

*20. $\{(x, y) \mid |x + y| \leq 2\} \cap \{(x, y) \mid |x| \leq 3\}$

PEDAGOGICAL CONSIDERATIONS

1. A student teacher ran into trouble when trying to teach a ninth grade algebra class how to draw the graph of the solution set for the in-

equality $2x - 3y + 6 > 0$. First he constructed the graph of the corresponding equality as in the following figure.

He then tested the point $(0, 0)$ in the upper half-plane, which gives this true statement of inequality:

$$0 - 0 + 6 > 0$$

Thus he concluded that the required half-plane includes the origin, and must be *above* the line. As a check, he then went on to write the given inequality in y-form:

$$2x - 3y + 6 > 0 \qquad -3y > -2x - 6 \qquad y < \tfrac{2}{3}x + 2$$

This last statement indicates that the required half-plane must be *below* the line! Where did he make the error?

2. Having made the error described above, state at least three different procedures the student teacher could have followed in terms of teaching techniques.

9-6

quadratic sentences   An expression of the form $ax^2 + bx + c$ is **quadratic** in x (see §8-6). A sentence that may be expressed in the form $y = ax^2 + bx + c$ is quadratic in the variable x and linear in y. In this section we are primarily concerned with graphical representations of quadratic sentences of this form.

We begin with a consideration of the statement $y = x^2$ and consider first a number of specific ordered pairs (x, y) such that $y = x^2$. Note that if $x = 0$, then $y = 0$; if $x = 1$ or -1, then $y = 1$; if $x = 2$ or -2, then $y = 4$. Several such pairs of numbers are summarized in the following table of values, and are then graphed and connected with a smooth curve to give the graph of $y = x^2$. This curve is known as a **parabola**.

x	−3	−2	−1	0	1	2	3
y	9	4	1	0	1	4	9

The symmetry of this parabola with respect to the y-axis arises from the fact that for each point (x, y) of the curve, there is also a point $(-x, y)$ of the curve. The y-axis is the **axis of symmetry** and the point $(0, 0)$ is the **vertex** of the parabola. Note that for $y = x^2$ the quantity squared is x, the axis of symmetry has equation $x = 0$, and the vertex $(0, 0)$ is the point of the axis that is also a point of the curve.

Using $y = x^2$ as a basis for comparison, we may now graph other sentences of the form $y = ax^2 + bx + c$. For example, the graph of $y = x^2 + 1$ has the same shape as $y = x^2$, has the y-axis ($x = 0$) as its axis of symmetry, and has the point $(0, 1)$ as its vertex. The graph of $y = x^2 + 1$ may be obtained by translating upward (raising) the graph of $y = x^2$ one unit. Similarly, the graph of $y = x^2 - 1$ has the y-axis ($x = 0$) as its axis of symmetry and has $(0, -1)$ as its vertex. The graph of $y = x^2 - 1$ may be obtained by translating downward (lowering) the graph of $y = x^2$ one unit.

$y = x^2 + 1$

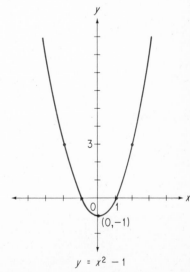

$y = x^2 - 1$

Next consider the graph of $y = (x + 1)^2$. A table of values may be used as a basis for drawing the graph.

x	-3	-2	-1	0	1	2	3
y	4	1	0	1	4	9	16

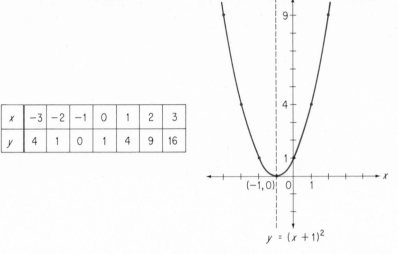

$y = (x + 1)^2$

This graph has the same shape as that of $y = x^2$ and indeed may be obtained by translating (shifting) the graph of $y = x^2$ one unit to the left. Note that the sentence involves the square of $x + 1$; the axis of symmetry of the graph has equation $x + 1 = 0$, that is, $x = -1$. The vertex is at $(-1, 0)$, and the graph may be obtained by translating the graph of $y = x^2$ one unit to the left, that is, -1 units.

Example 1: Graph $y = (x - 1)^2$ and identify the axis of symmetry and the vertex of the curve.

Solution: The sentence $y = (x - 1)^2$ involves the square of $x - 1$. The axis of symmetry has equation $x - 1 = 0$, that is, $x = 1$. The vertex is at $(1, 0)$. The graph may be obtained by translating the graph of $y = x^2$ one unit to the right.

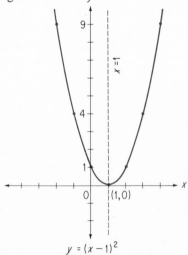

$y = (x - 1)^2$

A sentence such as $y = (x - 1)^2 + 1$ involves the square of $x - 1$. The graph of the sentence has axis of symmetry $x = 1$, has vertex $(1, 1)$, and may be obtained by translating the graph of $y = x^2$ one unit to the right and one unit upward. As in the figure it is often helpful to sketch the graph of $y = x^2$ as a dotted curve for reference.

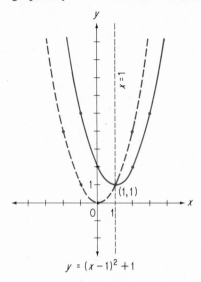

$$y = (x - 1)^2 + 1$$

Example 2: Graph $y = (x + 2)^2 - 1$.

Solution: The axis of symmetry is $x + 2 = 0$, that is, $x = -2$. The vertex is at $(-2, -1)$. The graph may be obtained by translating the graph of $y = x^2$ two units to the left and one unit down.

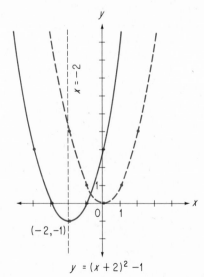

$$y = (x + 2)^2 - 1$$

Any quadratic sentence that can be expressed in the **standard form** $y = (x - h)^2 + k$ has $x - h = 0$ as its axis of symmetry, has (h, k) as its vertex, and may be graphed by translating the graph of $y = x^2$ a directed distance of h units along the x-axis and k units along the y-axis.

The method known as **completing the square** may be used to express any equation of the form

$$y = x^2 + bx + c$$

in the standard form

$$y = (x - h)^2 + k \quad \text{or} \quad y = x^2 - 2hx + h^2 + k.$$

For example, consider the sentence

$$y = x^2 - 6x + 5.$$

We wish to write this sentence in standard form. Compare the coefficients of x in these two forms, $-2h$ and -6. If $-2h = -6$, then $h = 3$ and $h^2 = 9$. The given sentence may then be written as

$$y = x^2 - 6x + 9 - 9 + 5.$$

(Add and subtract the square of one-half the coefficient of x.)

$$y = (x - 3)^2 - 4.$$

When the equation is considered in this form, we recognize the graph as a parabola with $x = 3$ as axis of symmetry, with $(3, -4)$ as vertex, and obtainable by translating the graph of $y = x^2$ three units to the right and four units down.

Note that for $x = 0, y = 5$. Also, when $y = 0$, we have

$$x^2 - 6x + 5 = 0$$
$$(x - 1)(x - 5) = 0$$
$$x = 1 \quad \text{or} \quad x = 5.$$

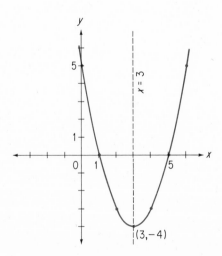

Example 3: Describe the graph of $y = x^2 + 4x + 5$.

Solution: First place the given equation in standard form:

$$y = x^2 + 4x + 5$$
$$y = x^2 + 4x + 4 - 4 + 5$$
(One-half of 4 is 2; $2^2 = 4$.)
$$y = (x + 2)^2 + 1$$
$$y = [x - (-2)]^2 + 1$$

The graph may be obtained by translating the graph of $y = x^2$ two units to the left and one unit up.

The general equation $y = ax^2 + bx + c$ may be expressed in the standard form $y = a(x - h)^2 + k$. The case $a = -1$ is considered in Example 4; other cases are considered in the exercises.

Example 4: Graph $y = -x^2$.

Solution: Each y value of x^2 for the points of the graph of $y = x^2$ is now graphed as its opposite, $y = -x^2$. Several ordered pairs are shown in the table of values. The graph of $y = -x^2$ may be obtained by rotating the graph of $y = x^2$ about the x-axis.

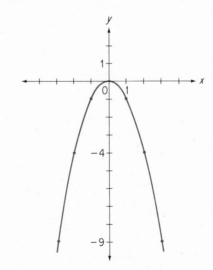

x	-3	-2	-1	0	1	2	3
y	-9	-4	-1	0	-1	-4	-9

Two other types of graphs are illustrated in the next two examples.

Example 5:   Graph $x = y^2$.

Solution: If $x = 0$, then $y = 0$; if $x = 1$, then $y^2 = 1$ and $y = 1$ or -1; if $x = 4$, then $y^2 = 4$ and $y = 2$ or -2, and so forth. This graph is a parabola with the x-axis as its axis of symmetry and $(0, 0)$ as its vertex. The graph of $x = y^2$ may be obtained by rotating the graph of $y = x^2$ clockwise $90°$ about the origin.

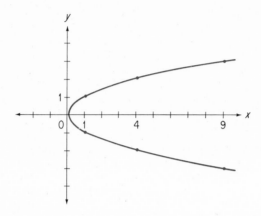

Example 6:   Graph $y \geq x^2$.

Solution: The graph consists of all the points on the graph of $y = x^2$, as well as the points in the shaded region above the graph.

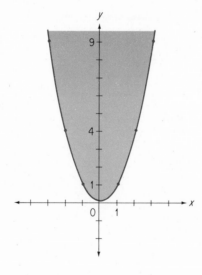

EXERCISES

Graph the solution set of each sentence.

1. $y = x^2 + 2$
2. $y = x^2 - 2$
3. $y = (x + 2)^2$
4. $y = (x - 2)^2$
5. $y = (x - 2)^2 + 2$
6. $y = (x + 2)^2 - 2$
7. $y = (x + 3)^2 - 1$
8. $y = (x - 3)^2 + 1$
9. $y \geq x^2 + 1$
10. $y \geq (x + 1)^2$
11. $y \leq (x - 1)^2$
12. $y \leq x^2 - 1$
13. $y \leq -x^2$
14. $y \geq -x^2$
15. $y = -x^2 + 1$
16. $y = -x^2 - 2$
17. $y = -(x - 1)^2$
18. $y = -(x + 1)^2$
19. $y = -(x - 2)^2 + 3$
20. $y = -(x + 3)^2 - 2$

In Exercises 21 through 28 place each equation in standard form. Then give the coordinates of the vertex and the equation of the axis of symmetry of the graph of each sentence.

21. $y = x^2 + 6x + 5$
22. $y = x^2 - 4x - 5$
23. $y = x^2 - 2x + 3$
24. $y = x^2 + 2x - 1$
25. $y = x^2 - 8x + 10$
26. $y = x^2 + 8x - 5$
27. $y = x^2 - 5x + 6$
28. $y = x^2 + 3x - 4$

*29. Graph $y = ax^2$ for (a) $a = 2$; (b) $a = \frac{1}{2}$; (c) $a = -3$; (d) $a = -\frac{1}{4}$.

*30. Assume that the graph of $y = ax^2$ has been previously graphed and describe a procedure for graphing the sentence (a) $y = a(x - h)^2$; (b) $y = a(x - h)^2 + k$.

9-7

relations and functions

A relation may be defined as any subset of $U \times U$; thus, a relation is a set of ordered pairs of numbers. It is most often defined by a rule. Consider, for example, $\{(x, y) \mid y > x - 1\}$, where $U = \{1, 2, 3\}$. This set, which *is* the relation, is $\{(1, 1), (1, 2), (1, 3), (2, 2), (2, 3), (3, 3)\}$. It may be graphed as in the accompanying figure.

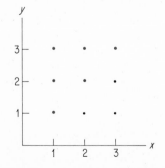

The relation may be defined by the rule, by the graph, or by a table of values for the variable:

x	1	1	1	2	2	3
y	1	2	3	2	3	3

Here is the table and graph for another relation:

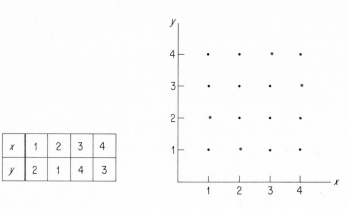

x	1	2	3	4
y	2	1	4	3

This relation is the set of ordered pairs {(1,2), (2, 1), (3, 4), (4, 3)}. Note that this second relation differs from the first in that, for any value of x, there is at most one value for y. We call this special type of relation a *function*.

A **function** is a set of ordered pairs (x, y) such that for each value of x there is at most one value of y; that is, no first element appears with more than one second element. You may think of the first element as the **independent variable** and the second element as the **dependent variable.** Each variable has a set of possible values. The set of all first elements of the ordered pairs of numbers is called the **domain** of the function; the set of all second elements is called the **range** of the function. In terms of its graph, any vertical line drawn meets the graph of a function in at most one point.

Here are the graphs of two relations that are also functions:

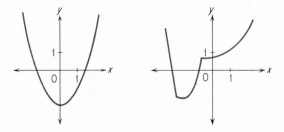

Notice that no vertical line intersects the graph in more than one point.
Here are the graphs of two relations that are not functions:

Notice that in each case there exists at least one vertical line that intersects the graph in two or more points.

Although formulas such as $y = x^2$ may *define* a function, they are not strictly functions. We have defined the function to be a set of ordered pairs (x, y), such as those obtained from the formula $y = x^2$ for the real variable x. Thus, a formula may provide a rule by which the function may be determined. In other words, a formula may provide a means for associating a unique element in the range with each element in the domain.

When a formula such as $y = x^2 - 2x + 3$ is used to define a function, it is customary to think of y as *a function of* x and to write the formula in **functional notation**:

$$y = f(x)$$

where

$$f(x) = x^2 - 2x + 3.$$

Then the value of y for any value b of x may be expressed as $f(b)$. For example,

$$f(2) = 2^2 - 2(2) + 3 = 3$$
$$f(-1) = (-1)^2 - 2(-1) + 3 = 6$$
$$f(0) = 0^2 - 2(0) + 3 = 3$$
$$f(1) = 1^2 - 2(1) + 3 = 2.$$

Notice that for the function $f(x) = (x - 1)^2 + 2$, the domain of the function is the set of all real numbers, and the range of the function is the set of real numbers greater than or equal to 2. The range may be determined from the graph of the function or from the observation that since $(x - 1)^2 \geq 0$ we must have $(x - 1)^2 + 2 \geq 2$. Other letters, as in $g(x)$ and $h(y)$, may be used in designating functions.

Example 1: Graph the function $y = x^2 - 1$ and identify the domain and range of this function.

Solution: The variable x may assume any real number as a value. Thus the domain is the set of real numbers. The range is the set of real numbers greater than or equal to -1, that is, $y \geq -1$, as in the graph at the top of the next page.

Example 2: Find the domain and range of $y = |x|$.

Solution: Domain: set of real numbers. Range: set of nonnegative real numbers, that is, $y \geq 0$.

Example 3: Which of these formulas define functions? (a) $y = x^3$; (b) $x = y^2$; (c) $x^2 + y^2 = 9$; (d) $y = (x + 1)^2$.

Solution: From the graphs we note that the formulas in (a) and (d) define functions.

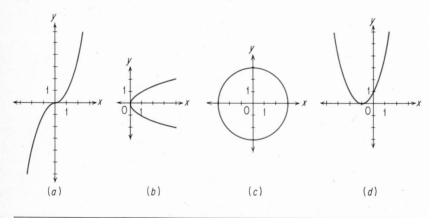

(a) (b) (c) (d)

Given any relation, the **inverse relation** is the set of ordered pairs obtained by interchanging the elements of each of the ordered pairs of the given relation. Thus, suppose we are given the relation

$$R = \{(2, 1), (3, 2), (4, 3), (4, 5)\}.$$

The inverse relation of R then consists of the set of ordered pairs

$$\{(1, 2), (2, 3), (3, 4), (5, 4)\}.$$

In a similar manner, we may find the inverse of a function, inasmuch as a function is nothing more than a special type of relation. That is, the inverse of a function is obtained by interchanging the elements of each of the ordered pairs of numbers that comprise the function. The inverse of a function is a relation; however, it may or may not be a function. It will be a function if and only if, for each y of the original function, we have at most one x.

Example 4: Find the inverse of the function

$$F = \{(1, 2), (2, 3), (3, 4)\}.$$

Solution: The inverse of this function is another function F', where

$$F' = \{(2, 1), (3, 2), (4, 3)\}.$$

Example 5: Find the inverse of the function

$$A = \{(1, 3), (2, 4), (3, 4), (4, 7)\}.$$

Solution: The inverse A' of this function is a relation, but it is not a function,

$$A' = \{(3, 1), (4, 2), (4, 3), (7, 4)\}.$$

If a relation is given by a formula, then a formula for its inverse may be obtained by interchanging the variables.

Example 6: Find a formula for the inverse of the function

$$F = \{(x, y) \mid y = 2x\}.$$

Solution: The inverse of F is $F' = \{(x, y) \mid x = 2y\}$. We may also write it as

$$F' = \left\{(x, y) \mid y = \frac{x}{2}\right\}.$$

Here the inverse of the given function is also a function.

Example 7: Find the inverse of $F = \{(x, y) \mid y = x^2\}$.

Solution: $\{(x, y) \mid x = y^2\}$. We may also write the solution as $\{(x, y) \mid y = \pm \sqrt{x}\}$. Note that the inverse of the given function is a relation but it is not a function. Explain why not.

---

1. If $f(x) = x - 2$, find **(a)** $f(1)$; **(b)** $f(2)$; **(c)** $f(17)$.
2. If $f(x) = x^2 - 4x + 5$, find **(a)** $f(0)$; **(b)** $f(2)$; **(c)** $f(-3)$.
3. If $g(x) = x^3 - 7$, find **(a)** $g(2)$; **(b)** $g(-1)$; **(c)** $g(-2)$.
4. If $f(x) = |x| + x$, find **(a)** $f(-2)$; **(b)** $f(0)$; **(c)** $f(3)$.

Graph each relation, identify the relations that are functions, and state the domain and range for each one that is a function. In each case $U = \{1, 2, 3\}$.

5. $y = x$
6. $y = x + 1$
7. $y < x$
8. $y < x + 1$

Proceed as in Exercises 5 through 8 for the universe of real numbers:

9. $y = x + 1$
10. $y = x - 1$
11. $y = |x - 1|$
12. $y = |x + 1|$
13. $y = |x| + 1$
14. $y = |x| - 1$
15. $y \geq x$
16. $y \leq x - 1$
17. $y = -x^2$
18. $y = (x + 1)^2$

Find the inverse of each function. Tell whether or not the inverse is a function.

19. $F = \{(1, 1), (2, 2), (3, 3)\}$
20. $F = \{(1, 2), (2, 2), (3, 2)\}$
21. $F = \{(1, 3), (2, 4), (3, 4)\}$
22. $F = \{(1, 3), (2, 2), (3, 1)\}$
23. $F = \{(x, y) \mid y = x + 1\}$
24. $F = \{(x, y) \mid y = x - 2\}$

Graph each function as a dotted curve and the inverse of the function as a solid curve. State whether or not the inverse of each function is a function.

25. $y = x + 2$
26. $y = x - 2$
27. $y = x$
28. $y = 2x$
29. $y = x^2$
30. $y = |x|$

PEDAGOGICAL CONSIDERATIONS

1. Many elementary texts now introduce the concept of function and relation on an intuitive basis through the use of arrow diagrams. For example, each of the following shows a correspondence (relation) between two sets. In each case the arrow indicates which members of the second set (range) correspond to the given elements of the first set (domain). Which of the relations are functions?

(a)

(b)

(c)

(d)

(e)

(f)

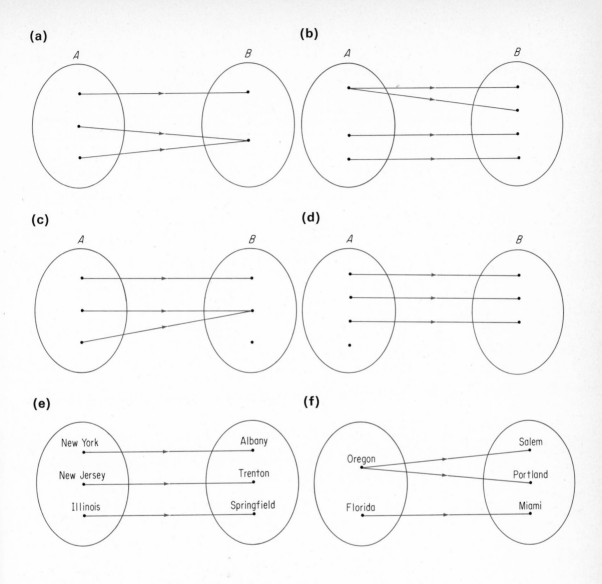

2. A mapping of a set *A* *onto* another set *B* is a correspondence where each element of *A* is mapped to a unique element of *B*, and each element of *B* is the image of at least one element in *A*. Draw several different arrow diagrams that illustrate such a mapping, both abstract and concrete as in Exercise 1.

3. A *one-to-one* mapping of a set *A* onto a set *B* is one where each element of set *A* corresponds to a unique element of set *B*, and each element of set *B* is the image of a unique element in set *A*. Draw an arrow diagram that illustrates such a mapping as in Exercise 1.

9-8

linear programming Graphs of linear statements in two variables provide a very important tool for solving modern problems. Although mathematicians are developing theories for statements that are not necessarily linear, we shall consider only the linear case. Thus we assume that the conditions of a problem have been represented by or approximated by linear statements. Then the solution of the problem depends upon the solution of a system (that is, a set) of linear statements. The usual method of solution is by graphing; that is, the method is geometrical. Accordingly, we first consider two examples of graphs of systems of linear statements in two real variables.

Example 1: Graph the system $x \geq 0$, $y \geq 0$, $y \geq x - 1$.

Solution: The solution set of the given system consists of the ordered pairs of real numbers (x, y) that satisfy all three of the statements. To graph the system, we graph each one of the three statements and then take the intersection of their graphs.

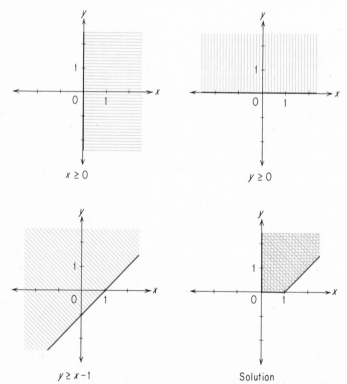

As in §9-3 we graph an inequality by first graphing the equality (using a solid line if its graph is part of the solution set, a dashed line otherwise).

Example 2: Graph the system

$$x \geq 0;\ y \geq 0;\ x \leq 6;\ y \leq 7;\ x + y \leq 10.$$

Solution:

$x \geq 0, \ x \leq 6$

$y \geq 0, \ y \leq 7$

$x + y \leq 10$

Solution

In Examples 1 and 2 the solutions of the systems of statements are known as **polygonal convex sets.**

In a linear programming problem we not only must set up our conditions, but we also must maximize or minimize an expression for profit, cost, or other quantity. Suppose the conditions for Example 2 represent the manufacture of x metal boxes and y glass jars in a given time, and $x + 2y$ represents the manufacturer's profit. If the manufacturer wishes to have a profit of $14, we then graph the equation $x + 2y = 14$ over the conditions of Example 2.

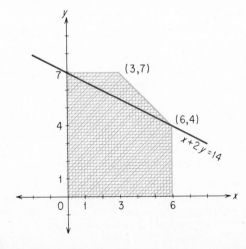

325

Note that there are now many ways in which the manufacturer can earn a profit of $14. In particular, consider the points $(0, 7)$ and $(6, 4)$. Thus he can earn $14 by manufacturing 0 metal boxes and 7 glass jars, or 6 metal boxes and 4 glass jars. Indeed, any point on the line $x + 2y = 14$ that is within the polygonal region represents an ordered pair (x, y) under the stated restrictions, and such that the profit is $14.

Next consider the same example, but this time we wish the profit to be k. For each value k the graph of $x + 2y = k$ is a straight line. As k takes on different values, we have a set of parallel lines. When several of these lines are graphed with the solution of the conditions in Example 2, we see that under these conditions k may have any value from 0 to 17 inclusive. The maximum (largest) value that is possible for k under these conditions is 17 and occurs at $(3, 7)$. The minimum (smallest) possible value for k is 0 and occurs at $(0, 0)$. Recall that the conditions are for the manufacture of x metal boxes and y glass jars in a given time and that $x + 2y$ represents the manufacturer's profit. Then if we assume that the manufacturer can sell all that he can make, he would make the most profit by manufacturing 3 boxes and 7 jars per unit of time.

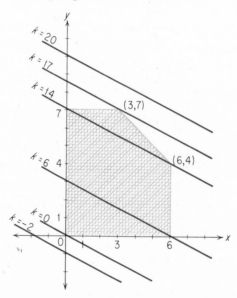

In any linear programming problem the maximum and minimum always occur at a vertex (possibly at two vertices, that is, along a side) of the polygonal region. Intuitively, the reason is that the region is convex and thus the lines of a set of parallel lines first intersect the region either by passing through a vertex, as in our example, or by passing along a side of the region. Accordingly, in our example, we could have found the maximum value of $x + 2y$ for points of the region by testing the values corresponding to the vertices $(0, 0)$, $(6, 0)$, $(6, 4)$, $(3, 7)$, and $(0, 7)$ of the region. The corresponding values of $x + 2y$ are 0, 6, 14, 17, and 14, respectively. Thus, as we observed before, the minimum value of $x + 2y$ is 0 and occurs as $(0, 0)$; the maximum value of $x + 2y$ is 17 and occurs at $(3, 7)$.

Example 3: Graph the system

$$x \geq 0, \qquad x \leq 3, \qquad y \leq 0, \qquad x - y \leq 5.$$

Solution:

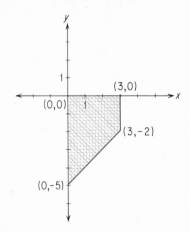

Example 4: Find the minimum and maximum values of the function, $2x + y$ defined over the system of Example 3.

Solution: Call the function f. Then at $(0, 0)$, $f = 0$; at $(0, -5)$, $f = -5$; at $(3, 0)$, $f = 6$; at $(3, -2)$, $f = 4$. The minimum value, -5, occurs at $(0, -5)$. The maximum value, 6, occurs at $(3, 0)$.

Example 5: A manufacturer produces gidgets and gadgets, and has his machines in operation 24 hours a day. To produce a gidget requires 2 hours of work on machine A and 6 hours of work on machine B. It takes 6 hours of work on machine A and 2 hours on machine B to produce a gadget. The manufacturer earns a profit of $5 on each gidget and $2 on each gadget. How many of each should he produce each day in order to earn the maximum profit possible?

Solution: If we let x represent the number of gidgets to be produced, and y the number of gadgets, then the conditions of the problem may be stated and graphed as follows:

$$x \geq 0$$
$$y \geq 0$$
$$2x + 6y \leq 24$$
$$6x + 2y \leq 24$$
$$\text{Profit} = P = 5x + 2y$$

327

We test the profit expression, $5x + 2y$, at the vertices of the polygon. At $(0, 0)$, $P = 0$; at $(0, 4)$, $P = 8$; at $(3, 3)$, $P = 21$; and at $(4, 0)$, $P = 20$. Thus to insure a maximum profit, the manufacturer should produce 3 gidgets and 3 gadgets daily.

EXERCISES

Graph each system.

1. $x \geq 2, x \leq 4, y \geq 0, x + y \leq 5$
2. $x \geq 0, y \geq 0, x + 3y \leq 9, 2x + y \leq 8$

Find the values of x and y such that under the set of conditions in the specified exercise the given expression has **(a)** a maximum value; **(b)** a minimum value.

3. $x + 2y$; Exercise 1.
4. $3x + y$; Exercise 1.
5. $x + y$; Exercise 2.
6. $x + 5y$; Exercise 2.

Use linear programming to solve these hypothetical problems.

7. A college is experimenting with a combination of teaching methods, using both teachers in the classroom and closed-circuit television. The college has facilities for handling five sections of a class at once using closed-circuit television. For the five sections of a class that meets three clock hours per week, the conditions appear to be as follows: The cost per minute of regular teaching is $5; the cost per minute of closed-circuit television is $3. For a certain week at most $750 can be spent on the instruction of these classes. Assume that the class hour can be spent in part with a teacher and in part with television. When neither the teacher nor the television is on, the students are free to discuss anything they wish. If the value to the students of x minutes of regular teaching and y minutes of closed-circuit television may be expressed as $3x + 2y$, how many minutes of regular teaching and how many minutes of closed-circuit television would be best for the students during the week?

8. Repeat Exercise 7 with the additional condition that the instructor must be present at least 30 minutes each week.

9. Use the conditions as for Exercise 8 and find the number of minutes of regular teaching and of closed-circuit television when the value to the students may be expressed as $x + 2y$.

10. Repeat Exercise 9 when the value to the students may be expressed as $2x + y$.

CHAPTER TEST

1. Find the solution set for the given universes:
   (a) $x + y = 5$, $U = \{1, 2, 3, 4\}$
   (b) $y < x - 2$, $U = \{-3, -2, -1, 0, 1, 2, 3\}$

2. Find the x-intercept and the y-intercept for the graph of each sentence:
   (a) $2x - y = 5$ \hspace{2cm} (b) $3x - 2y + 12 = 0$

In Exercises 3 through 7 graph for the universe of real numbers:

3. $\{(x, y) \mid x - 2y \geq 6\}$

4. $\{(x, y) \mid y \geq |x| - 2\}$

5. $\{(x, y) \mid x + y - 4 \geq 0\} \cap \{(x, y) \mid x - y + 1 \geq 0\}$

6. $\{(x, y) \mid x + y - 4 \geq 0\} \cup \{(x, y) \mid x - y + 1 \geq 0\}$

7. $y = (x + 2)^2 - 3$

8. Write the equation $y = x^2 + 4x - 1$ in standard form, then give the coordinates of the vertex and the equation of the axis of symmetry of the graph of the sentence.

9. If $f(x) = x^2 + 2x - 3$, find (a) $f(-2)$; (b) $f(1)$.

10. Graph each function and tell its domain and range:
    (a) $y = x - 1$ \hspace{2cm} (b) $y = |x - 1|$

10

Nonmetric Geometry

Geometry has evolved from a concern for earth measure (geo-metry), through the use of line segments and other figures to represent physical magnitudes, to a study of properties of sets of geometric elements. The figures serve as the elements of geometry. Relations among these elements and proofs of their properties from given sets of postulates are considered in more advanced courses.

All geometric figures are usually considered to be sets of points. Although other basic elements are considered in abstract geometry, we shall restrict our consideration to points. This is not a serious restriction, since points may be *interpreted* in many ways. For example, we usually think of a point as a position on a line, on a plane, or in space. We may also think of a point on a number scale in terms of its coordinate. We may even think of cities as points on a map, with the air routes joining them considered as lines. This freedom to interpret points in a variety of ways provides a basis for more abstract geometries.

The term **nonmetric geometry** is used for the study of geometric figures without using numbers as measures such as measures of angles, distance, area, and volume. In this chapter measures and coordinates are mentioned only when they are useful as an aid to our understanding of nonmetric geometry. All of the aspects of nonmetric geometry that we consider are also frequently considered in elementary school classes.

10-1

points, lines, and planes

We have an intuitive idea of what is meant by a point, but we have not and shall not *define* a point. In order to define a term, we need to distinguish it from other terms. In doing this, we must describe our term in words that are already known to us (that is, in simpler terms) in such a way that whenever the description is applicable, the term being defined is also appropriate.

The early Greeks described a point as "that which has no part." Today we recognize such a description as an aid to our interpretations, but we do not consider it a definition. We object to it as a formal definition because the terms used are not simpler terms and, indeed, simpler terms are not available.

Lines also are left undefined. In any logical system there must be some undefined terms. In other words, it is not possible to define everything; we must start with something. In geometry we start with points and lines. Then we define other figures in terms of these.

Even though lines are undefined, they do possess certain properties. Here are four of these properties:

1. *A line is a set of points.* Each point of the line is said to be *on* the line.

2. *Any two distinct points determine a unique line.* In other words, there is one and only one line *AB* through any two given points *A* and *B*. Thus a line may be named by any two of its points. If *C* and *D* are points on the line *AB*, then the line *CD* is the same as the line *AB*.

3. *Every line is a real number line.* Then as in §7-4 each line is a dense set of points and has infinitely many points.

4. *Any point A on a line separates the line into three parts: the point A and two half-lines, one on each side of A.* Each half-line is a set of points (§8-2). The point *A* is not a point (member) of either half-line.

> Half-line on the left   *A*   Half-line on the right

Three points *A, B, C* either lie on a line or do not lie on a line. Note that we think of (interpret) a "line" as a "straight line." If the points do not lie on the same straight line, then, since any two points determine a unique line, the three points determine three lines *AB, BC,* and *AC* as in the next figure.

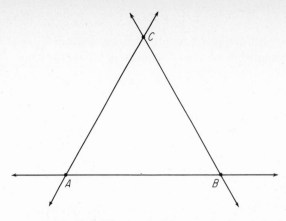

Each of the lines *AB*, *BC*, and *AC* is a set of points. Any two points on a line determine that line; for example, the points *D* and *E* on the line *BC* determine the line *BC*, and we say that *BC* and *DE* are two names for the same line. Any two points of the figure consisting of the lines *AB*, *BC*, and *AC* determine a line. If the points are not on the same one of these three lines, then the points determine a new line; for example, *D* and *F* determine a new line in the figure.

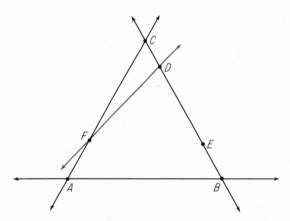

Many new lines can be determined in this way. A few such lines are shown in the next figure. The set of all points on such lines is the set of points of the plane determined by the three points *A*, *B*, and *C*.

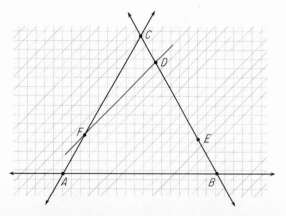

Planes are often accepted as undefined. However, all planes have certain basic properties. Here are five of these properties:

1. *A plane is a set of points.* Each point of the plane is said to be *on* the plane.

2. *Any three points that are not on the same line determine a unique plane.* In other words, there is one and only one plane through any three non-collinear points. For example, a piece of cardboard will balance firmly on the points of three thumbtacks; a three-legged stool is more stable than many four-legged chairs. We use this property when we name planes such as *ABC* by any three points that are on the plane but not on a straight line.

3. *If two distinct points of a line are on a plane, then every point of the line is on the plane.*

4. *Any line m on a plane separates the plane into three parts: the line m and two half-planes.* The points of the line *m* do not belong to either half-plane. However, the line *m* is often called the **edge** of both half-planes even though it is not a part of either.

Note that we may think of the points of the plane that are on the right of *m* as forming one half-plane and the points of the plane that are on the left of *m* as forming the other half-plane.

5. *Two distinct planes have either a line in common or no points in common.* In other words, any two distinct planes either intersect in a line or are parallel.

In ordinary geometry (often called **Euclidean geometry**) it is possible for two planes to fail to have a point in common. Two planes that do not have any point in common are said to be **parallel.** A line is parallel to a plane if it does not have any point in common with the plane.

Think of an ordinary classroom. On any given wall the line along the ceiling and the line along the floor do not appear to intersect, that is, do not have any point in common no matter how far they are extended. The line along the ceiling and one wall also does not appear to intersect the line of intersection of two other walls. Any distinction between these two situations must take into consideration the fact that in the first case the two lines were on the same wall (plane), whereas in the second case there could not be a single plane containing both lines. In general, two lines that are on the same plane and do not have any point in common are also said to be **parallel lines**; two lines that are not on the same plane are called **skew lines**; two distinct lines that have a point in common are called **intersecting lines**; lines that have all their points in common may be visualized as two names for the same line and are called **coincident lines**.

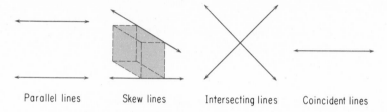

Parallel lines   Skew lines   Intersecting lines   Coincident lines

Note that the arrows indicate that the lines may be extended indefinitely, that is, that the figure involves representations for lines rather than actual lines. These representations are on the printed page and thus on a plane. Thus the figure for skew lines must be visualized as a "picture" on a plane of skew lines in space.

Points, lines, and planes are the building blocks of geometry. We use these elements and relations such as intersections, parallelism, coincidence, skewness, and separation as we define other geometric figures, that is, as we name other sets of points.

EXERCISES

1. Label three points *A*, *B*, and *C* that are not on the same line. Draw and name each of the lines that are determined by these points.

2. Repeat Exercise 1 for four points *A*, *B*, *C*, and *D* such that no three of the points are on the same line.

3. Assume that the four points in Exercise 2 are not on the same plane. Name each of the planes that is determined by these points.

4. Assume that all of the given points are on the same plane and no three of the given points are on the same line. State how many lines are determined by the given number of points: **(a)** 2; **(b)** 3; **(c)** 4; **(d)** 5; **(e)** 6; ***(f)** 10; ***(g)** *n*.

5. Assume that all of the given lines are on the same plane, that each line intersects each of the other lines, and that no three of the given lines are on the same point. State how many points are determined by the given number of lines: (a) 2; (b) 3; (c) 4; (d) 5; (e) 6; *(f) 10; *(g) n.

6. Consider the lines along the edges of the rectangular solid shown in the given figure and state whether the specified lines are parallel lines, skew lines, or intersecting lines:

(a) $\overleftrightarrow{AB}$ and $\overleftrightarrow{CD}$ (b) $\overleftrightarrow{AB}$ and $\overleftrightarrow{EF}$

(c) $\overleftrightarrow{AB}$ and $\overleftrightarrow{EH}$ (d) $\overleftrightarrow{CD}$ and $\overleftrightarrow{FG}$

Use pencils, table tops, etc. to represent lines and planes and determine whether each statement appears to be true or false in Euclidean geometry.

7. Given any line m and any point P that is not a point of m, there is exactly one line t that is parallel to m and contains P.

8. Given any line m and any point P that is not a point of m, there is exactly one line q such that q contains P and m and q are intersecting lines.

9. Given any line m and any point P that is not a point of m, there is exactly one line s such that s contains P and s and m are skew lines.

10. Given any line m and any point P that is not a point of m, there is exactly one plane that is parallel to m and contains P.

11. Given any plane ABC and any point P that is not a point of ABC, there is exactly one plane that contains P and intersects ABC.

12. Given any plane ABC and any point P that is not a point of ABC, there is exactly one plane that contains P and is parallel to ABC.

13. Given any plane ABC and any point P that is not a point of ABC, there is exactly one line that contains P and intersects ABC.

14. Given any plane ABC and any point P that is not a point of ABC, there is exactly one line that contains P and is parallel to ABC.

15. Given any plane ABC and any line m that is parallel to ABC, there is exactly one plane that contains m and intersects ABC.

16. Given any plane ABC and any line m that is parallel to ABC, there is exactly one plane that contains m and is parallel to ABC.

17. Given any plane ABC and any line m that is parallel to ABC, there is for any given point P that is not on m and not on ABC exactly one line that is parallel to m and also parallel to ABC.

Many interesting figures may be obtained by paper folding. Wax paper is especially well suited for these exercises.

18. Draw a line *m* and select a point *P* that is not on *m*. Fold the point *P* onto points of *m* in at least 20 different positions spaced as evenly as possible. Describe the pattern formed by the folds.

19. Draw a circle *C* and select a point *P* outside the circle. Fold the point *P* onto points of *C* in at least 20 different positions spaced as evenly as possible. Describe the pattern formed by the folds.

20. Repeat Exercise 19 for a point *P* that is inside the circle but not at the center of the circle.

21. Repeat Exercise 19 for the point *P* at the center of the circle.

22. Repeat Exercise 19 for a point *P* on the circle.

PEDAGOGICAL CONSIDERATIONS

Arithmetic, geometry, and algebra are basic parts of mathematics. The teaching of a topic from any of these three usually involves, at least informally, concepts from the other two. However, in the following items it is hoped that the reader will focus his attention on the geometric aspects of each exercise.

1. Read three articles from three different issues of *The Arithmetic Teacher* and write a brief report on each. Be sure that at least one of the three articles is concerned with the teaching of geometry in the elementary grades.

2. The famous French mathematician, René Descartes, concluded his book *La Géométrie* with this comment:

   I hope that posterity will judge me kindly, not only as to the things that I have explained, but also as to those which I have intentionally omitted so as to leave to others the pleasure of discovery.

Prepare a 15 minute lesson plan that attempts to present some geometric concept to an elementary class through a discovery approach.

3. List at least five specific techniques for motivating the learning of geometry at some specific elementary grade level.

4. Visit an elementary classroom and observe what provisions, if any, are made for individual differences of ability. See what you can find in the literature on this topic and prepare a report summarizing your findings.

5. In an attempt to provide for individual differences of ability in mathematics, some schools have adopted the Individually Prescribed Instruction (IPI) program developed at the University of Pittsburgh.

Others have attempted to make use of programmed instruction techniques. Find out what you can about either of these approaches, and prepare a report on your findings.

6. Dr. Jerome Brunner, Harvard psychologist, has said that one can teach almost anything to any child at any level in some intellectually honest form. Comment upon this statement, either in agreement or in opposition, as it pertains to the teaching of elementary geometry.

7. Beginning in 1970 *The Arithmetic Teacher* instituted a new section entitled "Ideas." This section consists of worksheets that can be used in elementary mathematics classrooms to allow students to engage in activities that lead to the formation of concepts. Study several of these and then develop two similar worksheets on nonmetric geometry for a specific grade level.

8. In recent years a great deal of attention has been given to the learning theories of Jean Piaget, a Swiss psychologist, concerning mathematics instruction. Find several references to Piaget's work in the literature and prepare a report on his findings concerning the learning of geometry.

10-2

rays,
line
segments,
and
angles

Suppose that we are given a line m and two points A and B on m. There is one and only one line m on both A and B. We call this the line AB and write it as $\overleftrightarrow{AB}$. As in §10-1 the point A separates the line m into two half-lines. Note that one of these half-lines contains the point B and one does not.

The set of points consisting of the half-line AB and the point A is called a **ray** (§8-2). The point A is the **endpoint** of the ray. There is one and only one ray with endpoint A and containing the point B. We call this the ray AB and write it as $\overrightarrow{AB}$. There is also a BA with B and containing A.

The **line segment** AB, written as $\overline{AB}$, consists of the points which $\overrightarrow{AB}$ and $\overrightarrow{BA}$ have in common. (See §8-2.) We may use the symbol ∩ for the intersection of two sets of elements (in this case, points) and write

$$\overline{AB} = \overrightarrow{AB} \cap \overrightarrow{BA}.$$

A line AB has no endpoints and may be named by any two of its points. A ray AB has one endpoint and may be named by that point A and any other point on the ray. A line segment AB has the two endpoints A and B and is named by those two points. However, for a line segment those two points may be stated in either order; that is, $\overline{AB} = \overline{BA}$.

When a direction, such as from A to B, is assigned to the points of a line segment, we obtain a **vector,** $\overrightarrow{AB}$. The vector AB has both direction and length. The direction is from A to B; the length is the same as the length of the line segment AB. Notice that a ray $\overrightarrow{AB}$ has only one endpoint A; a vector $\overrightarrow{AB}$ has two endpoints, an **initial point** A and a **terminal point** B.

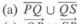
Vector AB

Any point of a line segment AB that is not an *endpoint* (that is, any point that is distinct from A and B) is called an **interior point** of $\overline{AB}$. If E is an interior point of AB, then E is **between** A and B. Thus the line segment AB consists of the endpoints A and B and the points of the line AB that are between A and B. The interior points of a line segment AB form the **open line segment** AB:

Open line segment AB

Example 1: Consider the given figure and identify each set of points.

(a) $\overline{PQ} \cup \overline{QS}$ (b) $\overline{PR} \cap \overline{QS}$
(c) $\overrightarrow{QR} \cup \overrightarrow{SR}$ (d) $\overrightarrow{PQ} \cap \overrightarrow{QS}$

Solution: (a) $\overline{PS}$; (b) $\overline{QR}$; (c) $\overleftrightarrow{PQ}$ (any one of several names for the line may be used); (d) $\overrightarrow{QS}$ (this ray may also be named $\overrightarrow{QR}$).

Any line AB is a set of points. If C is a point of the line such that A is between C and B, then the line is the union of the rays AB and AC. These two rays have a common endpoint A and are both on the line AB. Any figure formed by two rays that have a common endpoint is a **plane angle.** These rays may be on the same line but need not be on the same line. Each angle in the figure may be designated as $\angle BAC$; $\angle BAC = \overrightarrow{AB} \cup \overrightarrow{AC}$.

339

sec. 10-2

rays,

line

segments,

and angles

The rays *AB* and *AC* are called **sides** of ∠*BAC*. For the angle in the following figure there exist points *P* and *Q* of the angle such that the open line segment $\overset{\circ\!-\!\circ}{PQ}$ does not intersect the angle. All points of such open line segments are **interior points** of the angle.

Notice that an angle such as ∠*LMN* does not have any interior points, since $\overline{PQ} \subset ∠LMN$ for all points *P* and *Q* of the angle. Persons who have previously studied geometry may observe that any angle that has an interior has a measurement between 0° and 180°. For this reason some elementary books consider only angles whose measurements are greater than 0° and less than 180°.

If an angle has an interior, then the angle separates the plane into three disjoint sets: the points of the angle, the interior points of the angle, and the **exterior points** of the angle.

Example 2: Consider the given figure and identify each set of points.

(a) $\overleftrightarrow{AB} \cap \overleftrightarrow{DE}$

(b) $\overrightarrow{BC} \cup \overrightarrow{BD}$

(c) $∠CBD \cap \overrightarrow{ED}$

(d) $\overline{AF} \cap \overrightarrow{BD}$

(e) (Exterior ∠*ABE*) $\cap \overrightarrow{BD}$

(f) (Interior ∠*CBE*) $\cap \overleftrightarrow{AC}$

Solution: (a) The two lines intersect in the single point *B*; (b) ∠*CBD*;

(c) $\overrightarrow{BD}$; (d) ∅; (e) $\overset{\circ}{\overrightarrow{BD}}$; (f) ∅.

EXERCISES

Consider the given figure and identify each set of points.

A B C D

1. $\overline{AB} \cup \overline{BD}$ 2. $\overline{AB} \cap \overline{CD}$ 3. $\overline{AC} \cap \overline{BD}$
4. $\overline{AB} \cap \overline{BC}$ 5. $\overline{AD} \cap \overline{BC}$ 6. $\overline{AD} \cup \overline{BC}$
7. $\overleftrightarrow{AB} \cap \overline{BC}$ 8. $\overleftrightarrow{AB} \cap \overline{BC}$ 9. $\overrightarrow{BC} \cup \overrightarrow{CD}$
10. $\overrightarrow{BC} \cap \overrightarrow{CD}$ 11. $\overrightarrow{BC} \cap \overrightarrow{BA}$ 12. $\overrightarrow{DC} \cup \overrightarrow{CA}$

Consider the given figure and identify each set of points.

13. (Exterior $\angle ABC$) $\cap \overrightarrow{BD}$
14. (Interior $\angle ABC$) $\cap \overrightarrow{BD}$
15. (Interior $\angle ABC$) $\cap$ (exterior $\angle DBC$)
16. (Interior $\angle ABC$) $\cup$ (exterior $\angle ABC$) $\cup \angle ABC$
17. (Exterior $\angle DBC$) $\cap \overrightarrow{BA}$
18. (Exterior $\angle DBC$) $\cap \overline{AB}$

Consider the given figure and identify each set of points.

19. (Exterior $\angle BPC$) $\cap \overrightarrow{CP}$
20. (Exterior $\angle BPA$) $\cap \overleftrightarrow{AP}$
21. $\overrightarrow{PB} \cup \overrightarrow{PC}$
22. $\angle BPC \cap \angle PBA$
23. $\angle BPC \cup \angle BPA$
24. (Exterior $\angle BPC$) $\cap \overline{AC}$

SUPPLEMENTARY EXERCISES

Consider the given figure and identify each set of points.

1. $\overrightarrow{BC} \cap \overline{AD}$ 2. $\overline{BC} \cap \overleftrightarrow{AD}$ 3. $\overline{BC} \cap \overleftrightarrow{AD}$
4. $\overrightarrow{CA} \cup \overrightarrow{CD}$ 5. $\overline{CA} \cup \overline{CD}$ 6. $\overline{CA} \cap \overline{CD}$

341

sec. 10-2
rays,
line
segments,
and angles

7. $\overleftrightarrow{AB} \cap \overline{BD}$ 8. $\overleftrightarrow{AB} \cap \overline{BD}$ 9. $\overleftrightarrow{AB} \cup \overrightarrow{BD}$

10. $\overrightarrow{BD} \cup \overrightarrow{DB}$ 11. $\overrightarrow{BD} \cap \overrightarrow{DB}$ 12. $\overrightarrow{BD} \cap \overrightarrow{DB}$

Consider the given figure and identify each set of points.

13. $\angle ADB \cap \angle BDC$
14. (Exterior $\angle ADB$) $\cap$ (exterior $\angle BDC$)
15. (Interior $\angle ADB$) $\cap$ (interior $\angle BDC$)
16. $\angle ADB \cap \overrightarrow{DB}$
17. $\angle ADB \cap \overrightarrow{BD}$
18. $\angle ADB \cap$ (exterior $\angle BDC$)
19. $\overrightarrow{DB} \cup$ (interior $\angle ADB$) $\cup$ (interior $\angle BDC$)
20. (Exterior $\angle ADC$) $\cap \overrightarrow{DB}$
21. (Exterior $\angle ADB$) $\cup$ (exterior $\angle ADC$)

PEDAGOGICAL CONSIDERATIONS

Games may be used to present many mathematical concepts very effectively. Frequently slight modifications of the rules for a particular game lead to another very different game. The following games are described in terms of points and line segments on a piece of paper. However other procedures such as pegs and elastics on a peg board may be used instead.

1. Make several dots for points on a piece of paper as in the figure. Two players take turns drawing line segments with the given dots as endpoints. No line segment may cross another or contain more than two of the given points. The last person able to play is the winner.

2. As in Exercise 1 but with the added restriction that no point may be the endpoint of more than two line segments.

3. As in Exercise 1 but with the added restriction that no point may be the endpoint of more than three line segments.

4. As in Exercise 1, if you begin with three dots and your opponent has the first move, can you win?

10-3

plane
figures

We now continue our study of geometric figures on a plane; that is, sets of points on a plane. Consider the four figures in the accompanying diagram, where each is a union of line segments. Notice that in each case there seem to be three separate parts. A part may consist of one line segment or be the

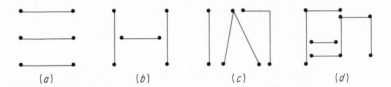

union of two or more line segments. However, you could pick out the three parts of each figure. You recognized that any two distinct parts were not "connected"; that you could not draw them both without removing the point of the pencil from the paper to move from one to the other. We now assume that the word "connected" is understood even though it has not been defined.

Each of the three figures in the next diagram is a union of line segments, and each figure is connected. Each may be drawn by starting at one of the points *A*, *B*, and ending at the other.

Each of the next five figures is a union of line segments, each is connected, each may be drawn as a continuous line by starting at one of the points *R*, *S* and ending at the other, and each differs from the figures *A*, *B* in the preceding set. See if you can identify that difference.

342

Any connected union of line segments is a broken line. Compare the three broken lines *AB* with the five broken lines *RS*. Each broken line *AB* may be drawn starting at *A* and ending at *B* without retracing any line segments or arriving at any point a second time. Each broken line *RS* may be drawn starting at *R* and ending at *S* without retracing any line segments, but for each broken line *RS* some point, possibly the starting point, must be used twice in the drawing. Briefly, we say that each broken line *RS* intersects itself. That is, each broken line *RS* contains at least one point that can be approached along at least three "paths." Such a point may appear, for example, as in any of these figures:

The word "simple" is used to distinguish between figures that intersect themselves and figures that do not intersect themselves. A figure such as one of the broken lines *AB* which does not intersect itself is a **simple figure**. Figures such as triangles, squares, circles, and rectangles are also simple figures. A figure such as one of the broken lines *RS* that does intersect itself is *not a simple figure*.

The broken lines *AB* and *RS* differ from each of the next two figures. Each of these next figures is simple; each is connected; each is a union of line segments. Each differs from the previous figures in that when it is drawn one starts at a point and returns to that point. We say that such a figure is **closed**.

Figures may be either simple or not simple; they may also be either closed or not closed. For example, the figure *LMNO* is simple and closed; the figure *PQRS* is closed and not simple; the broken lines *AB* are simple and not closed; the broken lines *RS* are neither simple nor closed.

A simple closed broken line on a plane is called a **polygon**. The line segments are called **sides** of the polygon; the endpoints of the line segments are called **vertices** of the polygon. A polygon is a **convex polygon** if every line segment *PQ* with points of the polygon as endpoints is either a subset of a side of the polygon or has only the points *P* and *Q* in common with the polygon. A polygon that is not a convex polygon is a **concave polygon**.

Convex polygon

Concave polygon

If P and Q are points of a convex polygon and $\overline{PQ}$ is not a subset of a side of the polygon, then the points of the open line segment PQ are **interior points** of the convex polygon. The union of the points of a convex polygon and its interior points is a **polygonal region**. The points of the plane of a convex polygon that are not points of the polygonal region are **exterior points** of the polygon.

Interior points   Polygonal region   Exterior points

A polygon with three sides is a **triangle**. For example, triangle ABC consists of the points of $\overline{AB} \cup \overline{BC} \cup \overline{CA}$. The **interior** of triangle ABC consists of the points in the intersection of the interiors of the three angles, $\angle ABC$, $\angle BCA$, and $\angle CAB$. The **exterior** of triangle ABC consists of the points of the plane ABC that are neither points of the triangle nor points of the interior of the triangle.

Triangle ABC Interior of triangle ABC Exterior of triangle ABC

In general, plane polygons are classified according to the number of sides. The common names are shown in this array:

Number of sides	Name of polygon
3	triangle
4	quadrilateral
5	pentagon
6	hexagon
7	heptagon
8	octagon
9	nonagon
10	decagon
12	dodecagon

Example: Consider the given figure and identify each set of points: (a) $\overrightarrow{AB} \cap \overrightarrow{ED}$; (b) $\angle BAE \cap \angle BCD$; (c) (interior $\triangle ACE$) $\cap$ (exterior $\angle EFD$); (d) three angles such that the intersection of their interior points is the interior of $\triangle ABF$.

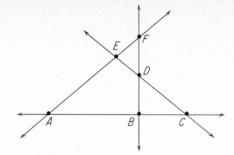

Solution: (a) $\{C\}$; (b) $\overleftrightarrow{AC} \cup E$; (c) interior $\triangle BCD$; (d) $\angle ABF$, $\angle BFA$, and $\angle FAB$.

EXERCISES

Sketch a union of line segments that is:

1. Connected.
2. Not connected.

Sketch a broken line that is:

3. Simple but not closed.
4. Closed but not simple.
5. Simple and closed.
6. Neither simple nor closed.
7. A convex polygon of five sides.
8. A concave polygon of six sides.

Consider the given figure and identify each set of points.

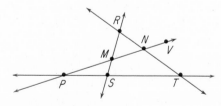

9. $\overleftrightarrow{MN} \cap \overrightarrow{TS}$
10. $\overrightarrow{SR} \cap \overrightarrow{ST}$
11. $\triangle RST \cap \overline{PM}$
12. (Exterior $\triangle RST$) $\cap \overline{MN}$
13. $\overrightarrow{TS} \cap \overline{PM}$
14. $\overrightarrow{PV} \cap \triangle RST$
15. (Interior $\triangle RST$) $\cap \overline{MN}$
16. $\overline{RS} \cap \overline{ST}$
17. $\overline{RM} \cup \overline{MN} \cup \overline{RN}$
18. (Exterior $\triangle RST$) $\cap \overrightarrow{NM}$

19. (Exterior $\triangle RST$) $\cap \overleftrightarrow{MN}$

20. $\overrightarrow{PM} \cup \overrightarrow{PS}$

21. $\overleftrightarrow{PV} \cap \triangle MRN$

22. $\angle SRT \cap \angle RTS$

23. (Interior $\triangle RST$) $\cap$ (interior $\triangle MRN$)

24. (Interior $\triangle RST$) $\cup$ (interior $\triangle MRN$)

25. (Interior $\angle VPT$) $\cap$ (interior $\triangle RST$)

26. (Interior $\angle SRT$) $\cap$ (interior $\angle RTS$)

27. (Interior $\angle RST$) $\cap$ (interior $\angle STR$) $\cap$ (interior $\angle TRS$)

28. (Interior $\triangle RMN$) $\cup$ (exterior $\triangle RMN$) $\cup$ $\triangle RMN$

SUPPLEMENTARY EXERCISES

Sketch a union of line segments that is:

1. Not a polygon.

2. A convex polygon of seven sides.

3. A concave polygon of seven sides.

4. Not simple, not closed, and not connected.

Consider the given figure and identify each set of points.

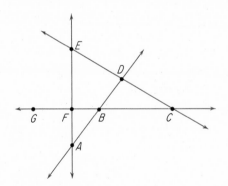

5. $\overrightarrow{GF} \cap \overrightarrow{DE}$

6. $\overrightarrow{CG} \cup \overrightarrow{CE}$

7. $\triangle CFE \cap \overrightarrow{AB}$

8. $\triangle ABF \cap \triangle BCD$

9. $\triangle ABF \cap \overrightarrow{CG}$

10. $\triangle CBD \cap \triangle CFE$

11. (Interior $\triangle CBD$) $\cap$ (interior $\triangle CFE$)

12. (Interior $\triangle ABF$) $\cap$ (exterior $\triangle CBD$)

13. (Exterior $\triangle ABF$) $\cap \overrightarrow{BF}$

14. (Exterior $\triangle ABF$) $\cap \overleftrightarrow{CD}$

15. (Interior $\triangle CFE$) $\cap \overleftrightarrow{BD}$

1. Study a contemporary series of elementary school mathematics books and identify the grade level(s) at which exercises equivalent to each of the exercises and supplementary exercises of this section are considered.

2. Name as many different types of convex polygons as possible with vertices at the indicated number of points on a plane:

(a) 2   **(b)** 3   **(c)** 4   **(d)** 5   **(e)** 6   **(f)** 10

10-4

space figures   Any two points A and B that are on a line but not on the same point determine a unique line segment AB. Any three points A, B, and C that are on a plane but not on the same line determine a unique triangle ABC. Any four points A, B, C, and D that are in space but not on the same plane determine a unique **tetrahedron**. The four points A, B, C, D are the **vertices** of the tetrahedron; the six line segments $\overline{AB}$, $\overline{AC}$, $\overline{AD}$, $\overline{BC}$, $\overline{BD}$, $\overline{CD}$ are the **edges** of the tetrahedron.

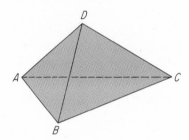

The triangular regions of the four triangles ABC, BCD, CDA, and ABD are the **faces** of the tetrahedron. The **interior points** of the tetrahedron are the points of the open line segments PQ, where P and Q are points of the tetrahedron (points of the union of the faces of the tetrahedron) and $\overset{\circ\,\,\circ}{PQ}$ does not intersect the tetrahedron. The points of the faces of the tetrahedron form a space figure that is also called a **triangular pyramid.** Any one of the triangular regions may be taken as the base of the pyramid with the remaining vertex as the **vertex** of the pyramid.

On a plane a polygon is a simple, closed union of line segments, that is, a union of line segments (sides of the polygon) such that exactly two sides have each endpoint of a line segment (vertex of the polygon) in common and no side contains an interior point of another side. In space a **polyhedron** is a simple, closed union of polygonal regions (**faces** of the polyhedron) such that exactly two faces have each side of a polygonal region (**edge** of the polyhedron) in common and no face contains an interior point of another face. The vertices of the polygonal regions are also called **vertices** of the polyhedron.

Each of these figures is a polyhedron. The **cube** has square faces and is a convex polyhedron. The rectangular box (**parallelepiped**) has rectangular faces and is a convex polyhedron. The third figure has two faces (top and bottom) that are not convex polygons and is not a convex polyhedron. The last figure appears to be formed by two triangular pyramids with their common base removed. This figure has six triangular faces and is a convex polyhedron. This last figure differs from the others in that each of the others has all of its vertices in two parallel planes, whereas this figure does not.

Prism

A polyhedron with all of its vertices in two parallel planes is a **prismatoid**. If the edges that are not in the parallel planes are parallel edges, then the polyhedron is a **prism**. The faces of the prism in the two parallel planes are the **bases** of the prism. The parallel edges joining the two bases are **lateral edges**. In more advanced courses it can be shown that the bases of a prism are congruent polygons. Notice in the sketches that the "hidden" edges are indicated by dotted line segments.

A polyhedron with all but one of its vertices on a plane is a **pyramid**. The vertices on a plane are the vertices of the **base** of the pyramid. Prisms and pyramids are often classified as triangular, rectangular, square, and so forth according to their bases.

Pyramid

EXERCISES

1. The tetrahedron *MNOP* is a triangular pyramid. Name its **(a)** vertices; **(b)** edges; **(c)** faces.

2. Repeat Exercise 1 for the **square pyramid** *PQRST*.

3. Repeat Exercise 1 for the cube *ABCDEFGH*.

4. Think of the lines along the edges of the cube in Exercise 3 and identify the lines that appear to be **(a)** parallel to $\overleftrightarrow{AB}$; **(b)** skew to $\overleftrightarrow{AB}$; **(c)** parallel to the plane *ABFE*.

5. A cube has all of its vertices on two parallel planes and has square faces in those planes. Accordingly, a cube is an example of a **square prism**. Repeat Exercise 1 for the **triangular prism** *JKLMNO*.

6. Let *V* be the number of vertices, *E* the number of edges, and *F* the number of faces. Then copy and complete the following table.

Figure	V	E	F
Triangular pyramid			
Square pyramid			
Cube			
Triangular prism			

In Exercises 7 through 13 extend the table in Exercise 6 for:

7. A pyramid with a pentagon as a base.

8. A prism with a pentagon as a base.

9. A prism with a hexagon as a base.

10. A pyramid with a hexagon as a base.

11. The given figure which represents two triangular pyramids with their common base removed.

12. The given figure *FGHIJK* which represents two square pyramids with their common base removed.

13. The given figure *ABCDEFGHI* which represents a cube with one face replaced by the four triangular faces of a square pyramid.

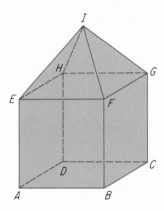

14. In the tables for Exercises 6 through 13 study the values of V, E, and F for each polyhedron. Compare the value of E with the value of $V + F$. Then conjecture a formula for E in terms of V and F. You should obtain the famous *Euler formula for polyhedrons*.

Consider the given figures and identify the:

15. Prisms. 16. Pyramids.

17. Triangular pyramids. 18. Square prisms.

(a)

(b)

(c)

(d)

(e)

(f)

(g)

(h)

350

Sketch:

19. A triangular pyramid.

20. A triangular prism.

21. A square pyramid.

22. A square prism.

23. A prismatoid with one square base and one triangular base.

24. A convex polyhedron that is not a prismatoid.

25. A polyhedron that is not convex.

26. A plane that intersects a cube in **(a)** a square; **(b)** a triangle; **(c)** a rectangle that is not a square.

27. A plane that intersects a square pyramid in **(a)** a square; **(b)** a triangle; **(c)** a trapezoid that is not a square.

28. A polyhedron with eight triangular regions as its faces.

29. A polyhedron with six triangular regions as its faces.

Consider cubes and their interior points.

30. A cube has two-inch edges and is painted red. Think of this two-inch cube as cut into eight one-inch cubes using three planes, one parallel to each pair of parallel faces of the cube. How many of the one-inch cubes have six painted faces? Five painted faces? Four painted faces? Three painted faces? Two painted faces? One painted face? No painted faces?

31. Repeat Exercise 30 for a painted three-inch cube divided into 27 one-inch cubes.

*32. Repeat Exercise 30 for a painted four-inch cube divided into 64 one-inch cubes.

*33. Repeat Exercise 30 for a painted five-inch cube divided into 125 one-inch cubes.

*34. Repeat Exercise 30 for a painted n-inch cube, $n \geq 2$, divided into n^3 one-inch cubes.

PEDAGOGICAL CONSIDERATIONS

1. Describe the treatment in a contemporary series of elementary school mathematics books of exercises like the ones in this section.

2. The "pictures" (sketches, drawings) of space figures on a sheet of paper never completely represent the figure and are often hard to visualize. Consider the accompanying pattern and make a model of a tetrahedron.

Tetrahedron

3. As in Exercise 2 consider the pattern and make a model of a cube.

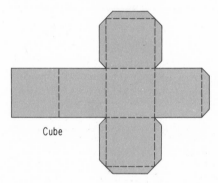

Cube

4. Name as many different types of polyhedrons as possible with vertices at the indicated number of points:

(a) 3   **(b)** 4   **(c)** 5   **(d)** 6

10-5

plane   Throughout this chapter we have considered points, lines, planes, rays, line
curves   segments, figures composed of line segments and rays, and regions of planes
determined by line segments, rays, and half-planes. We now consider plane
figures that may be approximated by broken lines and polygons; that is, we
consider **plane curves**. Each of the following figures is a **simple closed curve**.
Notice that a polygon is a special case of a simple closed curve.

Any two simple closed curves have a common property. Can you find that property? Size, shape, angles, and straight lines are certainly not common properties here. What is left to be considered? Some readers may consider the common property that we are seeking trivial because of its simplicity. We prefer to believe that the really basic properties of mathematics are inherently simple. The property that we have in mind may be stated very easily: Any simple closed curve has an inside and an outside.

In advanced mathematics we say that *any simple closed curve in a plane divides the plane into two regions.* This is the **Jordan curve theorem.** The curve is the common boundary of the two regions, and one cannot cross from one region to the other without crossing the curve. The Jordan curve theorem is a very powerful theorem and yet a very simple one. Notice that it is independent of the size or shape of the curve.

How can such a simple theorem have any significance? It provides a basis for Euclid's assumption that any line segment joining the center of a circle to a point outside the circle must contain a point of the circle. In more advanced courses it provides a basis for the existence of a zero of a polynomial on any interval on which the polynomial changes sign. It provides a basis for Euler diagrams and Venn diagrams in any two-valued logic. In §1-6 we assumed that all points of the universal set were either points of a circular region A or points of A', the complement of A.

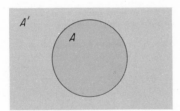

In §2-1 we considered statements p that are either true or false (in which case $\sim p$ is true) but not both true and false. Because of the Jordan curve theorem we were able to represent the situations under which statements were true (or false) by Euler diagrams and by Venn diagrams.

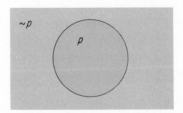

This simple theorem regarding the existence of an inside and an outside of any simple closed curve may also be used to answer questions raised by the next problem. This problem is a popular one, and many people have spent hours working on it. Consider three houses (×) in a row and three utilities (∘) in a second row on a plane surface. The problem is to join each house to each utility by an arc on the plane in such a way that no two arcs cross or pass through houses or utilities except at their endpoints. As in the figure on the left it is easy to designate paths from one house to each of the three utilities. One can also designate paths to each utility from the second house as in the second figure. Then one can designate two of the paths from the third house, but it is not possible on an ordinary plane to draw the path from the remaining house to the remaining utility. This assertion is based upon the fact that the simple closed curve indicated in the third figure divides the plane into two regions; the third house is inside the curve (shaded region), the remaining utility is outside the curve, and the two cannot be joined without crossing the curve.

As in the case of broken lines, plane curves may be either simple or not simple and either closed or not closed. Each of the figures that follow is a plane curve that may be approximated by a closed broken line that is not simple. Each curve is a closed curve that is not simple.

Closed curves that are not simple divide the plane into three or more regions. The points at which a curve intersects itself are **vertices**; also, any other points may be designated as vertices. The simple curves with vertices as endpoints and not containing any other vertices are **arcs**. Each of the following curves has three points designated as vertices. The numbers V of vertices, A of arcs, and R of regions into which the plane is divided are shown below each figure. The last figure leaves the plane as a single region, since, for any two points P and Q that are points of the plane but not points of the figure, there is a simple plane curve (arc) with endpoints P and Q that does not intersect the given figure.

V	3	3	3	3	3
A	3	3	4	5	2
R	2	2	3	4	1

The next figures may be approximated by simple broken lines that are not closed. They are called simple curves that are not closed. Notice that a line and a ray are special cases of simple curves that are not closed; thus in geometry curves do not have to be "crooked."

The next figures may be approximated by a broken line that is neither simple nor closed; each is called a curve that is neither simple nor closed.

EXERCISES

Sketch three curves that are different in appearance and are each:

1. Simple and closed.   2. Simple and not closed.
3. Closed and not simple.   4. Neither simple nor closed.

Sketch a broken line that closely approximates a given:

5. Simple closed curve.
6. Simple curve that is not closed.
7. Closed curve that is not simple.
8. Polygon.

In Exercises 9 through 14 find the numbers V of vertices, A of arcs, and R of regions for each figure.

9. (a)   (b)

10. (a)   (b)

11. (a)　　　　　　　　　　　(b)

12. (a)　　　　　　　　　　　(b)

13. (a)　　　　　　　　　　　(b)

14. (a)　　　　　　　　　　　(b)

15. Study the values of *V*, *A*, and *R* obtained in Exercises 9 through 14 and find an expression for *A* in terms of *V* and *R*. You should obtain a form of the *Euler formula for networks*.

PEDAGOGICAL CONSIDERATIONS

1. Describe the treatment in a contemporary series of elementary school mathematics books of exercises like the ones in this section.

2. Find a way of describing to third graders that the above curve is traversable. (*Hint:* Consider ways of drawing a "likeness" that is obviously traversable.)

10-6

networks

Any set of line segments (or arcs) forms a **network.** If the network can be drawn by tracing each line segment exactly once without removing the point of the pencil from the paper, the network is **traversable.** For example, any connected simple network is traversable. If such a network is closed, any point may be selected as a starting point for the drawing and this point will also be the terminating point of the drawing. If the connected simple network is not closed, the drawing must start at one of the endpoints and terminate at the other.

The study of the traversability of networks probably stemmed from a problem concerning the bridges in the city of Königsberg. There was a river flowing through the city, two islands in the river, and seven bridges as in the figure. The people of Königsberg loved a Sunday stroll and thought it would

be nice to take a walk that would take them across each bridge exactly once. They found that no matter where they started or what route they tried, they could not cross each bridge exactly once. Gradually it was observed that the basic problem was concerned with paths between the two sides of the river *A*, *B*, and the two islands *C*, *D* as in the figure. With this representation of

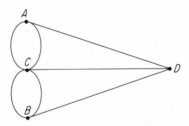

the problem by a network, it was no longer necessary to discuss the problem in terms of walking across the bridges. Instead one could discuss whether or not the network associated with the problem was traversable. The problem was solvable if and only if its network was traversable.

When is a network traversable? One can walk completely around any city block, and it is not necessary to start at any particular point to do so. In general, one may traverse any simple closed broken line in a single trip. We next consider walking around two blocks and down the street $\overline{BE}$ separating them. This problem is a bit more interesting in that it is necessary to start at B or E. Furthermore, if one starts at B, one ends at E, and conversely.

Note that it is permissible to pass through a vertex several times, but one may traverse a line segment only once. The peculiar property of the vertices B and E is based upon the fact that each of these endpoints is an endpoint of three line segments, whereas each of the other vertices is an endpoint of two line segments. A similar observation led a famous mathematician named Euler to devise a complete theory for traversable networks.

Euler classified the vertices of a network as odd or even. For example, in the given figure the vertex A is an endpoint of three arcs $\overline{AB}$, $\overline{AC}$, and $\overline{AD}$ and thus is an odd vertex; B is an endpoint of two arcs $\overline{BA}$ and $\overline{BC}$ and is an even vertex; C is an endpoint of four arcs $\overline{CB}$, $\overline{CA}$, $\overline{CD}$, and $\overline{CE}$ and is an even vertex; D is an endpoint of two arcs $\overline{DA}$ and $\overline{DC}$ and is an even vertex; E is an endpoint of one arc $\overline{EC}$ and is an odd vertex. Thus the figure has two odd vertices A and E and three even vertices B, C, and D. For any network a vertex that is an endpoint of an odd number of arcs is an **odd vertex**; a vertex that is an endpoint of an even number of arcs is an **even vertex**. Since each arc has two endpoints, there must be an even number of odd vertices in any network.

Any network that has only even vertices is traversable, and the trip may be started at any vertex. Furthermore, the trip will terminate at its starting point. If a network has exactly two odd vertices, it is traversable, but the trip must start at one of the odd vertices and will terminate at the other. If a network has more than two odd vertices, it is not traversable. In general, a network with $2k$ odd vertices may be traversed in k distinct trips. The network for the Königsberg bridge problem has four odd vertices and thus is not traversable in a single trip. Notice that the Königsberg bridge problem is independent of the size and shape of the river, bridges, or islands.

The study of simple closed curves, networks, and other plane curves that we have considered is a part of a particular geometry called **topology.**

EXERCISES

In Exercises 1 through 8 identify **(a)** *the number of even vertices;* **(b)** *the number of odd vertices;* **(c)** *whether or not the network is traversable and, if it is traversable, the vertices that are possible starting points.*

1.

2.

3.

4.

5.

6.

7. The network formed by the edges of a triangular pyramid.

8. The network formed by the edges of a cube.

9. Explain how a highway inspector can use a network to determine whether or not he can inspect each section of a highway without traversing the same section of the highway twice.

10. Use a related network to obtain and explain your answer to the following question: Is it possible to take a trip through a house with the floor plan indicated in the figure and pass through each doorway exactly once?

11. Explain whether or not it is possible to draw a simple connected broken line cutting each line segment of the given figure exactly once.

12. Consider the use of highway networks by a salesman who wants to visit each town exactly once. (a) Does the salesman always need to travel each highway at least once? (b) Consider each of the networks in Exercises 1 through 4 and indicate whether or not the salesman can visit each town exactly once without retracing any highway. (c) On the basis of your answer for part (b), does traversability of the network appear to be of interest to the salesman?

*13. Describe a modification that would make the Königsberg bridge problem possible.

*14. Describe a modification of the network for Exercise 11 such that the possibility of the desired construction is changed.

PEDAGOGICAL CONSIDERATIONS

There are many figures in topology that have very unusual properties. One of these is a surface that is one-sided. A fly can walk from any point on this surface to any other point without crossing an edge. Unlike a table top or a wall, it does not have a top and a bottom or a front and a back. This surface is called a **Möbius strip,** and it may be very easily constructed from a rectangular piece of paper such as a strip of gummed tape. Size is theoretically unimportant, but a strip an inch or two wide and about a foot long is easy to handle. We may construct a Möbius strip by twisting the strip of gummed tape just enough (one half-twist) to stick the gummed edge of one end to the gummed edge of the other end. If we cut across this strip, we again get a single strip similar to the one we started with. But if we start with a rectangular strip and cut around the center of the Möbius strip (see the dotted line in the second figure), we do not get two strips. Rather, we get one strip with two half-twists in it.

On one occasion one-sided surfaces of this sort were used as place cards at a seven-year-old's birthday party. While waiting for dessert, the youngsters were encouraged to cut the strip down the middle while guessing what the result would be. They were suitably impressed when they found only one piece, and were anxious to cut it again. Once more they were impressed when they found two pieces linked together. Almost a year after the party, one of the boys asked about the piece of paper that was in only one piece after it was cut in two.

When confronted with unusual properties such as those of the Möbius strip, both children and adults may ask questions that the nearest teacher cannot answer and that many college mathematics professors cannot answer. This is good for all concerned, since it impresses upon them that there is more to mathematics than formal algebraic manipulations and classical geometric constructions.

1. Construct a Möbius strip and cut around the center to obtain one strip with two half-twists in it.

2. Repeat Exercise 1 and then cut around the center again.

3. Construct a Möbius strip and cut along a path that is about one-third of the distance from one edge to the other.

4. Construct a Möbius strip, mark a point A on it, and draw an arc from A around the strip until you return to the point A.

5. Explain why a Möbius strip is called a one-sided surface.

6. Does a Möbius strip have one or two simple closed curves as its edge? Explain the reason for your answer.

CHAPTER TEST

1. Consider the given figure and identify each set of points: (a) $\overline{BD} \cap \overline{CA}$; (b) $\overline{CD} \cup \overline{CB}$.

2. Sketch and label two rays $\overrightarrow{PQ}$ and $\overrightarrow{RS}$ such that $\overrightarrow{PQ} \cap \overrightarrow{RS}$ is (a) a line segment; (b) a ray.

3. Sketch (a) a simple curve that is not closed; (b) a convex polygon with five sides.

4. Consider the six lines along the edges of a triangular pyramid $ABCD$ as in the figure. Identify the lines that **(a)** intersect $\overleftrightarrow{AB}$ in a single point; **(b)** form skew lines with $\overleftrightarrow{AB}$.

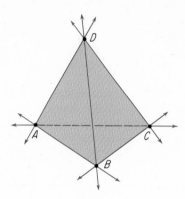

5. Consider the given figure and identify each set of points:

(a) $\angle ABC \cap \angle CDE$ **(b)** (interior $\triangle ABC$) $\cap \overleftrightarrow{DE}$

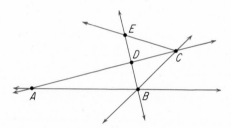

6. Think of a triangular prism and state the number of **(a)** vertices; **(b)** edges.

7. Sketch: **(a)** a curve that is neither simple nor closed; **(b)** a network with two odd vertices and two even vertices.

8. Consider the given figure and state whether or not each network is traversable. If a network is traversable, identify the vertices that are possible starting points.

(b)

(c)

(d)

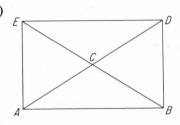

Identify as true (T, always true) or false (F, not always true).

9. **(a)** There is one and only one line that is perpendicular to a given plane and contains a given point.
   (b) There is one and only one plane that is perpendicular to a given plane and contains a given line.
10. **(a)** Any two lines intersect or are parallel.
    (b) Any two planes intersect or are parallel.

11

Metric
Geometry

Geometric figures may be studied in terms of the intersections of lines, rays, and planes as in nonmetric geometry (Chapter 10). Geometric figures may also be studied using lengths of line segments, measures of angles, area measures, and volume measures. The study of geometric figures making use of measures is often called **metric geometry**.

The three chapters on nonmetric geometry (Chapter 10), metric geometry (Chapter 11), and coordinate geometry (Chapter 12) are concerned with three basic approaches to the study of geometry. The chapter headings indicate the primary emphasis in each chapter. Other geometries are considered in each chapter whenever they add to the reader's understanding of the topic under consideration. In other words, our study of geometry is not strictly compartmentalized.

Elementary school students are using metric geometry when, for example, they speak of an acute angle (note dependence on the measure of the angle), an isosceles triangle (note dependence on measures of sides of the triangle), a rectangle, or a square. In this chapter metric geometry is used to introduce additional geometric figures and properties of geometric figures. Most of these figures and properties are already considered in elementary schools.

11-1

linear measures

Distances and lengths of objects are often called **linear measurements** since the shortest distance from one point to another is the length of the line segment joining them. Whenever distances or lengths are stated, some unit of measure needs to be given or assumed. For example, the length of a certain walking stick might be described by various people as

> a little more than a yard
> about 4 feet
> about 1 meter
> 1.15 meters
> 45 inches
> $1\frac{1}{4}$ yards
> 5 spans

A variety of units of length are considered in schools along with the relations among these units. Since very few people have difficulties with the study of units of linear measure, we simply assume that some unit has been selected and use a scale as on a number line.

Any line may be considered as a number line (§10-1). Then each point P_x on the line has a unique real number x as its *coordinate;* each real number x has a unique point P_x as its *graph.* Since each of the points of a number line may be identified by *one* real number, a line is often called a **one-space.**

Sets of points on a line form **line figures.** For example, any two points P_r and P_s on a number line are endpoints of a unique line segment $\overline{P_rP_s}$ with the absolute value of $s - r$ as its **linear measure.** We denote the linear measure of $\overline{P_rP_s}$ by $m(\overline{P_rP_s})$ (read as "measure of $\overline{P_rP_s}$") and write

$$m(\overline{P_rP_s}) = |s - r|.$$

Example 1: Find the linear measure of (a) $\overline{P_4P_1}$; (b) $\overline{P_{-2}P_3}$.

Solution: (a) $m(\overline{P_4P_1}) = |1 - 4| = |-3| = 3$;
  (b) $m(\overline{P_{-2}P_3}) = |3 - (-2)| = |5| = 5.$

Any two line segments with equal measures are called **congruent line segments.** Since any line segment has a unique linear measure, any line segment is congruent to itself; for example, $\overline{AB} \cong \overline{BA}$ (read as "line segment AB is congruent to line segment BA").

Measures of line segments on a number line are often used in the discussion of addition of whole numbers. In general, any line segment $\overline{AB}$ has a nonnegative real number $m(\overline{AB})$ as its linear measure. The **length** of the line segment is $m(\overline{AB})$ units for some unit of linear measure such as foot, inch, meter, mile, or kilometer. For example, a pencil with a linear measure of 3 in inches has length 3 inches.

A point has linear measure 0. In the figure $m(\overline{AB}) = 3$, $m(\overline{CD}) = 4$, $\overline{AB} \cap \overline{CD} = \overline{CB}$, $m(\overline{CB}) = 1$, $\overline{AB} \cup \overline{CD} = \overline{AD}$, $m(\overline{AD}) = 6$. Notice that $m(\overline{AB}) + m(\overline{CD}) \neq m(\overline{AD})$, since $m(\overline{AB} \cap \overline{CD}) \neq 0$.

$$m(\overline{AD}) = m(\overline{AB} \cup \overline{CD}) = m(\overline{AB}) + m(\overline{CD}) - m(\overline{AB} \cap \overline{CD})$$

In general, if the measure of the intersection of two line segments is 0, then the linear measure of the union of the two line segments is equal to the sum of their linear measures.

Example 2: Show the relationship among the linear measures of $\overline{P_1P_2}$, $\overline{P_2P_5}$, and $\overline{P_1P_5}$.

Solution: $m(\overline{P_1P_2}) = 1$, $m(\overline{P_2P_5}) = 3$,
$m(\overline{P_1P_2} \cap \overline{P_2P_5}) = m(\overline{P_2P_2}) = 0$,
$m(\overline{P_1P_5}) = 4 = m(\overline{P_1P_2}) + m(\overline{P_2P_5})$.

Example 3: Show the relationship among the linear measures of $\overline{P_{-2}P_3}$, $\overline{P_2P_5}$, and $\overline{P_{-2}P_5}$.

Solution: $m(\overline{P_{-2}P_3}) = 5$, $m(\overline{P_2P_5}) = 3$, $m(\overline{P_{-2}P_5}) = 7$,
$m(\overline{P_{-2}P_3} \cap \overline{P_2P_5}) = m(\overline{P_2P_3}) = 1$,
$m(\overline{P_{-2}P_5}) = m(\overline{P_{-2}P_3}) + m(\overline{P_2P_5}) - m(\overline{P_{-2}P_3} \cap \overline{P_2P_5})$.

The *linear measure* of a figure is a number n; the *length* of the figure is n units. The length of a simple closed curve, such as a polygon, is its **perimeter**. Since any two sides of a polygon have at most one point in common, the perimeter of the polygon is the sum of the lengths of its sides.

Example 4: Find the perimeter of the given polygon.

Solution: The perimeter in inches of polygon $ABCDE$ is

$$m(\overline{AB}) + m(\overline{BC}) + m(\overline{CD}) + m(\overline{DE}) + m(\overline{AE}) =$$
$$5 + 4 + 7 + 7 + 3 = 22.$$

Thus the perimeter is 22 inches.

The measurement of a line segment to determine its length is done with some standard scale such as a ruler marked off in inches. The reading of the linear measure from the ruler requires an estimation to the nearest unit used by the person making the measurement. For example, consider the measurement of $\overline{PQ}$ using the scale below it.

To the nearest unit $m(\overline{PQ}) = 6$. We write $m(\overline{PQ}) \approx 6$ (read as "the measure of the line segment PQ is approximately 6") to indicate that the length of $\overline{PQ}$ is about 6 units. Some people mentally subdivide the unit of length on a scale and *estimate* measures to smaller units. For example, different people might say:

> $m(\overline{PQ}) \approx 6$ to the nearest half-unit.
> $m(\overline{PQ}) \approx 6\frac{1}{4}$, that is, 6 and $\frac{1}{4}$ to the nearest quarter unit.
> $m(\overline{PQ}) \approx 6.2$, that is, 6 and $\frac{2}{10}$ to the nearest tenth of a unit.

Whatever subdivisions are used, the measure is based upon someone's "reading the scale" and thus is approximate. All linear measurements are approximate since they are based upon estimation from reading a scale. Indeed, all measurements are approximate. Even though a measurement is made with a very precise instrument, there is always a final estimation and thus an approximation.

If the length is 6 inches to the nearest inch, the greatest possible error is $\frac{1}{2}$ inch. If the length is $6\frac{1}{4}$ inches to the nearest quarter of an inch, the greatest possible error is one-half of the unit $\frac{1}{4}$ inch, and thus $\frac{1}{8}$ inch. If the length is 6.2 inches to the nearest tenth of an inch, the greatest possible error is one-half of the unit 0.1 inch, and thus 0.05 inch. For any measurement the **greatest possible error** is one-half of the smallest unit used.

There exist special techniques for computations with approximate numbers, that is, numbers that represent approximations. We shall leave such special techniques for more advanced courses. Thus we treat numbers obtained from measurements the same as other numbers and simply use the **approximation symbol** ≈ to indicate that the numbers are approximate rather than cardinal numbers of sets of individual elements such as 5 books or 72 people.

EXERCISES

On a number line find the linear measure of:

1. $\overline{P_1P_5}$

2. $\overline{P_{-1}P_4}$

3. $\overline{P_{-1}P_{-4}}$

4. $\overline{P_2P_{-3}}$

5. $\overline{P_{-7}P_{-1}}$

6. $\overline{P_3P_{-2}}$

Show the relationship among the linear measures of:

7. $\overline{P_2P_5}$, $\overline{P_5P_7}$, and $\overline{P_2P_7}$.

8. $\overline{P_{-3}P_2}$, $\overline{P_2P_5}$, and $\overline{P_{-3}P_5}$.

9. $\overline{P_1P_4}$, $\overline{P_2P_7}$, and $\overline{P_1P_7}$.

10. $\overline{P_{-4}P_0}$, $\overline{P_{-1}P_3}$, and $\overline{P_{-4}P_3}$.

11. $\overline{P_1P_{-3}}$, $\overline{P_{-2}P_4}$, and $\overline{P_{-3}P_4}$.

12. $\overline{P_5P_2}$, $\overline{P_5P_6}$, and $\overline{P_2P_6}$.

13. $\overline{P_4P_{-1}}$, $\overline{P_{-1}P_2}$, and $\overline{P_2P_4}$.

14. $\overline{P_{-1}P_{-5}}$, $\overline{P_{-5}P_{-2}}$, and $\overline{P_{-3}P_{-1}}$.

15. $\overline{P_2P_{-3}}$, $\overline{P_{-3}P_5}$, and $\overline{P_0P_5}$.

Find the perimeter of each given polygon.

16.

17.

18.

19.

Express the measure of $\overline{PQ}$:

20. To the nearest unit.

21. To the nearest half-unit.

22. To the nearest quarter unit.

23. Find the greatest possible error of the measurements in Exercises 20 through 22.

Which of these measurements is probably approximate?

24. I have 27 students in my class.

25. My breakfast has 627 calories.

26. Portland has a population of 67,500.

27. There are 527 pages in my history book.

28. Bobby is 18 months old.

PEDAGOGICAL CONSIDERATIONS

1. Select an elementary school grade level from 3 to 6. Then list as many uses of linear measures as you can think of for students at that grade level.

2. Look at mathematics and science books for the grade selected in Exercise 1 and try to extend your list. Be sure to include the word problems as you look through books.

3. Prepare a bulletin board display on the history of measurement.

4. Although a 12 inch ruler has each inch marked, not all of these divisions are necessary. For example, the 8 inch ruler below will measure any number of inches from 1 through 8. For example, the distance between the markings for 1 inch and 5 inches can be used to measure 4 inches.

Construct a drawing of a 12 inch ruler with the fewest possible markings that will allow you to measure any length from 1 inch to 12 inches.

5. Elementary school students can be introduced to the concept of the approximate nature of measurement through the use of arbitrary units of measure. For example, they can be asked to measure various line segments using lima beans as the unit of measure. Then consider other possible units, such as postage stamps. Prepare a 30 minute lesson plan on this topic for a particular grade level. As an example of this approach, see Unit 4 of Volume I of *Experiences In Mathematical Ideas*, published by the National Council of Teachers of Mathematics.

11-2

angular measures The union of any two rays $\overrightarrow{PQ}$ and $\overrightarrow{PR}$ with a common endpoint P is a *plane angle* with the rays as *sides* and the point P as *vertex*. We

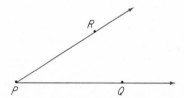

write $\angle QPR = \angle RPQ$ to indicate that "$\angle QPR$" and "$\angle RPQ$" are two names for the same angle. We use "$m\angle QPR$" to denote the measure of the angle in degrees or some other specified unit of angular measure.

The measure of an angle may be considered as an amount of rotation from one side to the other. The first side is called the **initial side**; the other side is called the **terminal side**. By convention the positive direction of rotation is taken counterclockwise. Unless otherwise indicated, the amount of rotation is assumed to be less than a complete rotation, that is, one revolution. The common units of angular measure are related as follows:

$$1 \text{ revolution} = 360 \text{ degrees} = 4 \text{ right angles}$$

Occasionally other units of angular measure are used; for example:

$$1 \text{ revolution} = 2 \text{ straight angles} = 2\pi \text{ radians}$$

Radians are used extensively in advanced work. One radian is about 57 degrees.

When an angle is represented but an initial side is not indicated, the measure is usually taken as a positive number of degrees and as small as possible. Thus in the figure $m\angle ABC = 60$ unless curved arrows are added to indicate the initial side and a direction of rotation that requires a different angular measure.

Protractors are used to measure angles just as rulers are used to measure line segments. One common type of protractor is shown in the figure.

Several common types of angles are defined in terms of their angular measures. An angle ABC is

an **acute angle** if $0 < m\angle ABC < 90$
a **right angle** if $m\angle ABC = 90$
an **obtuse angle** if $90 < m\angle ABC < 180$
a **straight angle** if $m\angle ABC = 180$.

Some textbooks postpone the use of the term straight angle and refer to the rays $\overrightarrow{BA}$ and $\overrightarrow{BC}$ as **opposite rays** when their union is a straight line.

Acute angle   Right angle   Obtuse angle   Straight angle or opposite rays

Any two angles with equal angular measures are called **congruent angles**. Since any angle has a unique angular measure, any angle is congruent to itself; for example $\angle QPR \cong \angle RPQ$ (read as "angle QPR is congruent to angle RPQ").

Congruent angles are used extensively in the study of geometric figures. For example, if two straight lines intersect, then four angles are formed. In the figure there are $\angle TRV$, $\angle VRS$, $\angle SRV$, and $\angle URT$. The sides of

$\angle TRV$ are opposite rays of the sides of $\angle SRV$ and the angles are **vertical angles**. In a formal course in geometry it can be proved that:

Any two vertical angles are congruent.

Example 1: Draw straight angles *ABC* and *DBE*. Then identify two pairs of vertical angles and two angles that are congruent to ∠*ABD*.

Solution: ∠*ABD* and ∠*CBE* are vertical angles; ∠*ABE* and ∠*CBD* are vertical angles; ∠*ABD* ≅ ∠*CBE*, ∠*ABE* ≅ ∠*CBD*.

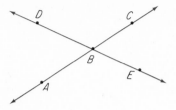

If the measure in degrees is known for any one of the angles formed when two straight lines intersect, then the measure of each of the four angles can be specified. Suppose in the figure for Example 1 that $m\angle EBC = 55$. Then $m\angle ABD = 55$ since vertical angles are congruent. Notice that ∠*EBD* is a straight angle and a rotation about *B* from the ray $\overrightarrow{BE}$ to the ray $\overrightarrow{BD}$ could be considered as a rotation from $\overrightarrow{BE}$ to $\overrightarrow{BC}$ followed by a rotation from $\overrightarrow{BC}$ to $\overrightarrow{BD}$. Therefore

$$m\angle EBC + m\angle CBD = m\angle EBD = 180.$$

Since $m\angle EBC = 55$,

$$55 + m\angle CBD = 180, \qquad m\angle CBD = 125.$$

Any two angles with measures that add up to 180 are called **supplementary angles**; ∠*EBC* and ∠*CBD* are supplementary angles. However, these two angles also have a special relationship since they have a common side $\overrightarrow{BC}$ and the other two sides form a straight angle. Any two such angles are **adjacent supplementary angles**. Since ∠*EBC* and ∠*EBA* are also adjacent supplementary angles, $m\angle EBA = 125$. This checks with the fact that ∠*EBA* and ∠*CBD* are vertical angles. Thus each angle of the figure can be labeled with its measurement.

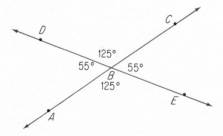

372

Next consider straight angles $\angle CAD$ and $\angle EBF$, where $A \neq B$ as in the figure. The line $\overleftrightarrow{AB}$ is a **transversal** of the lines $\overleftrightarrow{CD}$ and $\overleftrightarrow{EF}$.

Points H and G have been selected on $\overleftrightarrow{AB}$ such that $\angle BAH$ and $\angle ABG$ are straight angles. Then

$\angle DAH$ and $\angle FBA$ are **corresponding angles.**
$\angle CAH$ and $\angle EBA$ are corresponding angles.
$\angle CAB$ and $\angle EBG$ are corresponding angles.
$\angle DAB$ and $\angle FBG$ are corresponding angles.
$\angle DAH$ and $\angle EBG$ are **alternate exterior angles.**
$\angle HAC$ and $\angle GBF$ are alternate exterior angles.
$\angle CAB$ and $\angle FBA$ are **alternate interior angles.**
$\angle DAB$ and $\angle EBA$ are alternate interior angles.

In a formal course in geometry it can be proved that:

If the angles of any pair of corresponding angles, alternate interior angles, or alternate exterior angles are congruent, then the two given lines (in the above figure $\overleftrightarrow{CD}$ and $\overleftrightarrow{EF}$) are parallel.

If the two given lines are parallel, then the angles of each pair are congruent.

Example 2: In the figure $\angle DCE$ is a straight angle and $\overleftrightarrow{DE} \parallel \overleftrightarrow{AB}$. Name an angle with vertex C that is congruent to (a) $\angle BAC$; (b) $\angle ABC$.

Solution: The line $\overleftrightarrow{AC}$ is a transversal of the parallel lines $\overleftrightarrow{AB}$ and $\overleftrightarrow{DE}$. Since alternate interior angles are congruent, (a) $\angle BAC \cong \angle DCA$; (b) $\angle ABC \cong \angle ECB$.

EXERCISES

If possible, sketch and label. If not possible, explain why it is not possible.

1. An acute angle *ABC*.

2. Two acute angles *CDE* and *FGH* that are not congruent.

3. A right angle *IJK*.

4. Two right angles *LMN* and *OPQ* that are not congruent.

5. An obtuse angle *RST*.

6. Two obtuse angles *UVW* and *XYZ* that are not congruent.

7. Two straight angles *ABC* and *DEF* that are not congruent.

8. Vertical angles *FGH* and *IGJ*.

Consider the given figure and identify:

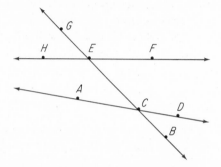

9. Four pairs of corresponding angles.

10. Two pairs of alternate interior angles.

11. Two pairs of alternate exterior angles.

Copy the given figure and label each angle with its measurement.

12.

13.

14.

15.

PEDAGOGICAL CONSIDERATIONS

It is important that elementary school students gain practice in estimating the measure of an angle. Worksheets can be prepared that show angles in various positions, with the student asked to estimate the measure of each angle in degrees to the nearest 5 or 10 degrees. Teachers also need to practice this skill. For the figures below, first estimate the size of each angle to the nearest 5 degrees. Then check your guess by using a protractor to measure each angle.

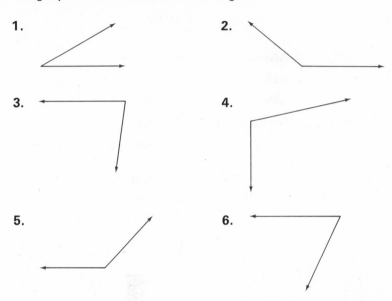

1.   2.

3.   4.

5.   6.

11-3

classifica-
tions of
polygons

Polygons were considered in §10-3 as simple closed unions of line segments on a plane. Then polygons were classified according to the number of sides as triangles, quadrilaterals, pentagons, hexagons, and so forth. Measures of line segments may be used to extend these classifications.

A triangle may be classified according to the lengths of its sides as a (an)

> **scalene triangle** if no two sides are congruent;
> **isosceles triangle** if at least two sides are congruent;
> **equilateral triangle** if all three sides are congruent.

Note that every equilateral triangle is an isosceles triangle, but not every isosceles triangle is an equilateral triangle.

A triangle may be classified according to the measures of its angles as a (an)

acute triangle if all three angles are acute angles;
right triangle if one angle is a right angle;
obtuse triangle if one angle is an obtuse angle;
equiangular triangle if all three angles are congruent.

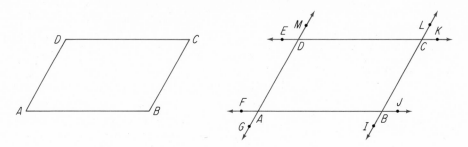

A quadrilateral $ABCD$ is a **parallelogram** if its sides are on two pairs of parallel lines, that is, $\overrightarrow{AB} \parallel \overleftrightarrow{CD}$ and $\overleftrightarrow{BC} \parallel \overrightarrow{AD}$. As in the second figure, $\overrightarrow{AD}$ is a transversal of $\overrightarrow{AB}$ and $\overleftrightarrow{CD}$ with corresponding angles $\angle BAD$ and $\angle CDM$; $\overleftrightarrow{CD}$ is a transversal of $\overrightarrow{AD}$ and $\overleftrightarrow{BC}$ with alternate interior angles $\angle CDM$ and $\angle BCD$; $\overleftrightarrow{BC}$ is a transversal of $\overrightarrow{AB}$ and $\overleftrightarrow{CD}$ with alternate interior angles $\angle BCD$ and $\angle CBJ$. If $m\angle BAD = 60$, then $m\angle CDM = 60$, $m\angle BCD = 60$, and $m\angle CBJ = 60$, as in §11-2, and each angle may be labeled with its measurement.

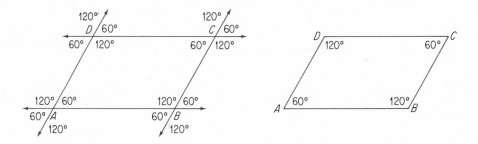

For the parallelogram $ABCD$ we may refer to the angles by their vertices and have $m\angle A = 60$, $m\angle B = 120$, $m\angle C = 60$, $m\angle D = 120$ as in the figure. Angles such as $\angle A$ and $\angle B$ with their sides on a common side $\overline{AB}$ of the quadrilateral are called **consecutive angles**. Angles of the quadrilateral such as $\angle A$ and $\angle C$ that are not consecutive angles are called **opposite angles**. The above procedure may be used for any parallelogram to prove that:

Any two opposite angles of a parallelogram are congruent angles.
Any two consecutive angles of a parallelogram are supplementary angles.

Two sides of a quadrilateral are **consecutive sides** if they have a common endpoint. Two sides of a quadrilateral that are not consecutive sides are called **opposite sides**. Thus for the parallelogram $ABCD$ the sides $\overline{AB}$ and $\overline{CD}$ are opposite sides, $\overline{AD}$ and $\overline{BC}$ are opposite sides, and in each case the opposite sides are parallel, that is, they are on parallel lines.

These definitions may be used to continue our classification of polygons. A quadrilateral is a **trapezoid** if at least one pair of its opposite sides are parallel. A quadrilateral is a **parallelogram** if each of its pairs of opposite sides are parallel. A parallelogram is a **rectangle** if at least one of its angles is a **right angle**. [Note that if one angle is a right angle, then we can show that all four angles are right angles. The definition only requires that at least one angle be a right angle since we want the explanations (proofs) of why parallelograms are rectangles to be as simple as possible.] A parallelogram is a **rhombus** if at least one pair of consecutive sides are congruent. (If at least one pair of consecutive sides are congruent, then it can be proved, as we shall see in Exercise 19 of §11-4, that all four sides are congruent.) A rectangle is a **square** if at least one pair of consecutive sides are congruent. (As in the case of the rhombus, all four sides can be proved to be congruent.) A rhombus is a *square* if at least one angle is a right angle. (As in the case of a rectangle, all four angles can be proved to be right angles.)

| Trapezoid | Parallelogram | Rectangle | Rhombus | Square |

EXERCISES

Sketch each figure and find its perimeter.

1. A square with an edge 5 inches long.

2. An equilateral triangle with a side 4 feet long.

3. A rectangle with sides 2 feet long and 7 feet long.

4. A triangle with sides 3 feet long, 5 feet long, and 6 feet long.

Find an expression for the linear measure of:

5. A square with an edge of linear measure e.

6. An equilateral triangle with a side of linear measure s.

7. A rectangle with sides with linear measures b and h.

8. A triangle with sides with linear measures a, b, and c.

Describe the change in the perimeter of:

9. A square when the measure of each of its edges is doubled.

10. A rectangle when the measure of each of its sides is multiplied by 3.

11. A triangle when the measure of each of its sides is multiplied by a constant k.

12. A polygon when the measure of each of its sides is multiplied by a constant h.

Consider the given figure and identify:

13. Two pairs of opposite sides.

14. Four pairs of consecutive sides.

15. Four pairs of consecutive angles.

16. Two pairs of opposite angles.

In Exercises 17 through 20 copy the given figure and label each angle with its measurement.

21. Explain why all four angles of a rectangle must be congruent.

SUPPLEMENTARY EXERCISES

If possible, sketch. If not possible, explain why it is not possible.

1. A scalene triangle.

2. An isosceles triangle.

3. An acute scalene triangle.

4. A right scalene triangle.

5. An obtuse scalene triangle.

6. An equiangular scalene triangle.

7. An acute isosceles triangle.

8. A right isosceles triangle.

9. An obtuse isosceles triangle.

10. An equiangular isosceles triangle.

11. An acute equilateral triangle.

12. An obtuse equilateral triangle.

13. A parallelogram that is not a rectangle.

14. A parallelogram that is not a rhombus.

15. A rhombus that is not a square.

16. A rhombus that is not a rectangle.

17. A rhombus that is not a parallelogram.

18. A trapezoid that is not a parallelogram.

19. A trapezoid that is not a rectangle.

20. A rectangle that is not a trapezoid.

PEDAGOGICAL CONSIDERATIONS

1. Reread the definition of a trapezoid as given in this section. Some elementary texts define a trapezoid as a quadrilateral with *exactly* one pair of its opposite sides as parallel. Compare these two definitions and discuss the advantage of using one versus the other.

2. Classroom explorations involving polygons can be used effectively with elementary school students to help promote discoveries. Consider, for example, the following outline that leads the student to discover that the diagonals of a rectangle are equal in measure. Then prepare a similar worksheet that leads to the discovery of some other fact about polygons.

   (a) In each figure draw line segments *AC* and *BD*. These segments are the diagonals of the quadrilateral.

(b) Measure, to the nearest eighth of an inch, the length of the two diagonals of each figure.

(c) Compare the lengths of $\overline{AC}$ and $\overline{BD}$ in each figure. For which figures do the lengths appear to be equal?

(d) Study your measurements and see if you can state a conclusion about the diagonals of a rectangle.

(e) Draw two more rectangles and check the conclusion that you reached.

11-4

triangles   Any triangle has three angles

$$\angle A = \angle BAC, \qquad \angle B = \angle ABC, \qquad \angle C = \angle ACB$$

and three sides $\overline{AB}$, $\overline{BC}$, and $\overline{AC}$. In the following figure there is a correspondence of vertices (A to D, B to E, and C to F) such that corresponding angles of the triangles are congruent:

$$\angle A \cong \angle D, \qquad \angle B \cong \angle E, \qquad \angle C \cong \angle F$$

The triangles ABC and DEF in the figure appear to be of the same shape but of different sizes. One triangle appears to be a "scale drawing" of the other with a correspondence of vertices A and D, B and E, C and F. We indicate this correspondence and "similarity" of the triangles by writing

$$\triangle ABC \sim \triangle DEF$$

(read as "triangle ABC is similar to triangle DEF"). We would not write $\triangle ABC \sim \triangle FDE$, since the correspondence of vertices A and F, B and D, C and E would not provide a correspondence of congruent angles. Thus as we name the triangles we list the vertices so that the order of the vertices indicates the correspondence between them. In general:

Two triangles are **similar** if there is a correspondence of the vertices of one triangle with the vertices of the other triangle so that corresponding angles are congruent.

For the triangles given in the figure notice that

$$\triangle ABC \sim \triangle DEF$$
$$\triangle ABC \sim \triangle ABC$$
$$\triangle ABC \not\sim \triangle BCA.$$

Example 1: The figure consists of four lines $\overleftrightarrow{AB}$, $\overleftrightarrow{AC}$, $\overleftrightarrow{BD}$, and $\overleftrightarrow{CD}$ with $\overleftrightarrow{AB} \parallel \overleftrightarrow{CD}$. Identify two similar triangles and explain why you expect them to be similar.

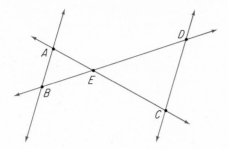

Solution: We appear to have $\triangle ABE \sim \triangle CDE$. When parallel lines are cut by a transversal, the alternate interior angles are congruent. Therefore, $\angle BAE \cong \angle DCE$ and $\angle ABE \cong \angle CDE$. Since vertical angles are congruent, $\angle AEB \cong \angle CED$. Thus under the correspondence of vertices A and C, B and D, E and E the corresponding angles of the triangles are congruent and the definition of similar triangles is satisfied.

Two line segments are *congruent* if they have equal linear measures. Two angles are *congruent* if they have equal angular measures. In general, two geometric figures are **congruent** if there is a correspondence between the figures such that all corresponding parts that have measures have equal measures. Thus *corresponding parts of congruent figures are congruent.*

Consider two triangles *ABC* and *RST*. In order to prove that the triangles are congruent, we need a correspondence of the vertices of one triangle with the vertices of the other. For example, from the appearance of the figures we might consider the correspondence of *A* and *R*, *B* and *T*, and *C* and *S*. This correspondence of vertices implies the following correspondence of angles and of sides of the triangle:

$$
\begin{array}{llll}
\angle A & \text{and} & \angle R & \quad \overline{AB} \ \text{and} \ \overline{RT} \\
\angle B & \text{and} & \angle T & \quad \overline{BC} \ \text{and} \ \overline{TS} \\
\angle C & \text{and} & \angle S & \quad \overline{AC} \ \text{and} \ \overline{RS}
\end{array}
$$

If the triangles are congruent under this correspondence, we write

$$\triangle ABC \cong \triangle RTS.$$

This statement implies:

$$\angle A \cong \angle R \qquad \overline{AB} \cong \overline{RT}$$
$$\angle B \cong \angle T \qquad \overline{BC} \cong \overline{TS}$$
$$\angle C \cong \angle S \qquad \overline{AC} \cong \overline{RS}$$

In general, if there is a correspondence between the vertices of a triangle *ABC* and the vertices of a triangle *RST* such that one of the following conditions is satisfied, then the triangles are congruent under that correspondence.

s.a.s. $\overset{\sim}{=}$ s.a.s: Two sides and the included angle of one triangle are congruent respectively to two sides and the included angle of the other triangle.

a.s.a. $\overset{\sim}{=}$ a.s.a: Two angles and the included side of one triangle are congruent respectively to two angles and the included side of the other triangle.

s.s.s. $\overset{\sim}{=}$ s.s.s.: The three sides of one triangle are congruent respectively to the three sides of the other triangle.

Example 2: Identify a pair of congruent triangles in the given parallelogram and state the congruence.

Solution: In the figure $\overline{AC} \cong \overline{AC}$ since any line segment is congruent to itself; $\angle BAC \cong \angle DCA$ and $\angle BCA \cong \angle DAC$ as alternate interior angles formed by two parallel lines and a transversal. Triangles *ABC* and *ADC* are congruent by a.s.a. $\cong$ a.s.a. The statement of the congruence of the triangles must show the correspondence of vertices *A* with *C*, *B* with *D*, and *C* with *A* of the respective triangles; for example,

$$\triangle ABC \cong \triangle CDA.$$

EXERCISES

In Exercises 1 through 5 identify a pair of congruent triangles and state the congruence.

1.

ABCD is a rectangle.

2.

EFGH is a parallelogram.

3.

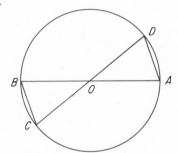

$\overline{AB}$ and $\overline{CD}$ are diameters of the circle.

4.

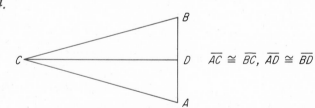

$\overline{AC} \cong \overline{BC}$, $\overline{AD} \cong \overline{BD}$

5.

The symbol ⌐ is used to denote a right angle.

6. In the figure for any triangle ABC a line $\overleftrightarrow{DE}$ has been drawn through C such that $\overleftrightarrow{DE} \parallel \overleftrightarrow{AB}$. Explain why in degrees the sum of the angle measures of any triangle ABC is 180.

7. For the given figure explain why (a) $\angle Q \cong \angle T$; (b) $\triangle PQR \sim \triangle STU$.

8. Explain why if two angles of one triangle are congruent respectively to two angles of another triangle, then the third angles of the triangles are congruent and the triangles are similar.

Identify a pair of similar triangles in each figure and state the similarity.

9.

$\overleftrightarrow{DE} \parallel \overleftrightarrow{AB}$

10.

$G = \overleftrightarrow{FH} \cap \overleftrightarrow{IJ}$

In Exercises 11 through 19 explain why each statement is true.

11. If $\triangle ABC \cong \triangle BAC$, then $\triangle ABC$ is isosceles.

12. If $\triangle ABC \cong \triangle BCA$, the $\triangle ABC$ is equilateral.

13. A parallelogram and one of its diagonals determine two congruent triangles.

14. The opposite sides of a parallelogram are congruent.

15. The base angles of an isosceles triangle are congruent.

16. If two angles of a triangle are congruent, then the triangle is isosceles.

17. Any two congruent triangles are similar.

18. All four sides of a square are congruent.

19. All four sides of a rhombus are congruent.

20. In the figure for any triangle ABC the point D is the midpoint of $\overline{AC}$, that is, $\overline{AD} \cong \overline{DC}$; E is the midpoint of $\overline{AB}$; F is the midpoint of $\overline{BC}$. Assume that a line that contains the midpoints of two sides of a triangle is parallel to the third side. Then name three pairs of parallel lines from the figure.

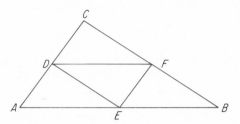

21. In the figure for Exercise 20, **(a)** explain why quadrilateral $AEFD$ is a parallelogram; **(b)** name two other parallelograms in the figure.

22. Use the figure for Exercise 20 and the statement in Exercise 14 to explain why the line segment joining the midpoints of two sides of a triangle is half the length of the third side.

23. Relative to the figure for Exercise 20 explain why

$$\triangle ABC \sim \triangle AED \sim \triangle EBF \sim \triangle DFC \sim \triangle FDE.$$

24. Relative to the figure for Exercise 20 explain why

$$\triangle AED \cong \triangle EBF \cong \triangle DFC \cong \triangle FDE.$$

PEDAGOGICAL CONSIDERATIONS

Many interesting geometric concepts can be developed intuitively through the media of paper folding. Below is an example of the sort of exploration that can be made. Wax paper is very useful for this kind of project. Follow the steps given, and then try to provide other examples of paper folding in geometry. A good source of ideas is contained in the booklet *Paper Folding in the Mathematics Class,* written by Donovan Johnson and published by the National Council of Teachers of Mathematics.

1. Cut out a triangular region ABC.

2. Fold the point B onto the line segment BC so that the crease $\overline{AD}$ passes through A. This determines the point D on $\overline{BC}$.

3. Fold B over to D, forming the crease EF, as in the figure on page 386.

4. Fold C over to D, forming the crease GH, as in the figure.

5. Fold *A* down to *D*, forming the crease *EG*.

Now notice that when all three points *A*, *B*, and *C* are folded over to the point *D*, the angles at *A*, *B*, and *C* appear to form an angle of 180°. Notice also that the length of $\overline{EG}$, the line segment joining the midpoints of two sides of △*ABC*, is equal to the length of $\overline{FH}$. Also, since $\overline{BF} \cong \overline{FD}$ and $\overline{DH} \cong \overline{HC}$, the length of $\overline{EG}$ is one-half the length of $\overline{BC}$. Thus we have a demonstration of the fact that the line segment joining the midpoints of two sides of a triangle is equal in length to one-half the third side of the triangle.

11-5

area measures　Area measure is a surface measure and is used for rectangular regions and other regions. The unit of area measure is the measure of a square region having an edge with length 1 unit of linear measure. For example, 1 square inch, 1 square foot, and 1 square mile may each be used as units of area measure.

1 square unit of area measure, that is, a unit square region

Consider a rectangular region *ABCD* with $m(\overline{AB}) = b$ and $m(\overline{BC}) = h$. When *b* and *h* are whole numbers, the rectangular region has the same area measure as *bh* unit square regions. In the second figure *b* = 3, *h* = 2, and the rectangular region *PQRS* has area measure 6; that is, the area of rectangular region *PQRS* is 6 square units.

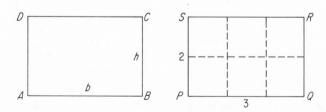

We use the symbol $\square ABCD$ for rectangle $ABCD$ and follow the custom of speaking of the area of the rectangle when we mean the rectangular region. Thus we write $\alpha(\square ABCD)$ for the **area measure** of rectangle $ABCD$. Then

$$\alpha(\square PQRS) = 6$$

and, in general,

$$\alpha(\square ABCD) = bh$$

where $m(\overline{AB}) = b$ and $m(\overline{BC}) = h$. We have illustrated the area formula for whole numbers b and h. We *define* the area measure of the rectangle to be bh in square units for all linear measures b and h.

A square is a special case of a rectangle, $b = h$. Thus the area of a square region with an edge e units long is e^2 square units. As in the case of rectangles we often speak of the areas of squares or other polygons and mean thereby the areas of the polygonal regions.

$$\alpha(\square\ LMNO) = e^2$$

A point has area measure 0 and a line segment has area measure 0. If two plane regions have area measures and the area measure of their intersection is 0, then the area measure of the union of the two regions is equal to the sum of their area measures.

$$\alpha(\square\ AGEF\ \cap\ \square\ GBCD) = \alpha(\overline{GE}) = 0$$
$$\alpha(\square\ AGEF\ \cup\ \square\ GBCD) = 9 + 10 = 19$$

In order to find the area measures of triangular regions, we assume that congruent figures have equal area measures. Then since a diagonal of a rectangle divides the rectangular region into congruent triangular regions, the area measure of a right triangle with legs of length b and h is $\frac{1}{2}bh$, as indicated in the next figure.

$$@(\triangle ABC) = @(\triangle CDA) = \tfrac{1}{2}\,bh$$

For any triangle ABC a side, such as $\overline{AB}$, may be selected as base. Then the line that contains C and is perpendicular to $\overleftrightarrow{AB}$ intersects $\overleftrightarrow{AB}$ at a point D. There are three cases to consider according as D is an interior point of $\overline{AB}$, an endpoint of $\overline{AB}$, or an exterior point of $\overline{AB}$.

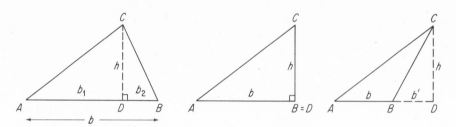

In the figure on the left D is an interior point of $\overline{AB}$, $m(\overline{AB}) = b$, $m(\overline{AD}) = b_1, m(\overline{DB}) = b_2, b_1 + b_2 = b, m(\overline{CD}) = h, @(\triangle ADC) = \tfrac{1}{2}b_1h$, $@(\triangle DBC) = \tfrac{1}{2}b_2h$, and

$$\begin{aligned}
@(\triangle ABC) &= @(\triangle ADC) + @(\triangle DBC) \\
&= \tfrac{1}{2}b_1h + \tfrac{1}{2}b_2h \\
&= \tfrac{1}{2}(b_1 + b_2)h = \tfrac{1}{2}bh.
\end{aligned}$$

In the center figure $B = D$, $m(\overline{AB}) = b$, $m(\overline{CD}) = h$, and $@(\triangle ABC) = \tfrac{1}{2}bh$.

In the figure on the right D is an exterior point of $\overline{AB}$, $m(\overline{AB}) = b$, $m(\overline{BD}) = b'$, $m(\overline{AD}) = b + b'$, $m(\overline{CD}) = h$, and

$$\begin{aligned}
@(\triangle ABC) &= @(\triangle ADC) - @(\triangle BDC) \\
&= \tfrac{1}{2}(b + b')h - \tfrac{1}{2}b'h \\
&= \tfrac{1}{2}[(b + b') - b']h \\
&= \tfrac{1}{2}[b + (b' - b')]h = \tfrac{1}{2}bh.
\end{aligned}$$

In all cases and thus for any triangle ABC, $@(\triangle ABC) = \tfrac{1}{2}bh$.

Areas of polygonal regions may be found, as in the supplementary exercises for this section, by subdividing the polygonal region into nonoverlapping rectangular and/or triangular regions, that is, regions with no common interior points. For example, any parallelogram $ABCD$ with base b and height h has area bh.

$$@(\square\, ABCD) = bh$$

Example: Find the area of the given region.

Solution: We subdivide the region into rectangular regions. The three rectangular regions indicated in the next figure have area measures 1, 12, and 4. Thus the area measure of the given region is 17 and the area is 17 square units.

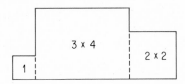

In the example notice that as for linear measure and length, the area measure is a number; the area is that number of square units.

EXERCISES

Find the area measurement of each region.

1.

2.

3.

4.

389

5. 12″ · 2″ · 6″ · 4″ · 8″

6. 7′ · 5′ · 10′

7. 7″ · 6″ · 2″ · 14″ · 16″

8. 12′ · 7′ · 6′ · 7′ · 7′ · 10′ · 7′ · 3′ · 5′

Find a formula for:

9. The surface area S of a cube with edge of linear measure e.

10. The surface area S of a rectangular parallelepiped with edges of linear measure b, d, and h.

Describe the change in the area of:

11. A square when the measure of each of its edges is doubled.

12. A square when the measure of each of its edges is multiplied by **(a)** 3; **(b)** $\frac{1}{2}$; **(c)** a constant k.

13. A rectangle when the measure of each of its edges is multiplied by **(a)** 2; **(b)** 3; **(c)** $\frac{1}{2}$; **(d)** a constant k.

14. A triangle when each of its edges is multiplied by **(a)** 2; **(b)** 3; **(c)** $\frac{1}{2}$; **(d)** a constant k.

15. A polygon when the measures of its angles are not changed and the measure of each of its edges is multiplied by **(a)** 2; **(b)** a constant k.

SUPPLEMENTARY EXERCISES

Any parallelogram has the same area as a rectangle that may be formed as in the figure. Give a reason for each of the following statements.

1. $\angle BAD \cong \angle FBC$.

2. Quadrilateral $EFCD$ is a rectangle.

3. $\overline{DE} \cong \overline{CF}$ and $\overline{DC} \cong \overline{EF}$.

4. $\overline{DC} \cong \overline{AB}$.

5. $\overline{AB} \cong \overline{EF}$.

6. $m(\overline{AE}) = m(\overline{BF})$.

7. $\triangle AED \cong \triangle BFC$.

8. $@(\square ABCD) = @(\square EFCD)$.

9. If $m(\overline{AB}) = b$, $\overleftrightarrow{DE}$ is perpendicular to $\overleftrightarrow{AB}$ with D on $\overleftrightarrow{AB}$, and $m(\overline{DE}) = h$, then $@(\square ABCD) = bh$.

Any trapezoid PQRS can be represented as in the figure. Let $m(\overline{PQ}) = b_1$, $m(\overline{SR}) = b_2$, $\overleftrightarrow{PQ} \parallel \overleftrightarrow{SR}$, $\overleftrightarrow{ST}$ be perpendicular to $\overleftrightarrow{PQ}$ with T on $\overleftrightarrow{PQ}$, $\overleftrightarrow{QU}$ be perpendicular to $\overleftrightarrow{SR}$ with U on $\overleftrightarrow{SR}$, and $m(\overline{ST}) = h$. Give a reason for each of the following statements.

10. $@(\triangle PQS) = \frac{1}{2}b_1 h$.

11. $@(\triangle QRS) = \frac{1}{2}b_2 h$.

12. $@(\square PQRS) = \frac{1}{2}(b_1 + b_2)h$.

PEDAGOGICAL CONSIDERATIONS

1. It is important that elementary school students understand that the measure of the area of a figure depends upon a standard unit of measure. One way to develop an appreciation of this concept is to ask them to find the measure of various geometric figures using different units of area measure. For example, estimate the measure of each of the following figures in terms of dimes. That is, estimate the whole number of dimes that can be enclosed within each figure. Then check estimates by actually using dimes within each figure.

2. Examine a set of elementary mathematics textbooks for grades 1 through 6. List as many different formulas as you can find that are presented there for the areas of geometric figures.

11-6

volume measures

Volume measure is a space measure and is used for cubic and other regions. As for area measure it is customary to speak of the volume of a cube or other polyhedron and to mean thereby the volume of the polyhedral region. The unit of volume measure is the measure of a cubic region having an edge with length 1 unit of linear measure. For example, 1 cubic inch, 1 cubic centimeter, and 1 cubic foot may each be used as units of volume measure.

1 cubic unit of volume measure, that is, a unit volume measure

Volumes are considered in a manner very similar to areas. We define any rectangular parallelepiped with edges of linear measure a, b, and c on a common endpoint to have volume measure abc:

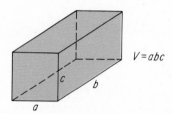

$V = abc$

Points, line segments, and plane regions have volume measure 0. Any pyramid with a base of area B square units and height to that base h linear units, can be proved in more advanced courses to have volume measure $\frac{1}{3}Bh$:

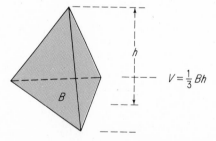

$V = \frac{1}{3}Bh$

Volumes of many other regions may be obtained by subdividing the region into pyramids and/or rectangular parallelepipeds, or by using the prismoidal formula (Exercise 12). As in the case of length and area, the volume measure is a number; the volume is that number of cubic units.

EXERCISES

Sketch and find **(a)** *the surface area, and* **(b)** *the volume, of a cube with edge of length:*

1. 2 feet   **2.** 3 yards   **3.** 4 centimeters   **4.** 5 inches

Find **(a)** *the surface area, and* **(b)** *the volume, of each rectangular parallelepiped.*

5.

*6.

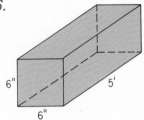

Find the volume of each pyramid.

7.

ABCD is a square

8.

$\triangle PQR$
has a
right
angle
at Q.

In Exercises 9 through 11 describe the change in the volume of:

9. A cube when the measure of each of its edges is doubled.

10. A cube when the measure of each of its edges is multiplied by **(a)** 3; **(b)** $\frac{1}{2}$; **(c)** a constant k.

11. A rectangular parallelepiped when the measure of each of its edges is multiplied by **(a)** 2; **(b)** 3; **(c)** $\frac{1}{2}$; **(d)** a constant k.

*12. The most general common formula for volume measures is the **prismoidal formula**

$$V = \frac{h}{6}(B_1 + 4M + B_2).$$

This formula may be used, for example, for any prismatoid where B_1 and B_2 are the area measures of the bases, M is the area measure of the intersection of the prismatoid with the plane halfway between the bases, and the bases are h units apart; that is, the altitude to the bases has linear measure h. The plane section halfway between the bases is often called the **midsection**. Then M is the area measure of the midsection.

Show that for a cube with an edge of linear measure e, $B_1 = e^2$, $M = e^2$, $B_2 = e^2$, and the prismoidal formula gives the usual expression for the volume.

13. Think of a sphere as in the figure and use the prismoidal formula to find an expression for the volume of a sphere of radius r. (The area measure of a circle of radius r is πr^2.)

14. For a square pyramid with height h and a base with an edge of linear measure e the midsection is a square with an edge $\frac{1}{2}e$. Use the prismoidal formula to find the volume of the square pyramid.

15. Use the prismoidal formula to find the volume of a circular cone with height h if the base has radius r and the midsection has radius $\frac{1}{2}r$.

PEDAGOGICAL CONSIDERATIONS

1. Examine a set of elementary mathematics textbooks for grades 1 through 6 and list all of the formulas presented therein for the volumes of geometric solids.

2. An interesting way to introduce elementary school students to the concept of volume is to ask such questions as "Do you think that it is possible to fit one million pennies into this room?" "Can one million ping-pong balls fit into this room?" Answer these questions for one of your classrooms, and then develop some procedure to help you verify your answers.

3. A contest to guess the number of beans, or other such similar objects, in a jar is another interesting approach to the concept of volume for elementary school students. Try to list 5 specific suggestions for activities that help develop the idea of volume.

4. Find formulas for the area of a circle and the volume of a right circular cylinder. Prepare a short lesson plan to illustrate how these formulas might be presented to upper elementary school students, using discovery or laboratory type procedures.

1. On a number line find the linear measure of: (a) $\overline{P_2P_{-7}}$; (b) $\overline{P_{-8}P_{-3}}$.

2. Copy the given figure and label each angle with its measurement.

3. In the figure for Exercise 2 identify **(a)** four pairs of corresponding angles; **(b)** two pairs of vertical angles with vertex at Q.

4. Identify a pair of similar triangles in the given figure and state the similarity.

$\overleftrightarrow{AB} \parallel \overleftrightarrow{DE}$

5. Sketch **(a)** an obtuse scalene triangle; **(b)** a rhombus that is not a square.

6. Find the area of the given region.

7. Find the surface area of a cube with an edge of length 3 inches.

8. Find the surface area of a rectangular parallelepiped with edges of lengths 3 inches, 5 inches, and 7 inches.

9. Find the volume of **(a)** a cube with an edge of length 6 feet; **(b)** a rectangular parallelepiped with edges of lengths 3 inches, 4 inches, and 5 inches.

10. Describe the change in the **(a)** surface area and **(b)** volume of a cube when the measure of each of its edges is multiplied by 5.

12

Coordinate Geometry

On a line, on a plane, or in space, points may be identified by coordinates. The study of sets of points in terms of coordinates of the points is called **coordinate geometry** or **analytic geometry**. Coordinate geometry is based upon the work of René Descartes, who pioneered this unification of algebra and geometry. The importance of Descartes' work is often recognized by referring to a coordinate plane as a **Cartesian plane**.

The use of coordinates as an aid to the study of geometry is not yet common in elementary schools. However, coordinate geometry is important for prospective elementary school teachers as a means of increasing their own understanding of the geometric figures (sets of points) considered in elementary schools. Coordinate geometry also serves as a sample of one way in which elementary school students will later use the geometric concepts that are introduced in elementary school classes. Teachers of junior high school mathematics, grades 7 through 9, will find much of the material considered in this chapter already in the curriculum of these grades.

12-1
coordinates on a plane

Any plane may be considered as a coordinate plane as in §9-3. Any two points $A: (x_1, y_1)$ and $B: (x_2, y_2)$ determine a line $\overleftrightarrow{AB}$.

Example 1: Graph A: (2, 1) and B: (2, 5). Draw the line $\overleftrightarrow{AB}$ and find the length of the line segment $\overline{AB}$.

Solution: The given points may be graphed as in the figure. The line $\overleftrightarrow{AB}$ is parallel to the y-axis and has equation $x = 2$ since A and B both have x-coordinate 2.

To find the length of $\overline{AB}$ we consider the two points D: (0, 1) and C: (0, 5) on the y-axis, note that the opposite sides $\overline{AB}$ and $\overline{CD}$ of the rectangle $ABCD$ are congruent (have the same length), and note that on the y-axis $m(\overline{CD}) = |5 - 1|$, that is, 4. Therefore, the length of $\overline{AB}$ is 4 units.

For any points $A: (x_1, y_1)$ and $B: (x_1, y_2)$ the line $\overleftrightarrow{AB}$ is parallel to the y-axis and the linear measure of the line segment $\overline{AB}$ is $|y_2 - y_1|$. Similarly, for any points $A: (x_1, y_1)$ and $C: (x_2, y_1)$ the line $\overleftrightarrow{AC}$ is parallel to the x-axis and $m(\overline{AC}) = |x_2 - x_1|$. In general, the linear measure of any line segment is a number; the length of the line segment is that number of units of length.

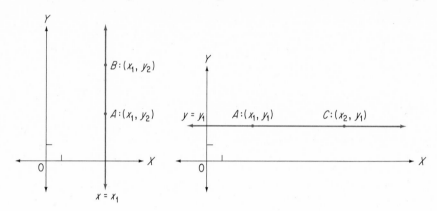

Consider the points P_x with coordinates x on a number line. We define the **directed distance** from P_r to P_s to be $s - r$. Then the directed distance from P_1 to P_4 is $4 - 1$, that is, 3; the directed distance from P_4 to P_1 is $1 - 4$, that is, -3; the directed distance from P_{-1} to P_1 is $1 - (-1)$, that is, 2. We use directed distances to develop a formula for the coordinate of the midpoint of a line segment. To find the midpoint M of a line segment $\overline{P_rP_s}$ we start at P_r and proceed halfway from P_r to P_s. Therefore, the coordinate of M is $r + \frac{1}{2}(s - r)$, that is, $\frac{1}{2}(r + s)$.

Example 2: Find the linear measure and the coordinate of the midpoint of $\overline{AB}$ for $A: (1, 5)$ and $B: (7, 5)$.

Solution: Consider the points $C: (1, 0)$ and $D: (7, 0)$ and the rectangle $CDBA$ as in the figure;

$$m(\overline{AB}) = m(\overline{CD}) = |7 - 1| = 6.$$

The y-coordinate of the midpoint M of $\overline{AB}$ must be 5, since every point of $\overleftrightarrow{AB}$ has y-coordinate 5. The midpoint E of $\overline{DC}$ has x-coordinate $\frac{1}{2}(1 + 7)$, that is, 4. The lines, $\overleftrightarrow{AC}$, $x = 4$, and $\overleftrightarrow{BD}$ are parallel and intercept (cut off) congruent segments $\overline{CE}$ and $\overline{ED}$ on the x-axis. Therefore these lines intercept congruent segments on $\overline{AB}$, M must be on the line $x = 4$, and M has coordinates $(4, 5)$.

In general, any two distinct points $A: (x_1, y_1)$ and $C: (x_2, y_1)$ on a line $y = y_1$ parallel to the x-axis are endpoints of a line segment $\overline{AC}$ with linear measure $|x_2 - x_1|$ and midpoint $(\frac{1}{2}(x_1 + x_2), y_1)$. Similarly, any two distinct points $A: (x_1, y_1)$ and $B: (x_1, y_2)$ on a line $x = x_1$ parallel to the y-axis are endpoints of a line segment $\overline{AB}$ with linear measure $|y_2 - y_1|$ and midpoint $(x_1, \frac{1}{2}(y_1 + y_2))$.

Example 3: Find the coordinates of the midpoints of the line segments with endpoints (a) $(1, 0)$ and $(-3, 0)$; (b) $(2, 1)$ and $(2, 7)$.

Solution:

(a) $\left(\dfrac{1 + (-3)}{2}, 0 \right) = (-1, 0)$

(b) $\left(2, \dfrac{1 + 7}{2} \right) = (2, 4)$

For two points $A: (x_1, y_1)$ and $B: (x_2, y_2)$ such that $\overline{AB}$ is not parallel to a coordinate axis we consider also the point $C: (x_2, y_1)$. Then the midpoints of $\overline{AC}$ and $\overline{AB}$ are on the line $x = \frac{1}{2}(x_1 + x_2)$, and the midpoints of $\overline{CB}$ and $\overline{AB}$ are on the line $y = \frac{1}{2}(y_1 + y_2)$. Thus, as in the next figure, the midpoint of $\overline{AB}$ has coordinates

$$\left(\frac{x_1 + x_2}{2}, \frac{y_1 + y_2}{2} \right).$$

399

This is the **midpoint formula** for points on a coordinate plane. Notice that the formula holds whether or not $\overline{AB}$ is parallel to a coordinate axis.

Example 4: Find the midpoint of $\overline{AB}$ for (a) A: $(6, 1)$ and B: $(-2, 5)$; (b) A: $(-3, 4)$ and B: $(1, -2)$.

Solution:

(a) $\left(\dfrac{6 + (-2)}{2}, \dfrac{1 + 5}{2} \right) = (2, 3)$

(b) $\left(\dfrac{-3 + 1}{2}, \dfrac{4 + (-2)}{2} \right) = (-1, 1)$

EXERCISES

1. Find the linear measure and the coordinates of the midpoint of the line segment with endpoints:

   (a) $(0, 3)$ and $(4, 3)$ (b) $(2, 1)$ and $(5, 1)$

   (c) $(2, 1)$ and $(2, 5)$ (d) $(5, -2)$ and $(5, 7)$

   (e) $(-3, 2)$ and $(-3, 7)$ (f) $(-3, 2)$ and $(-7, 2)$

2. Find formulas for the lengths of line segments with endpoints:

   (a) $(0, 0$ and $(x, 0)$ (b) $(0, 0$ and $(0, y)$

   (c) $(x_1, 0)$ and $(x_2, 0)$ (d) $(0, y_1)$ and $(0, y_2)$

   (e) (x_1, y_1) and (x_2, y_1) (f) (x_1, y_1) and (x_1, y_2)

Find the coordinates of:

3. The midpoint of $\overline{AB}$ for **(a)** A: $(1, -6)$ and B: $(7, 2)$; **(b)** A: $(5, 11)$ and B: $(-1, -3)$; **(c)** A: $(-2, -7)$ and B: $(-4, -5)$.

4. The endpoint B if A: $(1, 3)$ and C: $(2, 5)$ is the midpoint of $\overline{AB}$.

5. The endpoint B if A: $(3, 0)$ and C: $(-1, 4)$ is the midpoint of $\overline{AB}$.

6. The endpoint A if B: $(-5, 6)$ and C: $(-3, -1)$ is the midpoint of $\overline{AB}$.

7. The vertex C of a square $ABCD$ with A: $(0, 0)$, B: $(a, 0)$, and D: $(0, a)$.

8. The vertex S of a rectangle $QRST$ with Q: $(0, 0)$, R: $(a, 0)$, and T: $(0, b)$.

9. The vertex S of an isosceles trapezoid $QRST$ with Q: $(0, 0)$, R: $(a, 0)$, and T: (b, c), where $2b < a$.

Draw coordinate axes, sketch each set of points, and label each line with its equation.

10. The points 5 units from the y-axis.

11. The points 2 units from the x-axis.

12. The points 3 units from the line $x = 2$.

13. The points 4 units from the line $y = -1$.

properties of figures on a coordinate plane and prove:

14. If A: $(0, 0)$, B: $(a, 0)$, and C: (b, c) are the vertices of a triangle, then the midpoints of $\overline{AC}$ and $\overline{BC}$ are endpoints of a line segment that is parallel to $\overline{AB}$ and half as long as $\overline{AB}$.

*15. The line segments with the midpoints of opposite sides of a quadrilateral as endpoints bisect each other. [*Hint:* Use vertices A: $(0, 0)$, B: $(a, 0)$, C: (b, c), and D: (d, e).]

PEDAGOGICAL CONSIDERATIONS

Elementary school students can be given practice with coordinates in a variety of ways. One procedure is to have them find the locus (location) of all points that satisfy certain conditions. The use of graph paper for such activities is an excellent aid. Thus consider the following exercises.

1. On a sheet of graph paper:
   (a) Select a point A at an intersection of the lines of the graph paper.
   (b) Along one of the lines through A select a point B six units from A. (Consider the edge of each small square on the paper as one unit.)
   (c) Mark at least 20 points P that are two units from $\overline{AB}$. (Remember that the distance from a point P to a figure such as $\overline{AB}$ is the length of the shortest possible line segment PQ where Q is a point of $\overline{AB}$.)

(d) Is it necessary for all of the points P to be on the same side of $\overline{AB}$? Can some of the points P be at opposite ends of $\overline{AB}$?

(e) Try to sketch a curve consisting of all the points P that are exactly two units from $\overline{AB}$.

2. Repeat Exercise 1 using a square with sides three units long instead of $\overline{AB}$ as in the original figure.

3. Prepare a worksheet appropriate for an elementary school class with an exploration similar to the one presented in Exercise 1.

4. Try to describe the figure obtained in Exercise 1 if all points in space are considered, rather than just those in the plane of the graph paper.

12-2

lines on a plane

Any equation of the form

$$ax + by + c = 0$$

where a and b are not both zero and each letter represents a real number, has a line as its *graph* as in §9-3; any line on a coordinate plane has an equation of this *standard form*. Several other common forms for equations of lines involve the concept of slope.

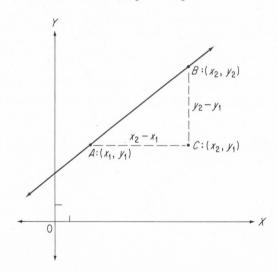

For any points A: (x_1, y_1) and B: (x_2, y_2), where $x_2 \neq x_1$, the line $\overleftrightarrow{AB}$ is not parallel to the y-axis, and the **slope** m of $\overleftrightarrow{AB}$ is defined by the equation

$$m = \frac{y_2 - y_1}{x_2 - x_1}.$$

Slope is not defined for lines parallel to the y-axis since for any line parallel to the y-axis, $x_1 = x_2$, $x_2 - x_1 = 0$, and division by zero is not allowed. For example, for C: $(2, 3)$ and D: $(2, 5)$, $x_2 - x_1 = 2 - 2 = 0$, and the slope of $\overleftrightarrow{CD}$ is not defined.

Example 1: Find the midpoint C of $\overleftrightarrow{AB}$ for A: $(1, -3)$ and B: $(7, 5)$. Then find the slope of $\overleftrightarrow{AB}$, $\overleftrightarrow{AC}$, and $\overleftrightarrow{BC}$.

Solution: By the above midpoint formula we have C: $(4, 1)$, since $\frac{1}{2}(1 + 7) = 4$ and $\frac{1}{2}(-3 + 5) = 1$. By the definition of slope, the slope of $\overleftrightarrow{AB}$ is

$$\frac{5 - (-3)}{7 - 1} = \frac{8}{6} = \frac{4}{3}$$

the slope of $\overleftrightarrow{AC}$ is

$$\frac{1 - (-3)}{4 - 1} = \frac{4}{3}$$

and the slope of $\overleftrightarrow{BC}$ is

$$\frac{5 - 1}{7 - 4} = \frac{4}{3}.$$

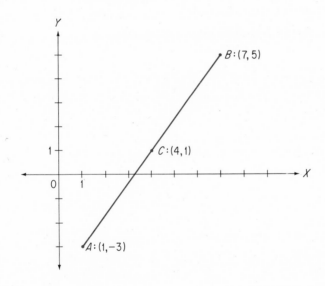

Notice in Example 1 that $\overleftrightarrow{AB}$, $\overleftrightarrow{AC}$, and $\overleftrightarrow{BC}$ have the same slope. This result illustrates the fact that the coordinates of any two points of a line may be used to find the slope of the line. Thus for any points A: (x_1, y_1)

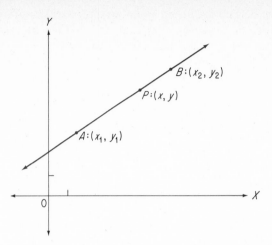

and B: (x_2, y_2), where $x_2 \neq x_1$, a point P: (x, y) is on the line $\overleftrightarrow{AB}$ if and only if the slopes of $\overleftrightarrow{AP}$ and $\overleftrightarrow{AB}$ are equal; that is,

$$\frac{y - y_1}{x - x_1} = \frac{y_2 - y_1}{x_2 - x_1}.$$

This equation is the **two-point form** of the equation of the line $\overleftrightarrow{AB}$.

Example 2: Find an equation for $\overleftrightarrow{AB}$ for A: $(1, -3)$ and B: $(7, 5)$.

Solution:

$$\frac{y - (-3)}{x - 1} = \frac{5 - (-3)}{7 - 1}, \qquad \frac{y + 3}{x - 1} = \frac{4}{3}$$

Several other equations of the line $\overleftrightarrow{AB}$ in Example 2 can be found from the two-point form of the equation given in the solution. For example, we can obtain

$$y + 3 = \tfrac{4}{3}(x - 1)$$

which involves the point $(1, -3)$ and the slope $\tfrac{4}{3}$. This equation is in the **point-slope form**

$$y - y_1 = m(x - x_1).$$

We can obtain the equation

$$y = \tfrac{4}{3}x - 4\tfrac{1}{3}$$

which involves the slope $\tfrac{4}{3}$ and the y-intercept $-4\tfrac{1}{3}$. This equation is in the **slope-intercept form**

$$y = mx + b.$$

Finally, we can obtain

$$3y = 4x - 13$$
$$4x - 3y = 13$$
$$\frac{x}{\frac{13}{4}} + \frac{y}{-\frac{13}{3}} = 1$$

which involves the x-intercept $\frac{13}{4}$ and the y-intercept $-\frac{13}{3}$. This equation is in the **intercept form**

$$\frac{x}{a} + \frac{y}{b} = 1.$$

Example 3: Consider the line that contains the points $(1, 5)$ and $(3, -2)$ and find the equation of the line in (a) two-point form; (b) point-slope form; and (c) slope-intercept form.

Solution: The slope of the line is

$$\frac{-2 - 5}{3 - 1} = \frac{-7}{2}.$$

(a) $\dfrac{y - 5}{x - 1} = \dfrac{-2 - 5}{3 - 1}$. Any other point of the line may be used in place of $(1, 5)$ in the left member of this equation; likewise in part (b) any other point of the line may be used.

(b) $y - 5 = -\frac{7}{2}(x - 1)$.

(c) $y = -\frac{7}{2}x + 8\frac{1}{2}$.

In analytic geometry two lines are *parallel* if and only if they have the same slope or they are both parallel to the y-axis. This definition provides the basis for the proofs of some common theorems.

Example 4: Prove that the quadrilateral $ABCD$ with vertices $A: (0, 0)$, $B: (a, 0)$, $C: (a + b, c)$, and $D: (b, c)$ is a parallelogram.

Solution: $\overleftrightarrow{AB} \parallel \overleftrightarrow{DC}$, since both lines are parallel to the x-axis. If $b = 0$, $\overleftrightarrow{AD}$ and $\overleftrightarrow{BC}$ are both parallel to the y-axis and thus to each other. If $b \neq 0$, $\overleftrightarrow{AD}$ has slope c/b, $\overleftrightarrow{BC}$ has slope c/b, and $\overleftrightarrow{AD} \parallel \overleftrightarrow{BC}$, since they have the same slope. The quadrilateral $ABCD$ is a parallelogram, since in any case $\overleftrightarrow{AB} \parallel \overleftrightarrow{DC}$ and $\overleftrightarrow{AD} \parallel \overleftrightarrow{BC}$.

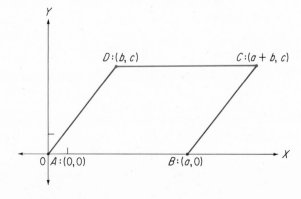

405

Find the slope of the line that contains each pair of points:

1. (1, 1) and (4, 7)   *2.* (2, 1) and (5, −2)

3. (3, 5) and (6, 5)   *4.* (−2, 6) and (5, 3)

Find (a) a two-point form, (b) a point-slope form, and (c) the slope-intercept form of the equation of the line determined by the points given in:

5. Exercise 1.   *6.* Exercise 2.

7. Exercise 3.   *8.* Exercise 4.

Prove that the quadrilateral ABCD is a parallelogram for the given points.

9. A: (1, −2), B: (6, 0), C: (6, 4), D: (1, 2)

10. A: (−2, −1), B: (4, −2), C: (6, 2), D: (0, 3)

11. A: (3, 1), B: (7, −3), C: (5, 2), D: (1, 6)

12. A: (−4, 7), B: (1, −1), C: (4, 3), D: (−1, 11)

Prove for figures on a coordinate plane:

**13.* The midpoints of the sides of any quadrilateral are the vertices of a parallelogram.

**14.* If a line m intersects one of two parallel lines, then the line m intersects the other parallel line also.

PEDAGOGICAL CONSIDERATIONS

The squares of graph paper can be used to help elementary school students draw a line parallel to any given line. For example, suppose that a line, *AB*, and a point *P* are given as in the figure.

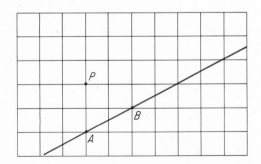

1. The point P is two units above A. Let T be the point two units above B. Draw the line PT. Does $\overleftrightarrow{PT}$ appear to be parallel to $\overleftrightarrow{AB}$? (Compare the slopes of the two lines.) What sort of figure is the quadrilateral $ABTP$?

2. Copy the given figure. Describe the manner of going from A to B by counting squares to the right and then up on the graph paper. Go from P to a point T using the same procedure as from A to B. Repeat the rest of Exercise 1 for this point T.

3. Can you find another line through P that is parallel to $\overleftrightarrow{AB}$?

12-3

distances on a plane

The distance formula for points on a coordinate plane is based upon the *Pythagorean theorem* (§7-1):

> For any right triangle the area of the square upon the hypotenuse is equal to the sum of the areas of the squares upon the other two sides.

This is essentially the form in which the early Greeks thought of the Pythagorean theorem before algebraic notations had been developed. We now usually select the right angle at the vertex C of a right triangle ABC, let c be the linear measure of the hypotenuse (the side opposite the right angle), let a and b be the linear measures of the other two sides (legs), and write

$$c^2 = a^2 + b^2.$$

This is an algebraic statement of the equivalence of the areas of the squares considered by the early Greeks.

The Pythagorean theorem may be used to obtain a general formula for the linear measure of any line segment on a coordinate plane. Let the endpoints of the line segment be $A: (x_1, y_1)$ and $B: (x_2, y_2)$. As in §12-1, either the line segment $\overline{AB}$ is parallel to a coordinate axis, or a right triangle may be formed with the vertex of the right angle at $C: (x_2, y_1)$. The linear measures of the legs of the right triangle are $|x_2 - x_1|$ and $|y_2 - y_1|$. By the Pythagorean theorem the linear measure of the hypotenuse is

$$m(\overline{AB}) = \sqrt{(x_2 - x_1)^2 + (y_2 - y_1)^2}.$$

This is the **distance formula** on a plane. The formula holds whether or not the line segment is parallel to a coordinate axis.

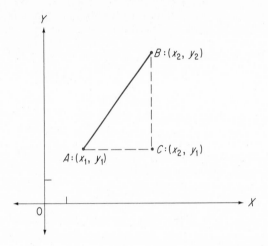

Example 1: Find the linear measure of the line segment with endpoints $A: (2, -3)$ and $B: (5, 1)$.

Solution:

$$m(\overline{AB}) = \sqrt{(5 - 2)^2 + [1 - (-3)]^2} = \sqrt{3^2 + 4^2} = 5$$

The circle with center at $C: (h, k)$ and radius r is the set of points at a distance r from the point C. The distance of any point $P: (x, y)$ from $C: (h, k)$ is

$$\sqrt{(x - h)^2 + (y - k)^2}.$$

Therefore, the circle with center $C: (h, k)$ and radius r has equation

$$(x - h)^2 + (y - k)^2 = r^2.$$

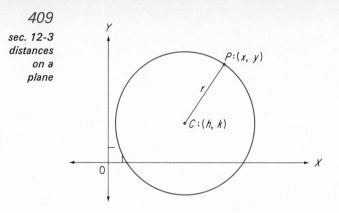

Example 2: Describe the set of points (x, y) such that
(a) $x^2 + y^2 = 4$.
(b) $(x - 1)^2 + (y - 3)^2 = 1$.
(c) $(x + 2)^2 + (y + 1)^2 = 9$.

Solution: (a) Since $x = x - 0$ and $y = y - 0$, the given equation can be expressed as

$$(x - 0)^2 + (y - 0)^2 = 2^2.$$

Each point of the set is 2 units from the origin. Thus the set of points is a circle with center $(0, 0)$ and radius 2. (b) The set of points is a circle with center $(1, 3)$ and radius 1. (c) Since $x + 2 = x - (-2)$ and $y + 1 = y - (-1)$, the given equation can be expressed as

$$[x - (-2)]^2 + [y - (-1)]^2 = 3^2$$

and the set of points is a circle with center $(-2, -1)$ and radius 3.

Example 3: Give an algebraic representation for the points that are on a coordinate plane and are within 2 units of $(3, -5)$.

Solution: The points form a **circular region** (the points of a circle and the interior points of the circle) with center $(3, -5)$ and radius 2. Thus we have

$$(x - 3)^2 + (y + 5)^2 \leq 4.$$

EXERCISES

Find the linear measures of the line segments with endpoints:

1. A: $(1, 2)$ and B: $(4, 6)$

2. C: $(-1, 6)$ and D: $(4, -6)$

3. E: $(2, 16)$ and F: $(9, -8)$

4. G: $(-2, 5)$ and H: $(7, 11)$

5. I: $(-1, -3)$ and J: $(2, 0)$

6. L: $(4, -5)$ and M: $(-7, -11)$

Describe the set of points (x, y) such that:

7. $x^2 + y^2 = 9$ 8. $x^2 + y^2 \le 36$

9. $x^2 + y^2 > 64$ 10. $(x - 1)^2 + (y - 2)^2 = 4$

11. $(x - 1)^2 + (y - 2)^2 \le 4$ 12. $(x + 3)^2 + (y - 1)^2 < 4$

13. $(x + 1)^2 + (y + 2)^2 \le 25$ 14. $(x + 3)^2 + (y + 1)^2 > 1$

Consider the points on a coordinate plane and give an algebraic representation for:

15. The points of the circle with center $(2, 5)$ and radius 3.

16. The points of the circle with center $(-3, 4)$ and radius 2.

17. The points of the circular region with center $(-3, -1)$ and radius 4.

18. The interior points of the circle with center $(6, -4)$ and radius 5.

19. The exterior points of the circle with center $(-4, 3)$ and radius 4.

20. The exterior points of the circle with center $(-3, -2)$ and radius 6.

Graph the solution set of each sentence:

*21. $\{(x, y) \mid x^2 + y^2 \le 9 \text{ and } x - y \ge 0\}$.

*22. $\{(x, y) \mid x^2 + y^2 \le 4 \text{ and } x - y - 1 < 0\}$.

PEDAGOGICAL CONSIDERATIONS

1. Many proofs (several hundred, in fact) of the Pythagorean theorem are in existence. Find at least two different proofs of this theorem.

2. Prepare a report on the life of Pythagoras that is appropriate for presentation to a sixth grade class.

3. Read about the Egyptian rope-stretchers in a history of mathematics text. Then prepare a 10-minute lesson on this topic suitable for an elementary school mathematics class.

12-4

**parabolas,
ellipses,
hyperbolas**

Distances on a plane are used to define several common figures. In this section we introduce three types of curves (the conic sections) that are used extensively in many ways including the study of the paths of space probes and the orbits of satellites.

The conic sections arise as the intersections of planes and cones. A mathematical cone is double-ended relative to the common "ice cream" variety. For example, think of a circle and the line t that is perpendicular to the plane of the circle at the center of the circle. Select a point P that is on the line t but not on the plane of the circle. The set of all lines m that contain P and a point of the circle form a **cone**. Usually we think of a single such line m being "held" at the point P and rotated to trace the points Q of the circle. The rays PQ generate one half (nappe) of the cone and the opposite rays generate the other half of the cone with **vertex** P. Since a circle has been used and the vertex P is on the line t that is perpendicular to the plane of the circle at the center of the circle, the cone that we have considered is a special type of cone called a **right circular cone**.

Any plane parallel to the plane of the circle used to generate a right circular cone must intersect the cone at the vertex P or in a circle. A plane that is not parallel to the plane of the circle and does not contain the vertex P may intersect the cone in an ellipse, a parabola, or a hyperbola (note the two branches of the hyperbola) as indicated in the following figures. Circles are often considered as special cases of ellipses.

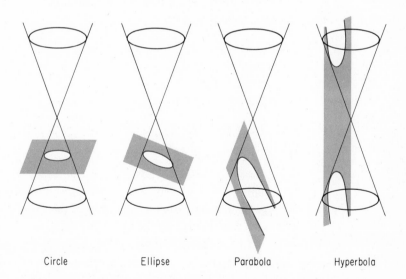

Circle   Ellipse   Parabola   Hyperbola

The three types of conic sections (parabolas, ellipses, and hyperbolas) may also be defined in terms of distances on a coordinate plane. The distance definition for a parabola is illustrated in Example 1.

Example 1: On a coordinate plane give an algebraic representation for the points that are equidistant from the line $y = -1$ and the point F: $(0, 1)$.

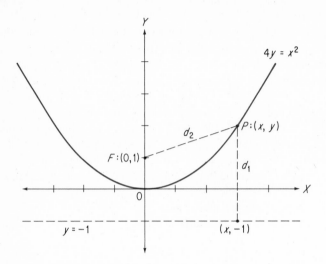

Solution: If d_1 is the distance of a point P: (x, y) from the line $y = -1$, then $d_1 = |y - (-1)|$ and $d_1^2 = (y + 1)^2$. If d_2 is the distance of a point P: (x, y) from F: $(0, 1)$, then $d_2^2 = (x - 0)^2 + (y - 1)^2$. Thus the condition that a point P: (x, y) be equidistant from the line $y = -1$ and the point F: $(0, 1)$ may be expressed as

$$d_1^2 = d_2^2$$
$$(y + 1)^2 = x^2 + (y - 1)^2$$
$$y^2 + 2y + 1 = x^2 + y^2 - 2y + 1$$
$$4y = x^2.$$

The equation $4y = x^2$ provides an algebraic representation for the given set of points.

The equation $4y = x^2$ obtained in Example 1 may be expressed as $y = \frac{1}{4}x^2$. In general, as in §9-6, any equation of the form

$$y = ax^2 + bx + c, \qquad a \neq 0$$

has a *parabola* as its graph. The graph in Example 1 is a parabola, and any parabola may (although we shall not study the procedures for doing so) be defined as a set of points that are equidistant from a given line and a given point that is not on the given line. The given line is the **directrix** and the given point is the **focus** of the parabola.

On a plane the set of points P such that the *sum* of the distances of P from two given points **(foci)** is a constant is an **ellipse**. If the foci are

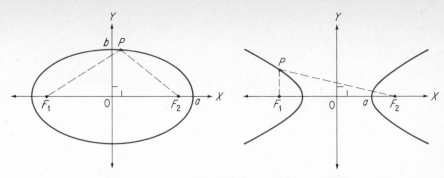

F_1: $(-c, 0)$ and F_2: $(c, 0)$, then for $m(\overline{PF_1}) + m(\overline{PF_2}) = 2a$ and for $b = \sqrt{a^2 - c^2}$, the ellipse has an equation of the form

$$\frac{x^2}{a^2} + \frac{y^2}{b^2} = 1.$$

The set of points P such that a *difference* of the distances of P from two given points (**foci**) is a constant is a **hyperbola**. If the foci are F_1: $(-c, 0)$ and F_2: $(c, 0)$, then for $|m(\overline{PF_1}) - m(\overline{PF_2})| = 2a$ and $b = \sqrt{c^2 - a^2}$, the hyperbola has an equation of the form

$$\frac{x^2}{a^2} - \frac{y^2}{b^2} = 1.$$

We shall use the general forms of the equations only to identify the graphs of such sentences. We shall not attempt to simplify equations of ellipses or hyperbolas to obtain equations in a general form.

Example 2: Without sketching the graph, identify the graph of each sentence as a circle, ellipse, hyperbola, or parabola.

(a) $\dfrac{x^2}{1} + \dfrac{y^2}{4} = 1$ (b) $\dfrac{x^2}{2} + \dfrac{y^2}{2} = 1$

(c) $x^2 - y = 3$ (d) $x^2 - y^2 = 1$

Solution: (a) Ellipse; (b) circle; (c) parabola; (d) hyperbola.

Notice that an ellipse with $a = b$ in its general equation is a circle. In other words, a circle is a special case of an ellipse. However, in making identifications we identify the circles as circles.

Example 3: Write an equation for the set of points P: (x, y) such that the sum of the distances of P from the points $(2, 1)$ and $(6, 1)$ is 8.

Solution: $\sqrt{(x - 2)^2 + (y - 1)^2} + \sqrt{(x - 6)^2 + (y - 1)^2} = 8$
Notice that simplification of the equation to a standard form is *not* requested.

EXERCISES

Without sketching the graph, identify the graph of each sentence as a circle, ellipse, hyperbola, or parabola.

1. $x^2 + y^2 = 4$
2. $x^2 - y^2 = 4$
3. $x^2 - 4y^2 = 4$
4. $x^2 + 4y^2 = 4$
5. $y = x^2 + 1$
6. $y^2 = x^2 + 1$

On a coordinate plane give an algebraic representation for:

7. The points that are equidistant from the line $y = -2$ and the point $(0, 4)$.
8. The points that are equidistant from the line $y = -p$ and the point $(0, p)$.
9. The points that are equidistant from the line $x = -3$ and the point $(3, 0)$.
10. The points that are equidistant from the line $x = -g$ and the point $(g, 0)$.
11. The points P: (x, y) such that the sum of the distances of P from F_1: $(-3, 0)$ and F_2: $(3, 0)$ is 10 units.
12. The points P: (x, y) such that a difference of the distances of P from the points F_1: $(-5, 0)$ and F_2: $(5, 0)$ is 8 units.

PEDAGOGICAL CONSIDERATIONS

1. There are many interesting applications of the conic sections that are suitable for presentation to elementary school mathematics classes. Some of these deal with orbits of planets, trajectories of missiles, parabolic reflectors in telescopes, etc. Prepare a 15 minute talk, with illustrations, that could be used as an enrichment lesson.

2. Curves from straight lines is an interesting supplementary topic for elementary school classes. Note the appearance of a parabola derived from straight lines only in the following figure. The endpoints of the line segments are equally spaced on the sides of the angle.

Prepare additional figures of your own. Then try to repeat the figure shown, using needle and thread for construction of the line segments.

414

3. Try to construct an ellipse in the following manner. Use a piece of string and two suction cups on the blackboard. (Chewing gum at the ends of the string is a suitable, albeit unhygienic alternative!) Place a piece of chalk at some point P on the string and pull it taut. Then trace a figure in this manner as shown.

12-5

coordinates in space

We may think of the xy-plane in space with a line $\overleftrightarrow{OZ}$ perpendicular to the xy-plane at the origin. It is customary to select the coordinate axes $\overleftrightarrow{OX}$, $\overleftrightarrow{OY}$, and $\overleftrightarrow{OZ}$ as in the figure, where the tick marks indicate the unit points on each axis.

In space the locus of points 1 unit from the xy-plane consists of two planes; that is, the plane $z = 1$ unit above the xy-plane and the plane $z = -1$ one unit below the xy-plane. Any plane parallel to the xy-plane has an equation of the form

$$z = t.$$

Similarly, any plane parallel to the *xz*-plane has an equation of the form

$$y = n.$$

Any plane parallel to the *yz*-plane has an equation of the form

$$x = k.$$

We locate points (k, n, t) as the intersection of these loci. By convention the *x*-coordinate is always listed first, the *y*-coordinate second, and the *z*-coordinate third in specifying the coordinates of a point in space as an **ordered triple** of real numbers (k, n, t).

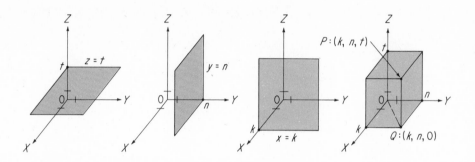

Each point in space has a unique ordered triple of real numbers as its *coordinates* with respect to a given set of coordinate axes; each ordered triple of real numbers has a unique point in space as its *graph*. When k, n, and t are each different from zero, we frequently can visualize the position of the point P: (k, n, t) after sketching a rectangular parallelepiped with a vertex at the origin, three edges on the coordinate axes, and a vertex at P.

EXERCISES

In the last figure in this section a rectangular box (parallelepiped) is sketched in coordinate space to assist in the identification of the position of the given point P: (k, n, t). The rectangular box in the figure has $m(\overline{AB}) = 2$, $m(\overline{AF}) = 4$, and $m(\overline{AD}) = 3$.

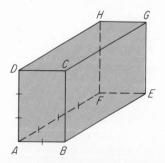

Copy the box ABCDEFGH and give the coordinates of each vertex if:

1. The origin is at F with $\overrightarrow{OX} = \overrightarrow{FA}$, $\overrightarrow{OY} = \overrightarrow{FE}$, and $\overrightarrow{OZ} = \overrightarrow{FH}$.

2. The origin is at A with $\overrightarrow{OX}$ opposite $\overrightarrow{AF}$, $\overrightarrow{OY} = \overrightarrow{AB}$, and $\overrightarrow{OZ} = \overrightarrow{AD}$.

3. The origin is at E with $\overrightarrow{OX} = \overrightarrow{EB}$, $\overrightarrow{OY}$ opposite $\overrightarrow{EF}$, and $\overrightarrow{OZ} = \overrightarrow{EG}$.

4. The origin is at H with $\overrightarrow{OX} = \overrightarrow{HD}$, $\overrightarrow{OY} = \overrightarrow{HG}$, and $\overrightarrow{OZ}$ opposite $\overrightarrow{HF}$.

5–8. Repeat Exercises 1–4 for the next box with $m(\overline{AB}) = 5$, $m(\overline{AF}) = 7$, and $m(\overline{AD}) = 2$.

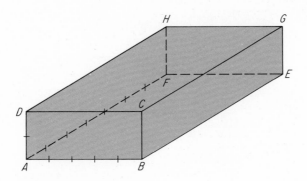

Graph each equation.

9. $z = 3$ 10. $x = 5$

11. $y = -1$ 12. $z = -2$

Graph each point.

13. $(3, 2, 1)$ 14. $(2, 1, 3)$

15. $(1, 2, 2)$ 16. $(3, -1, 2)$

Describe the set of points such that:

17. $x = 0$ 18. $y = 0$

19. $z = 0$ 20. $x = a$

21. $y = b$ 22. $z = c$

23. $\begin{cases} x = 0 \\ y = 0 \end{cases}$ 24. $\begin{cases} y = 0 \\ z = 0 \end{cases}$

25. $\begin{cases} x = 0 \\ z = 0 \end{cases}$ 26. $\begin{cases} x = 1 \\ z = 0 \end{cases}$

Describe each set of points in space by an equation, inequality, or system of such statements.

27. The points 5 units above the *xy*-plane.

28. The points above the *xy*-plane.

29. The points at least 2 units above the *xy*-plane.

30. The points at least 1 and at most 3 units above the *xy*-plane.

*31. The points within 2 units of the *yz*-plane.

*32. The points within 3 units of the *xz*-plane.

12-6

distances in space

For any point $P: (k, n, t)$ that is in a coordinate space but not on a coordinate plane, we consider also $Q: (k, n, 0)$. Then for $R: (k, 0, 0)$ there is a right triangle ORQ and by the Pythagorean theorem

$$m(\overline{OQ}) = \sqrt{k^2 + n^2}.$$

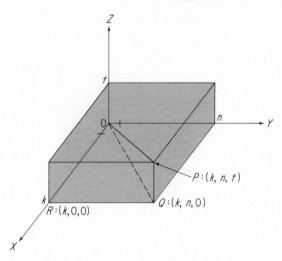

Also there is a right triangle OQP and by the Pythagorean theorem

$$m(\overline{OP}) = \sqrt{[m(\overline{OQ})]^2 + t^2} = \sqrt{k^2 + n^2 + t^2}.$$

This formula holds whether or not P is on a coordinate plane.

Example 1: Find $m(\overline{OP})$ for $P: (1, 5, 7)$.

Solution:

$$m(\overline{OP}) = \sqrt{1^2 + 5^2 + 7^2} = \sqrt{75} = 5\sqrt{3}$$

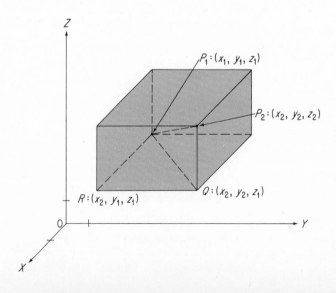

Let P_1: (x_1, y_1, z_1) and P_2: (x_2, y_2, z_2) be any two points that are not on a plane that is parallel to a coordinate plane. Then, as in the figure, we can identify the vertices of a rectangular parallelepiped with P_1 and P_2 as vertices and with its edges parallel to the coordinate axes. For Q: (x_2, y_2, z_1) we have

$$m(\overline{P_1Q}) = \sqrt{(x_2 - x_1)^2 + (y_2 - y_1)^2}$$

and from the right triangle P_1QP_2

$$m(\overline{P_1P_2}) = \sqrt{[m(\overline{P_1Q})]^2 + [m(\overline{P_2Q})]^2}$$
$$m(\overline{P_1P_2}) = \sqrt{(x_2 - x_1)^2 + (y_2 - y_1)^2 + (z_2 - z_1)^2}.$$

This **distance formula** holds for any points P_1 and P_2 in space.

Example 2: Find $m(\overline{AB})$ for A: $(2, -3, 5)$ and B: $(7, 1, 11)$.

Solution:

$$\begin{aligned} m(\overline{AB}) &= \sqrt{(7 - 2)^2 + [1 - (-3)]^2 + (11 - 5)^2} \\ &= \sqrt{5^2 + 4^2 + 6^2} \\ &= \sqrt{25 + 16 + 36} = \sqrt{77} \end{aligned}$$

The sphere with center (a, b, c) and radius r is the set of points at a distance of r from the point (a, b, c). The distance of any point P: (x, y, z) from the point (a, b, c) is

$$\sqrt{(x - a)^2 + (y - b)^2 + (z - c)^2}.$$

Therefore, the sphere with center (a, b, c) and radius r has equation

$$(x - a)^2 + (y - b)^2 + (z - c)^2 = r^2.$$

Example 3: Find an equation for the sphere with center $(0, 0, 0)$ and radius 3.

Solution: From the formula, $(x - 0)^2 + (y - 0)^2 + (z - 0)^2 = 9$, that is, $x^2 + y^2 + z^2 = 9$.

Example 4: Find an equation for the sphere with center $(5, -7, 11)$ and radius 4.

Solution:

$$(x - 5)^2 + (y + 7)^2 + (z - 11)^2 = 16$$

Example 5: Describe the set of points at most 2 units from the point $(1, -2, 3)$ by an inequality.

Solution:

$$(x - 1)^2 + (y + 2)^2 + (z - 3)^2 \leq 4$$

EXERCISES

Find the linear measure of $\overline{AB}$ for:

1. A: (1, 0, 2) and B: (1, 4, 5).
2. A: (2, 7, −3) and B: (1, 5, 0).
3. A: (2, −1, 0) and B: (3, 1, 2).
4. A: (1, 5, −6) and B: (−2, 3, −1).

Find an equation for the sphere with:

5. Center A: (1, −2, 5) and radius 2.
6. Center B: (2, 3, −4) and radius 5.
7. Center C: (−1, −2, 3) and radius 4.
8. Center D: (5, −7, 11) and radius 6.

Describe the set of points such that:

9. $x^2 + y^2 + z^2 = 5$
10. $x^2 + y^2 + z^2 = r^2$
11. $(x - 1)^2 + (y - 2)^2 + (z - 3)^2 = 4$
12. $(x - 2)^2 + (y + 1)^2 + (z + 2)^2 = 9$
13. $(x + 3)^2 + (y - 1)^2 + (z + 1)^2 = 1$

Describe each set of points by an equation, inequality, or system of such statements.

14. The points 3 units from the origin.
15. The points 2 units from (1, −2, 3).
16. The points at most 2 units from (2, 3, 4).
17. The points at least 2 units from (2, 3, 4).
18. The points at most 3 units from (1, 2, −3).
*19. The points at least 2 and at most 5 units from (1, 6, 4).
*20. The points at least 5 and at most 7 units from (2, −3, 5).

CHAPTER TEST

On a coordinate plane:

1. Find the linear measure and the coordinates of the midpoint of the line segment with endpoints (2, −5) and (7, 7).
2. Find the coordinates of the endpoint B of a line segment $\overline{AB}$ with endpoint A: (2, 6) and midpoint C: (5, 1).

3. Find the slope intercept form of the equation of the line determined by the points $(1, 5)$ and $(3, 2)$.

4. Describe the set of points such that:
   (a) $x^2 + y^2 \leq 81$ (b) $(x - 1)^2 + (y + 2)^2 > 4$

5. Find an algebraic representation for the circle with center $(5, -2)$ and radius 3.

6. Find an algebraic representation for the points that are equidistant from the line $x = -5$ and the point $(5, 0)$.

7. Without sketching the graph, identify the graph of each sentence as a circle, ellipse, hyperbola, line, or parabola.
   (a) $x^2 + 4y^2 = 1$ (b) $4x^2 + 4y^2 = 1$
   (c) $4x^2 + y = 1$ (d) $4x^2 - 4y^2 = 1$
   (e) $4x - y = 1$

In coordinate space:

8. Give an algebraic representation for the y-axis.

9. Find $m(\overline{AB})$ for A: $(1, 5, 3)$ and B: $(7, 1, 5)$.

10. Find an equation for a sphere with center $(2, -3, 1)$ and radius 5.

13

An Introduction to Statistics

The average consumer is besieged daily with statistical data that is presented over radio and television, in newspapers, and in various other media. He is urged to buy a certain commodity because of statistical evidence presented to show its superiority over other brands. He is cautioned *not* to consume a particular item because of some other statistical study carried out to show its danger. He is told to watch certain programs, read certain magazines, see certain movies, and eat certain foods because of evidence produced to indicate the desirability of these acts as based upon data gathered concerning the habits of others.

Almost every issue of the daily newspaper presents data in graphical form to help persuade the consumer to follow certain courses of action. Thus we are told what the "average" citizen eats, what he reads, what he earns, and even what he does with his leisure time.

Because of the widespread use of statistics for the consumer, we find that even elementary school children are ready to quote facts that they read or hear, often without real understanding. Unfortunately, this is equally true for the adult. Therefore in this chapter we shall make an effort to acquaint the reader with a sufficient number of basic concepts so that he may better understand and interpret statistical data.

Statistical data are often presented incorrectly to impress or mislead the reader. We shall consider several imaginary misuses as samples of the sort of abuses of statistics that occur daily. Consider the statement:

Brush your teeth with GLUB and you will have fewer cavities.

The basic question that this statement fails to answer is: "Fewer cavities than what?" Regardless of the merits of GLUB, it is probably safe to say the following of every brand of toothpaste: "Brush your teeth with XXXX and you will have fewer cavities than *if you never brush your teeth at all.*" In other words, the example given is misleading in that it implies the superiority of GLUB without presenting sufficient data to warrant comparisons. Here is another example:

There is a high degree of correlation between expenditures on education and alcohol. Education leads one to drink.

It may well be that some students face examinations in school with such dread that alcohol becomes their only salvation. Hopefully, this book will not drive anyone to drink. There is no doubt that expenditures on education and alcohol are correlated; that is, both items have seen increases in recent years. This does not mean, however, that the two items are related or that one necessarily affects the other. Thus the conclusion drawn from the given fact is unwarranted and misleading.

Many types of statements must be examined very carefully to separate their statistical implications from the impressions they are intended to make. Consider this statement:

More people are killed in automobile accidents than in wars. Therefore, it is safer to be on the battlefield.

Although the first statement may be true, the second one does not necessarily follow. Inasmuch as there are generally many more people at home driving cars than there are in combat, the *per cent* of deaths on the battlefield is actually much higher than in automobiles. Consider this advertisement:

95% of mothers who use BOON milk for their babies do so on advice of their doctors.

This statement is made to impress the consumer. It is hoped that he will then conclude, incorrectly of course, that 95% of all doctors recommend BOON milk for children. Actually the claim being made, if true, is still not very impressive. Most modern mothers follow doctor's recommendations concerning their babies' formula. Therefore, this statement could undoubtedly be made not only of BOON milk, but also of every other milk used for babies.

Frequently graphs are used to mislead readers:

Why not subscribe to MATH magazine? The next graph shows our growth of new subscribers over the past five years.

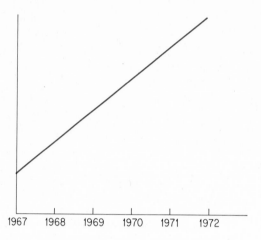

The visual impression given by the graph is that of tremendous growth over a five-year period of time. However without any scale on the vertical axis it is not possible to draw any conclusion other than that there has been a change in the circulation. If the vertical scale is oriented upward and there has been an increase, this increase may be quite small, since we have no assurance that the vertical axis begins with 0. For example, the actual facts may be as in the next graph, where there has been an increase, but only a very modest one of 400 readers (an increase of 0.4%).

Notice that in the preceding graph the scale on the vertical axis did not begin at 0, as indicated by the broken line near the base. This is what

425

*sec. 13-1
uses
and
misuses
of
statistics*

the graph of the same data would look like if the horizontal scale were set to begin at 0. This time the graph shows the correct picture, namely a very gradual and almost imperceptible increase over the five-year period.

At times data is presented graphically to purposefully mislead a reader. In the following graph, let us assume that the data presented is correct. Note, however, how the picture graph distorts the data by showing the box representing KORNIES to be not only three times as high as the OATIES picture, but also three times as wide. Thus the area of the KORNIES box is nine times as large as the OATIES box, instead of three times as large as indicated by the data. With the numerical facts absent, as is often the case, the picture would be completely dishonest.

The appearance of a contrast between the sales of KORNIES and OATIES is even more striking when three-dimensional boxes are pictured. (See page 426.) Then the volumes of the boxes are visualized and the volume of the KORNIES box is twenty-seven times that of the OATIES box.

The collecting of actual examples of uses of statistics to mislead readers can be a very interesting activity. In particular, an analysis of advertising claims can be rewarding in the insight it brings to an understanding of basic statistical concepts.

The intelligent consumer needs to ask basic questions about statistical statements that he hears or reads. Unfortunately too many people tend to accept as true any statement that contains numerical data, probably because such statements sound impressive and authentic. Consider, for example, this statement:

Two out of three adults no longer smoke.

Before accepting and repeating such a statement, one should ask such questions as the following:

Does the statement sound reasonable?
What is the source of this data?
Can the facts be verified?
Do the facts appear to conform with your own observations?

The reader will undoubtedly think of other questions to raise. The important thing is that one raises these questions, and does not accept all statistical facts as true in general. Furthermore this is not to imply that all such facts are biased or untrue. Rather, as we have seen and shall see later in this chapter, even true statements can be presented in many ways so as to provide various impressions and interpretations.

Finally, one should examine all statements to search for ambiguous words or meaningless comparisons. Thus, consider this assertion:

If you take vitamin C daily, your health will be better.

In addition to the questions one might raise about the scientific basis for this statement, note that the word "better" is ambiguous. We need to ask the question: "Better than what?" That is, we need to know the basis for comparison. As an extreme case, we certainly can say that your health will be better than one who is desperately ill! The exercises that follow will provide the reader with an opportunity to test his powers of reasoning and questioning.

EXERCISES

Discuss each of the following examples and tell what possible misuse or misinterpretation of statistics each one involves.

1. Most automobile accidents occur near home. Therefore one is safer taking long trips than short ones.

2. Over 95% of the doctors interviewed endorse SMOOTHIES as a safe cigarette to smoke. Therefore it is safe to smoke SMOOTHIES.

3. More college students are now studying mathematics than ever before. Therefore mathematics must be a very popular subject.

4. Professor X gave out more A's last semester than Professor Y. Therefore, one should try to enroll in X's class next semester rather than in Y's class.

5. At a certain college 100% of the students bought *Elements of Mathematics* last year. Therefore this must be a very popular book at that school.

6. Most accidents occur in the home. Therefore it is safer to be out of the house as much as possible.

7. Over 75% of the people surveyed favor the Democratic candidate. Therefore he is almost certain to win the election.

8. A psychologist reported that the American girl kisses an average of 79 men before getting married.

9. In July, 1960 a magazine reported the presence of 200,000 stray cats in New York City.

10. A recent report cited the statistic of the presence of 9,000,000 rats in New York City.

11. Arizona has the highest death rate in the nation for asthma. Therefore if you have asthma, you should not go to Arizona.

12. During World War II more people died at home than on the battlefield. Therefore it was safer to be on the battlefield than to be at home.

13. In a pre-election poll, 60% of the people interviewed were registered Democrats. Thus the Democratic candidate will surely win the election.

14. Over 90% of the passengers that fly to a certain city do so with Airline X. It follows that most people prefer Airline X to other airlines.

15. A newspaper exposé reported that 50% of the children in the local school system were below average in arithmetic skills.

427

Assume that each of the following statements are made. For each one, list two or three questions that you would wish to raise before accepting the statement as true. Identify words, if any, that are ambiguous or misleading.

16. Most people can swim.

17. Short men are more aggressive than tall men.

18. If you walk three miles a day you will live longer.

19. Brand *A* aspirin is twice as effective as Brand *B*.

20. Teen-agers with long hair are happier individuals.

21. People who swim have fewer heart attacks.

22. If you sleep 8 hours the night before an exam you will do better.

23. Most students have success with this textbook.

24. Teachers tend to be less conscious about social ills.

25. About 3 out of 4 college students marry within one year of graduation.

Use the following graph for Exercises 26–27.

Average monthly temperature

26. Disregarding the scale, what visual impression do you get from this graph about the fluctuation of temperatures during the year?

27. In what ways, if any, is the graph misleading?

Use the following bar graph to answer Exercises 28–30.

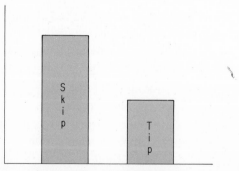

Circulation of two leading magazines

429

sec. 13-1

*uses
and
misuses
of
statistics*

28. Can one deduce from the graph that the circulation of **SKIP** is greater than the circulation of **TIP**?

29. Can one deduce from the graph that the circulation of **SKIP** is approximately twice as great as the circulation of **TIP**?

30. What, if anything, could possibly be misleading or deceiving about the graph?

PEDAGOGICAL CONSIDERATIONS

1. Read *How to Lie With Statistics* by Darrell Huff, New York: W. W. Norton and Co., Inc., 1954.

Begin a collection from newspapers, magazines, and other media of examples of misuses of statistics. In particular look for examples with one or more of the following:

2. Scales that have been cut off to mislead the reader.

3. Areas or volumes used to exaggerate comparisons.

4. Parts of the figure magnified or enlarged to provide undue emphasis.

Discuss the following classic quotes:

5. Facts are facts.

6. Figures don't lie but liars figure.

"Looks better this way"

Cartoon by Lurie, LIFE Magazine © Time Inc.
Reproduced by permission.

13-2

collecting and presenting data

Although the typical consumer needs to be able to interpret rather than to present data, a discussion of methods of presentation should serve to help develop skills of interpretation. As an illustration of such methods, let us assume that four coins have been tossed 32 times and that each time the number of heads showing has been recorded. This **raw data** is listed in the order of the occurrence of the events.

Number of heads

1	2	3	2	4	2	2	1
0	2	4	2	2	1	3	3
2	3	1	1	2	2	3	0
4	1	3	2	4	2	1	2

The information is somewhat more meaningful when the number of heads are tallied and summarized in tabular form as **frequency distribution.**

Number of heads	Tally	Frequency
0	II	2
1	ℕ II	7
2	ℕ ℕ III	13
3	ℕ I	6
4	IIII	4

This information can now be treated graphically in a number of different ways. One common form of presenting data is by means of a **histogram,** a bar graph without spaces between the bars.

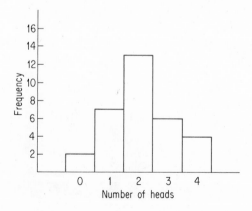

Frequently a histogram is used to construct a **frequency polygon** or **line graph.** Thus the histogram can be approximated by a line graph by

430

431

sec. 13-2
collecting
and
presenting
data

connecting the midpoints of the tops of each bar. The graph is then extended to the base line as in the figure.

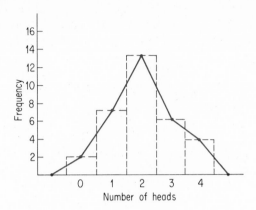

Line graphs appear to associate values with all points on a continuous interval rather than just with points with integers as coordinates. Thus such line graphs are more appropriate for **continuous data** (all intermediate values have meaning) rather than with **discrete data** (only isolated values have meaning). For example, consider a graph of the temperatures in a particular city. Although it was 40° at 5 P.M. and 35° at 6 P.M., the temperature must have passed through *all* possible values between 40° and 35° in one hour. This, then, is an example of continuous data. Temperatures may be plotted and connected to produce a line graph as in the figure. Since gradual changes are expected, we "smooth out" the graph as a curve without "sharp turns."

Hour	6 A.M.	7	8	9	10	11	12	1	2	3	4	5	6
Temp.	32	35	35	38	40	42	45	46	45	42	41	40	35

There are many different types of graphs used to present statistical data. One very popular type is the **circle graph,** which is especially effective when one wishes to show how an entire quantity is divided into parts. Local and federal government documents frequently use this type of graph to show the distribution of tax money, and various budget distributions. As an example, consider a family that plans to distribute their income according to the following guide:

Food:	40%
Household:	25%
Recreation:	5%
Savings:	10%
Miscellaneous:	20%
Total:	100%

To draw a circle graph for this data, we first recognize that there are 360° in a circle. Then we find each of the given per cents of 360, and with the aid of a protractor construct central angles of the appropriate sizes as in the figure.

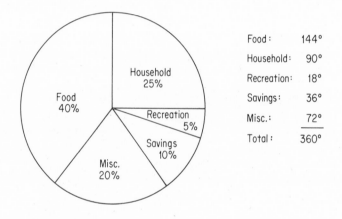

Food:	144°
Household:	90°
Recreation:	18°
Savings:	36°
Misc.:	72°
Total:	360°

The same data can also be presented in the form of a divided bar graph. Here we select an arbitrary unit of length to represent 100%, and divide this in accordance with the given per cents.

Food 40%	Household 25%	Rec. 5%	Savings 10%	Misc. 20%

There is an almost endless supply of graphs that one could use to illustrate this method of presenting data. Indeed, the reader will note from observation that he is constantly subjected to graphical presentations in the daily newspapers and in other periodicals as well. Other examples of graphs are given in the exercises that follow.

EXERCISES

1. Toss four coins simultaneously and record the number of heads obtained. Repeat this for a total of 32 tosses of four coins. Present your data in the form of a frequency distribution.

2. Present the data for Exercise 1 in the form of a bar graph.

3. Present the data for Exercise 1 in the form of a line graph.

4. Repeat Exercise 1 for 32 tosses of a set of five coins.

5. Present the data for Exercise 4 in the form of a bar graph.

6. Present the data for Exercise 4 in the form of a line graph.

7. Here is the theoretical distribution of heads when four coins are tossed for a total of 64 times:

Number of heads	Frequency
0	4
1	16
2	24
3	16
4	4

Present this data in the form of a bar graph.

8. Here is the theoretical distribution of heads when five coins are tossed for a total of 64 times:

Number of heads	Frequency
0	2
1	10
2	20
3	20
4	10
5	2

Present this data in the form of a bar graph.

9. Construct a circle graph to show this distribution of time spent by one student:

Sleep:	8 hours
School:	6 hours
Homework:	4 hours
Eating:	2 hours
Recreation:	4 hours
Total:	24 hours

10. Construct a divided bar graph for the data of Exercise 9.

Use the following graph to answer Exercises 11–17.

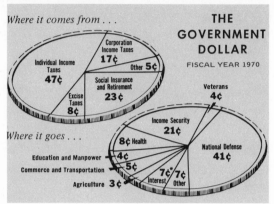

Where it comes from . . .

THE
GOVERNMENT
DOLLAR

FISCAL YEAR 1970

Corporation
Income Taxes
17¢

Other **5¢**

Individual Income
Taxes
47¢

Social Insurance
and Retirement
23¢

Excise
Taxes
8¢

Veterans
4¢

Where it goes . . .

8¢ Health

Income Security
21¢

National Defense
41¢

Education and Manpower **4¢**

Commerce and Transportation **5¢**

7¢ Interest **7¢** Other

Agriculture **3¢**

Executive Office of the President/Office of Management and Budget

*Excludes Interest Paid to Trust Funds

11. What per cent of the government dollar comes from individual income taxes?

12. What per cent of the government dollar goes to national defense?

13. Can you tell from the graphs **(a)** what per cent of the government expenditures were for health matters? **(b)** how many dollars were spent on health matters? Explain your answers.

14. What is the size of the central angle in the graph that shows the amount spent for commerce and transportation?

15. To the nearest degree, what is the size of the central angle that shows the amount obtained from individual income taxes?

16. Of every million dollars collected, how many dollars were collected from excise taxes?

17. Of every million dollars spent, how many dollars were spent for health?

Insurance companies compile data on deaths, and use this to make predictions of life expectancy and to determine insurance rates. Use the following two graphs to answer Exercises 18–22.

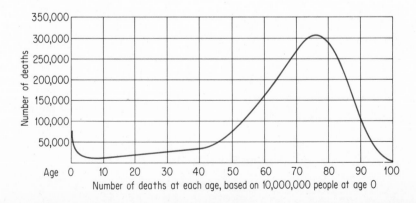

Number of deaths at each age, based on 10,000,000 people at age 0

435

sec. 13-2
collecting
and
presenting
data

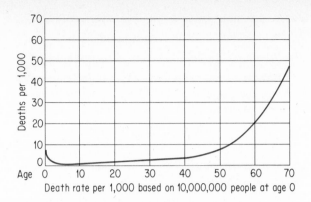

Death rate per 1,000 based on 10,000,000 people at age 0

18. At approximately what age do the largest number of deaths occur?

19. At approximately what ages are there 150,000 deaths?

20. Explain why both graphs show an initial decrease before both start to rise again.

21. Both graphs have the same shape at first. The latter portions, however, are quite different. Explain why this is so.

22. The graph of the death rate per 1000 stops at age 70. What do you think happens to this curve after age 70?

Use the following graph for Exercises 23–30.

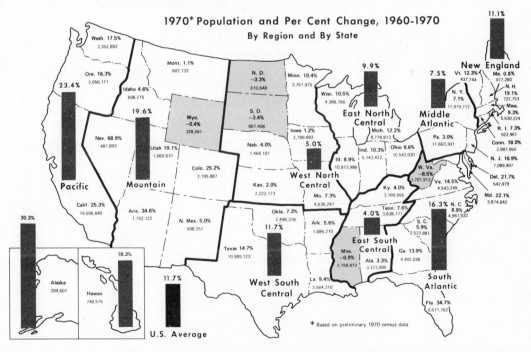

1970* Population and Per Cent Change, 1960-1970
By Region and By State

© 1970 the National Industrial Conference Board. Reproduced by permission.

23. Which states showed a decrease in population during the past decade?

24. Which state showed the largest per cent increase in population?

25. Which state has the largest population?

26. Which state showed the greatest per cent decrease in population?

27. Alaska shows a greater per cent increase in population than Hawaii. Yet the population of Hawaii is more than twice as great as that of Alaska. How can you explain this apparent paradox?

28. Within the continental U.S., which region showed the greatest per cent increase?

29. Within the continental U.S., which region showed the greatest per cent decrease?

30. What general conclusions can you form on the basis of this graph?

PEDAGOGICAL CONSIDERATIONS

1. Select an elementary mathematics textbook from grades 4, 5, or 6. Determine the amount of attention given to the construction and interpretation of graphs. Ascertain the number of different types of graphs presented.

2. Make a collection of graphs found in newspapers and magazines. See how many different types of graphs you can find.

3. Write for a copy of the booklet "Sets, Probability, and Statistics" by Clifford, Keiffer, and Sobel. This can be obtained free of charge from The Institute of Life Insurance, 277 Park Avenue, New York, New York, 10017.

13-3

measures of central tendency

Most of us have neither the ability nor the desire to digest large quantities of statistical data. Rather, we prefer to see a graphical presentation of such data, or some figure cited as representative of the entire collection. Thus we are often faced by such statements as:

"On the average, 9 out of 10 doctors recommend H_2O."

"The average family earns $4800 per year and has 2.3 children."

"The average college teacher gives out 21% A's each year."

The word "average" is used loosely and may have a number of different meanings. However, in each case it is used in an effort to provide a capsule summary of a collection of data by means of a single number.

Consider the following set of data, which represents the number of dates which each of 10 girls in Alpha Beta Gamma sorority had last semester:

$$6, \ 7, \ 8, \ 9, \ 10, \ 12, \ 15, \ 15, \ 20, \ 28$$

Each of the following statements is now correct:

1. *The average sorority member had 13 dates last semester.*

In this case the average is computed by finding the sum of the given numbers and dividing the sum by 10, the number of girls. This is the most commonly used type of average, is referred to as the **arithmetic mean,** or simply the **mean,** of a distribution, and is generally denoted by the symbol $\overline{X}$.

6	
7	
8	
9	$\overline{X} = \frac{130}{10} = 13$
10	
12	Arithmetic mean: 13
15	
15	
20	
28	
130	

2. *The average sorority member had 11 dates last semester.*

In this case the word *average* is used to denote the number which divides the data so that half of the scores are above this number and half are below it.

$$
\left.
\begin{array}{l}
6 \\
7 \\
8 \\
9 \\
10
\end{array}
\right\} \text{Five scores below 11}
$$

— Median: 11

$$
\left.
\begin{array}{l}
12 \\
15 \\
15 \\
20 \\
28
\end{array}
\right\} \text{Five scores above 11}
$$

The number that divides a set of scores in this way is called the **median** of the distribution. In this case it is determined as the number midway between 10 and 12. Note that the data has to be arranged in order of size before the median can be found.

3. *The average sorority member had 15 dates last semester.*

In this case, the average is determined as the number that appears most frequently in the distribution. That is, it is correct to say that more girls had 15 dates last semester than any other number of dates. Such an average is referred to as the **mode** of a set of data. Obviously, in this case, the mode is the best average to use if the sorority wishes to impress others with the success of their members. Actually, however, the median is more representtive of the number of dates of the typical member of the sorority.

Each of the three types of averages that we have discussed is known as a **measure of central tendency**. That is, each average is an attempt to describe a set of data by means of a single representative number. Note, however, that not every distribution has a mode, whereas some may have more than one mode. The set of scores {8, 12, 15, 17, 20} has no mode. The set {8, 10, 10, 12, 15, 15, 17} is **bimodal** and has both 10 and 15 as modes.

Example 1: Find (a) the mean; (b) the median; and (c) the mode for this set of test scores:

$$72, 80, 80, 82, 88, 90, 96$$

Solution: (a) The sum of the seven scores is 588.

$$\overline{X} = \frac{588}{7} = 84$$

(b) The median is 82. This is the middle score; there are three scores below 82 and three above it.

(c) The mode is 80, since this score appears more frequently than any other one.

Example 2: Repeat Example 1 for these scores:

$$30, 80, 80, 82, 88, 90, 96$$

Solution: (a) $\overline{X} = \frac{546}{7} = 78$. (b) The median is still 82. (c) The mode is still 80.

A comparison of Examples 1 and 2 indicates that the test scores are the same except for the first one. Note that the arithmetic mean is the only one of the three averages that is affected by the one low score. In general, the mean is affected by extreme scores, whereas the median and mode are not. Thus the arithmetic mean should be used as a representative of a set of data when extreme (high or low) scores should be reflected in the average. Otherwise, the arithmetic mean should not be used to describe a set of data that contains extreme scores unless deception is one's major objective.

Consider this example of the earnings of the employees of a small business run by a foreman and three other employees. The foreman earns $19,000 per year. The others earn $4800, $4500, and $4500, respectively. To impress the union, the owner claims that he pays his employees an average salary of $8200. He selects the arithmetic mean as the representative salary.

$$19{,}000 + 4800 + 4500 + 4500 = 32{,}800$$
$$32{,}800 \div 4 = 8200$$

Actually, the median, $4650, would present a fairer picture of the average, or typical, salary.

As a more extreme example, consider a group of 49 people with a mean income of $5000. Let us see what happens to the average income of the group when an additional person with an income of $100,000 joins the group.

$$49 \times 5000 = 245,000 \quad \text{(total income of the 49 people)}$$
$$1 \times 100,000 = 100,000$$
$$\text{Sum} = 345,000 \quad \text{(total income of the 50 people)}$$
$$\overline{X} = \frac{345,000}{50} = 6900$$

The average (arithmetic mean) salary is now $6900. The addition of this one extreme score has raised the mean $1900, whereas the median and the mode of the salaries have not been affected.

In reading or hearing advertisements, the consumer must always raise the question of the type of average being used. Furthermore, he should question the source and plausibility of the data. For example, one might wonder how a psychologist can say that the "average" American girl kisses 79 men before getting married!

EXERCISES

Find **(a)** *the mean;* **(b)** *the median; and* **(c)** *the mode for each set of data.*

1. 73, 79, 80, 82, 85, 85, 97

2. 62, 63, 67, 67, 72, 75, 75, 81, 86

3. 7, 12, 15, 16, 18, 21, 23, 24

4. 6, 7, 10, 10, 10, 13, 14

5. 87, 63, 70, 75, 93, 70, 95

In Exercises 6 through 8 tell which one, if any, of the three measures of central tendency seems most appropriate to represent the data that is described.

6. The average salary in a shop staffed by the owner and five employees.

7. The average salary of the workers in a factory that employs 100 people.

8. The average number of cups of coffee ordered by individual diners in a restaurant.

9. What relationship does the mode have to the use of this word in everyday language?

10. The mean score on a set of 15 tests is 74. What is the sum of the 15 test scores?

11. The mean score on a set of 30 tests is 82. What is the sum of the test scores?

12. The mean score on a set of 40 tests is 76. What is the sum of the test scores?

13. The mean score on nine of a set of ten tests is 80. The tenth score is 50. What is the sum of the test scores?

14. The mean score on 25 of a set of 27 tests is 78. The other two scores are 30 and 35. What is the sum of the scores?

15. Two sections of a course took the same test. In one section the mean score on the 25 tests was 80. In the other section the mean score on the 20 tests was 70. **(a)** What is the sum of the 45 test scores? **(b)** To the nearest integer what is the mean of the 45 test scores?

16. The mean score on five tests is 85. The mean on three other tests is 75. What is the mean score on all eight tests? (*Hint:* The answer is *not* 80.)

17. A student has a mean score of 85 on seven tests taken to date. What score must she achieve on the eighth test in order to have a mean score of 90 on all eight tests? Comment on your answer.

*18. An interesting property of the arithmetic mean is that the sum of the deviations (considered as signed numbers) of each score from the mean is 0. Show that this is true for the set of scores: 9, 10, 13, 17, 21.

PEDAGOGICAL CONSIDERATIONS

The term **percentile rank** is frequently found in educational literature, especially as it applies to scores on standardized tests. For example, when we say that someone's test score has a percentile rank of 80 we mean that 80% of the scores on this test fall below this particular grade. Of course this also means that 20% of all the scores lie above the given grade. A score with a percentile rank of 80 is also said to be at the eightieth percentile.

1. What is the percentile rank of the median of a distribution?

2. A student scores a grade of 127 on an aptitude test. What can you say about his aptitude as a result of this test score? Suppose you are then given the additional information that the score of 127 is at the ninety-fifth percentile in a distribution of scores for all students who have taken this particular test. What can you conclude then about his aptitude?

13-4

measures of dispersion

A measure of central tendency describes a set of data through the use of a single number. However, as in the examples considered in §13-3, a single number without other information can be misleading. Some information can be obtained by comparing the arithmetic mean and the median, since the mean is affected by extreme scores and the median is relatively stable. In this section we consider other ways of obtaining information about sets of data without considering all of the elements of the data individually.

A **measure of dispersion** is a number that provides some information on the variability of a set of data. The simplest such measure to use for a set of numbers is the **range**, the difference between the largest and the smallest number in the set.

Consider the test scores of these two students

$$\text{Betty:} \quad \{68, 69, 70, 71, 72\}, \qquad \overline{X} = 70$$
$$\text{Jane:} \quad \{40, 42, 70, 98, 100\}, \qquad \overline{X} = 70$$

The range for Betty's test scores is $72 - 68$, or 4; the range for Jane's test scores is $100 - 40$, or 60. The averages and ranges provide a more meaningful picture of the two students' test results than the averages alone:

$$\text{Betty:} \quad \overline{X} = 70, \qquad \text{range, } 4$$
$$\text{Jane:} \quad \overline{X} = 70, \qquad \text{range, } 60$$

Although the range is an easy measure of dispersion to use, it has the disadvantage of relying on only two extreme scores—the lowest and the highest scores in a distribution. For example, consider Bob's test scores:

$$\{40, 90, 95, 95, 100\}$$

Bob is quite consistent in his work, and there may well be some good explanation for his one low grade. However, in summarizing his test scores one would report a range of 60, since $100 - 40 = 60$.

There is another measure of dispersion that is widely used in describing statistical data. This measure is known as the **standard deviation**, and is denoted by the Greek letter σ (sigma). The standard deviation is relatively difficult to compute but can be very informative. Fortunately, the average consumer of statistics needs to be able to understand and interpret, rather than to compute standard deviations. This need is particularly acute for teachers since they need to interpret the performance of their students on standardized tests.

To understand the significance of standard deviations as measures of dispersion, we first turn our attention to a discussion of **normal distributions.** The line graphs of these distributions are the familiar bell-shaped curves that are used to describe distributions for so many physical phenomena. For example, the distribution of intelligence quotient (I.Q.) scores in the entire population of the United States can be pictured by a **normal curve.**

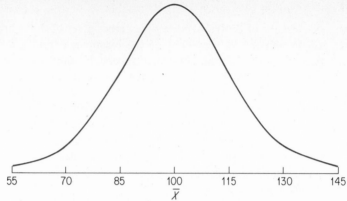

According to psychologists, a score over 130 is considered to be very superior. At the other end of the scale, a score below 70 generally indicates some degree of retardation.

A normal-appearing curve should be expected only for data that involve a large number of elements and are based upon a general population. For example, the I.Q. scores of the honor society members in a school should not be expected to fit the distribution for the school as a whole.

In a normal distribution the mean, the median, and the mode all have the same value. This common value is associated with the axis of symmetry of the normal curve.

If three standard deviations are added to and subtracted from the mean of a normal distribution, practically all (99.7%) of the data will fall on the interval from $\bar{X} - 3\sigma$ to $\bar{X} + 3\sigma$. If an interval of two standard deviations from the mean is considered, approximately 95% of all data is included. An interval of one standard deviation about the mean includes approximately 68% of all data in a normal distribution. We may summarize these statements as follows:

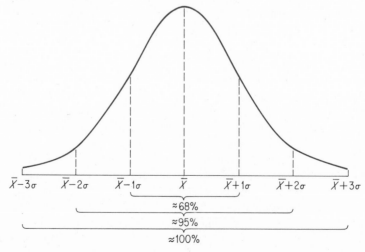

Let us now return to the graph of I.Q. scores. Suppose we are told that for this distribution $\sigma = 15$. We may then show these standard deviations on the base line of the graph. According to our prior discussion, we may now say that approximately 68% of the population have I.Q. scores between 85

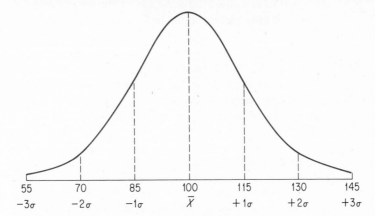

and 115, that is, on the interval $100 \pm 1\sigma$. Approximately 95% of the population have I.Q. scores between 70 and 130, that is, on the interval $100 \pm 2\sigma$. Finally, almost everyone has an I.Q. score between 55 and 145, that is, on the interval $100 \pm 3\sigma$. "Almost everyone" really means 99.7%; that is, we might expect 0.3% (3 in 1000) of the population to have scores below 55 or above 145.

Example 1: What per cent of the population have I.Q. scores less than 115?

Solution: We know that 50% of the population have I.Q. scores below 100, the mean. Also, 68% have scores on the interval $\overline{X} \pm 1\sigma$. Then by the symmetry of the normal curve, 34% have scores between 100 and 115, and, as in the figure, 84% have scores less than 115.

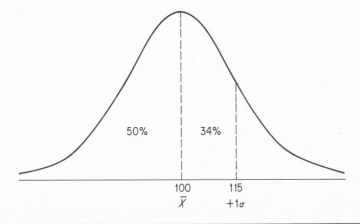

Notice that when considering questions such as that in Example 1 we do not concern ourselves with the number of people who make any particular score such as 100 or 115. The scale $-3, -2, -1, 0, 1, 2, 3$ in standard deviations enables us to use the properties of a normal distribution, such as the 68% within one standard deviation of the mean, for a wide variety of situations. For example, we found in Example 1 that 84% of the population have I.Q. scores less than 115, that is, less than one standard deviation above the mean. This same result can be used in many other situations. Scholastic Aptitude Test (S.A.T.) scores have a mean of 500 and a standard deviation of 100. Thus 84% of the S.A.T. scores on a particular test are below 600. College Entrance Examination Board (C.E.E.B.) scores also have a mean of 500, a standard deviation of 100, and 84% of the scores below 600.

Example 2: What per cent of the population have I.Q. scores greater than 130?

Solution: The interval $\overline{X} \pm 2\sigma$ includes the scores of 95% of the population, so 47.5% of the population have scores on the interval from 100 to 130. Thus 97.5% of the population have scores below 130, and 2.5% of the population have I.Q. scores above 130.

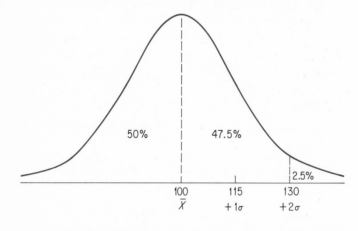

The result obtained in Example 2 as applied to S.A.T. and C.E.E.B. scores shows that 97.5% of the scores are below 700 and 2.5% of the scores are over 700.

Note that we have discussed only normal distributions. However, the standard deviation can be computed for other distributions as well. In general, many sets of data tend to be approximately normal, and the standard deviation is a very useful measure of dispersion.

As stated earlier, the standard deviation is relatively difficult to compute, especially for data grouped in a table. However, it may be instructive to note the computation of σ for a small set of scores. For such data, we generally use the formula

$$\sigma = \sqrt{\frac{\Sigma\, d^2}{n}}$$

where $\Sigma\, d^2$ represents the sum of the squares of the deviations of each score from the mean, and n is the number of scores.

Consider the set of scores

$$\{8,\ 10,\ 12,\ 16,\ 19\}.$$

We first compute the mean as 13. In the column headed d we find the differences when the mean $\overline{X}$ is subtracted from each score. The sum of these deviations is 0 as noted in §13-3, Exercise 18. Thus to obtain a meaningful average of these deviations we may use either absolute values or the squares of the deviations. The standard deviation is based upon the squares of the deviations, as in the column headed d^2.

Scores	d	d^2	
8	-5	25	$\sigma = \sqrt{\frac{80}{5}} = \sqrt{16}$
10	-3	9	$\sigma = 4$
12	-1	1	
16	3	9	
19	6	36	
5$\overline{)65}$		80	
$\overline{X} = 13$			

Recall that σ is a measure of dispersion. In Example 3 we have a set of scores with the same mean as the above set, but with a much smaller range. Note that the standard deviation is correspondingly smaller.

Example 3: Compute σ: $\{10, 11, 13, 14, 17\}$.

Solution:

Scores	d	d^2	
10	-3	9	$\sigma = \sqrt{\frac{30}{5}} = \sqrt{6}$
11	-2	4	$\sigma \approx 2.5$
13	0	0	
14	1	1	
17	4	16	
5$\overline{)65}$		30	
$\overline{X} = 13$			

These two sets of five scores for which standard deviations have been computed are much too small to expect a close match with a normal distribution. In each of these two cases all scores are within one and a half standard deviations of the mean.

1. Find the range for the set of scores:

$$\{55, 60, 80, 85, 90, 92, 98\}$$

2. Give a set of seven scores with the same mean as in Exercise 1 but with a smaller range.

3. Repeat Exercise 2 for a larger range.

4. Give a set of seven scores with the same range as in Exercise 1 but with a smaller mean.

5. Repeat Exercise 4 for a larger mean.

In Exercises 6 through 8 tell which of the distributions can be expected to be approximately normal in shape.

6. The scores of graduating high school seniors on a particular college board examination.

7. The weights of all college freshmen boys.

8. The number of heads obtained if 100 coins are tossed by each college graduate in the country.

9. For a normal distribution of 10,000 test scores the mean is found to be 500 and the standard deviation is 100.
   (a) What per cent of the scores will be above 600?
   (b) What per cent of the scores will be below 300?
   (c) About how many scores will be above 700?
   (d) About how many scores will be below 400?
   (e) About how many scores will be between 400 and 700?

10. If 100 coins are tossed repeatedly, the distribution of the number of heads is a normal one with a mean of 50 and a standard deviation of 5. What per cent of the number of heads will be: **(a)** greater than 60? **(b)** less than 45? **(c)** between 40 and 60?

11. Compute the standard deviation for the set of scores:

$$\{7, 9, 10, 11, 13\}$$

12. Compute the standard deviation for the set of scores:

$$\{11, 12, 13, 15, 20, 20, 21\}$$

PEDAGOGICAL CONSIDERATIONS

In reading educational literature one frequently comes across the term **coefficient of correlation,** usually denoted by the letter *r*. The coefficient of correlation is given as a decimal from −1.00 to +1.00 and provides an indication of how two variables are related.

A perfect positive correlation of 1.00 indicates that two sets of data are related so that as one increases, so does the other, and does so uniformly. For example, here is a set of ages and weights for a group of six individuals. Note that for each increase of 5 years of age there is a corresponding increase of 10 pounds of weight.

Age	Weight	
20	130	
25	140	
30	150	The coefficient of
35	160	correlation is 1.00.
40	170	
45	180	

Now consider the following table for six other individuals. For each increase of 5 years of age, there is a corresponding decrease of 10 pounds of weight.

Age	Weight	
20	180	
25	170	
30	160	The coefficient of
35	150	correlation is −1.00.
40	140	
45	130	

A coefficient of correlation of 0 indicates no uniform change of either variable with respect to the other. In general practice one would not expect to find these extreme cases occur. Furthermore one has to read the literature accompanying any particular test or research study to determine whether any particular coefficient of correlation can be considered as significant.

Finally, it is important to note that correlation does not imply causation. For example, there is a positive correlation between size of shoe that one wears and handwriting ability. Now this does not mean that big feet improves one's handwriting. Rather the explanation is probably that as one grows older, one's feet tend to get larger and also at the same time one tends to write more legibly.

1. Obtain a copy of a standardized test and read the instructions to see what references, if any, are made to correlation.

2. There has been a high positive correlation between expenditures for alcohol and for higher education in recent years. Does this mean that drinking alcohol provides one with the thirst for knowledge? Does it mean that education leads one to drink? How can you explain this high correlation?

447

3. Estimate whether there is a high or a low correlation between the following pairs of data, and attempt to justify your answer:
  (a) I.Q. scores and scores on college entrance examinations.
  (b) Scores made by elementary school students in reading and in arithmetic.
  (c) Age and physical ability.

13-5

binomial distributions

Any distribution of data for which there are only *two* possible events is a **binomial distribution.** The distribution of heads in tossing coins is a binomial distribution since each coin is assumed to have exactly one of two possible outcomes, heads or tails. A "fair" coin can be expected to land heads about half the time. Accordingly, we say that the *probability* (§14-2) of heads is $\frac{1}{2}$ on each toss of the coin; that is, the ratio of the number of heads to the total number of tosses is expected to be about $\frac{1}{2}$. If the experiment of tossing a coin n times is repeated over and over, the expected mean of the distribution of heads on these trials of the experiment is $\frac{1}{2}n$. The result of tossing one coin n times is the same as that of tossing n coins once. For example, if one coin is tossed 100 times or 100 coins are tossed once, we have probability $p = \frac{1}{2}$ and $n = 100$. Therefore, the mean or average number of heads expected is $(\frac{1}{2})(100) = 50$.

The standard deviation of a binomial distribution is easily found by the formula

$$\sigma = \sqrt{p \times (1 - p) \times n}$$

where p is the probability of success and n is the number of trials. For heads on 100 coins, we have $p = \frac{1}{2}$, $1 - p = \frac{1}{2}$, $n = 100$, and

$$\sigma = \sqrt{(\tfrac{1}{2})(\tfrac{1}{2})(100)} = \sqrt{25} = 5.$$

It has been shown for a normal distribution that if three standard deviations are added to and subtracted from the mean, we have the limits within which almost all of the data will fall. For the example of 100 coins, the mean is 50, and three times the standard deviation is 15, thus giving the limits 35 to 65. This is frequently stated in terms of **confidence limits**: we may say with "*almost* 100%" confidence that the number of heads will be between 35 and 65. Furthermore, we may say with approximately 95% confidence that the number of heads will be between 40 and 60, that is, within two standard deviations of the mean. We may say with approximately 68% confidence that the number of heads will be between 45 and 55, that is, within one standard deviation of the mean.

A **control chart** is a graph on which lines are drawn to represent the limits within which all data are expected to fall a given per cent of the time. In industry, for example, a chart like the following may be drawn on which data from samples are plotted.

The middle horizontal line represents the **mean** ($\overline{X}$) or average around which these data are expected to lie. The other horizontal lines, called the **upper control limit (UCL)** and **lower control limit (LCL)** represent the limits within which all the data are expected to fall most of the time. Occasionally data may fall outside of these limits; as long as most of the samples tested produce data which lie between these lines, the process is said to be "in control."

For the case of 100 coins, the theoretical average is 50. The lower control limit would be $\overline{X} - 3\sigma$, or 35. The upper control limit would be $\overline{X} + 3\sigma$, or 65. We would suspect that something was wrong if a toss of 100 coins were to produce fewer than 35 or more than 65 heads. Such an event *can* take place just by chance, but it would be very unusual, occurring less than 3 times in 1000.

Example 1: What is the mean number of fives to be expected in 180 tosses of a single die?

Solution: When dice (singular die) are tossed, the possible results are

$$1, \quad 2, \quad 3, \quad 4, \quad 5, \quad 6.$$

For a "fair" die these six possibilities are equally likely and we say that the probability of a five is one out of six that is, $\frac{1}{6}$. Thus we have $p = \frac{1}{6}$ and $n = 180$. Therefore, the average number of fives expected is $(\frac{1}{6})(180) = 30$. Theoretically we can expect one out of every six tosses to produce a five and expect 30 fives in 180 tosses.

Example 2: What is the standard deviation for the distribution of Example 1?

Solution: The probability of tossing a five is $\frac{1}{6}$. Thus $p = \frac{1}{6}$, $(1 - p) = \frac{5}{6}$, $n = 180$, and $\sigma = \sqrt{(\frac{1}{6})(\frac{5}{6})(180)} = \sqrt{25} = 5$.

The results of Examples 1 and 2 can be shown in terms of a graph as on the next page.

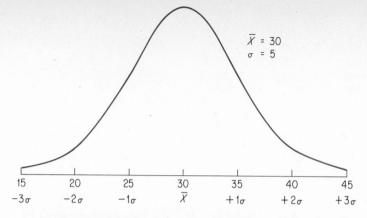

Thus, in a set of 180 tosses of a die, we would expect with almost 100% confidence to have between 15 and 45 fives. With 95% confidence we would expect to have between 20 and 40 fives; and with 68% confidence we would expect between 25 and 35 fives.

These examples should serve to show the reader that fluctuation is normal. That is, although the mean of the distribution is 30, we expect there to be variability. Not every set of 180 tosses will produce exactly 30 fives—in fact, very few will. However, on the other hand, we expect this fluctuation to be within limits that can be described by our knowledge of the behavior of normal distributions. Someone could conceivably toss a die 180 times and produce 100 fives. However, an understanding of the principles of this chapter would dictate that you challenge the tosser of the die and examine the die quite carefully for evidence of foul play. In other words, although it is possible to obtain 100 fives, it is far from probable that this would happen.

EXERCISES

In Exercises 1 through 5 consider a single die that is tossed 720 times.

1. What is the mean number of sixes to be expected?

2. What is the standard deviation for the distribution of sixes?

3. What is the maximum number of sixes we can expect to obtain with almost 100% confidence?

4. What is the minimum number of sixes we can expect to obtain with almost 100% confidence?

5. Within what limits can we expect the number of sixes to fall approximately 95% of the time?

6. Toss 100 coins 10 times and record the results on a control chart.

7. A coin is tossed 64 times. Within what limits can we say, with 95% confidence, that we will find the total number of heads?

In Exercises 8 through 11 consider a single die that is tossed 180 times.

8. What is the mean number of threes to be expected?

9. What is the maximum number of threes we can expect to obtain with almost 100% confidence?

10. What is the minimum number of threes we can expect to obtain with almost 100% confidence?

11. Within what limits can we expect the number of threes to fall approximately 95% of the time?

12. Every student in a class is asked to toss a coin 64 times and record the number of heads obtained. One student claims to have tossed 45 heads. React to his claim, basing your reaction on the statistical concepts presented in this section.

PEDAGOGICAL CONSIDERATIONS

1. Use the concepts of this section to establish the limits within which the total number of heads can be expected to fall when a coin is tossed 64 times, using the 95% confidence limits. Then test these results by tossing a coin 64 times, and asking as many other people as possible to repeat this experiment. (Note that instead of tossing a single coin for 64 tosses, you may also toss four coins for 16 tosses, sixteen coins for 4 tosses, etc.)

2. Repeat Exercise 1 for 180 tosses of a single die, counting the number of sixes that appear.

"It's made a lot of stupid mistakes."

1. Tell what possible misuse or misinterpretation of statistics each of the following statements involves:

   (a) Mathville's population increased by 100% last year. This is a greater per cent increase than for any other city in the state. Therefore the population of Mathville must be the largest of any city in that state.

   (b) Statistics show that college students drink more.

2. Draw a circle graph to show the distribution of grades given by Professor X last semester.

Grade	A	B	C	D	F
Per cent	20	25	40	10	5

3. Draw a histogram for the following data of the number of heads obtained on 32 tosses of five coins.

Number of heads	0	1	2	3	4	5
Frequency	1	4	11	9	5	2

Use the following graphs to answer Exercises 4–6.

© 1969 Wide World Photos, Inc. Reproduced by permission.

4. Assume that the world population in 1980 will be six billion. How many figures would have to be placed on top of the world map to represent this figure?

5. The number of births per year in North America is approximately what per cent of the population of North America?

6. What is the size of the central angle that represents the population of Africa?

7. Find **(a)** the mean; **(b)** the median; **(c)** the mode for this set of scores:

$$55, 62, 70, 74, 74, 79$$

8. Find the range for each of the following sets of scores:
   (a) 75, 82, 64, 98, 79   **(b)** 100, 45, 79, 82, 96, 68

9. A collection of 1000 scores forms a normal distribution with a mean of 50 and a standard deviation of 10.
   (a) What per cent of the scores will be above 70?
   (b) What per cent will be below 40?

10. A coin is tossed 64 times.
   (a) Within what limits can we say, with almost 100% confidence, that we will find the total number of heads?
   (b) If the experiment is repeated many times, what per cent of the times should be expected to produce fewer than 36 heads?

14

An Introduction to Probability

We make frequent reference to probability in everyday language. For example, we say, "It probably will rain," "The odds are in his favor," and make similar comments about many things.

Ever since the fifteenth century, mathematicians have been exploring the topic of probability. Interestingly enough, the subject is said to have had its foundation in the realm of gambling and to have arisen from a discussion of the distribution of stakes in an unfinished game. Probabilities are still used to understand games of chance such as the tossing of coins, throwing of dice, or drawing of lottery tickets. Probabilities are also used extensively by persons who are concerned with the cost of insurance (based upon rates of mortality), the construction of various polls such as the Gallup Poll to appraise public opinion, and numerous other types of statistical studies. The topics in this chapter provide a basis for understanding probabilities and their applications to real life situations.

14-1
counting problems

Many problems depend for their solution upon an enumeration of all possible outcomes. Thus, the simple task of counting becomes an important one in the study of probability. To illustrate various problems in this chapter, we shall invent a fictitious club consisting of a set *M* of members:

$$M = \{\text{Betty, Doris, Ellen, John, Tom}\}$$

Let us form a committee that is to consist of one boy and one girl, each selected from the set *M* of club members. How many such committees are possible? One way to answer this question is by means of a **tree diagram**, which helps list each of the possibilities.

For each of the two possible choices of a boy, there are three possible choices of a girl. Thus the following six distinct possible committees can be formed and can be read from the tree diagram:

John–Betty   Tom–Betty

John–Doris   Tom–Doris

John–Ellen   Tom–Ellen

Suppose that we had selected a girl first. Then the tree diagram would be as shown in the next figure and there would still be six possibilities, the same six committees as before.

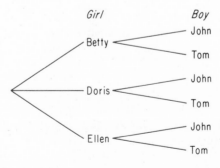

Example 1: In how many different ways can two officers, a president and a vice-president, be elected from the set *M* of club members?

Solution: Let us select the officers in two stages. There are five possible choices for the office of president. Each of these five selections may be paired with any one of the remaining four members. Thus there are 20 possible choices in all, which can be read from the following diagram.

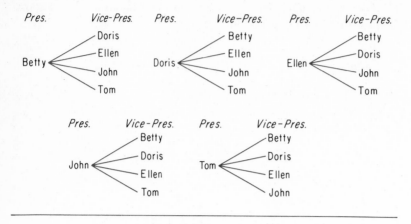

In general, if one task can be performed in *m* different ways and a second task can be performed in *n* different ways, then the first and second tasks together can be performed in $m \times n$ different ways. This general principle can be extended if there are additional tasks:

$$m \times n \times r \times \cdots \times t$$

Example 2: The club members *M* must send a delegate to a meeting tomorrow and also a delegate to a different meeting next week. In how many ways may these delegates be selected if any member of the club may serve as a delegate to each meeting?

Solution: There are five possible choices of a delegate to the first meeting. Since no restriction is made, we assume that the same member may attend each of the two meetings. Thus, there are five choices for the delegate to next week's meeting. In all, there are 5×5, that is, 25 choices.

Example 3: How many three-letter "words" may be formed from the set of vowels $V = \{a, e, i, o, u\}$ if no letter may be used more than once? (A word in this sense is any arrangement of three letters, such as aeo, iou, etc.)

Solution: There are five choices for the first letter, four for the second, and three for the third. In all, there are $5 \times 4 \times 3$, that is, 60 possible "words." Notice that 125 words would be possible if repetitions of letters were permitted.

EXERCISES

1. How many three-letter "words" may be formed from the given set $R = \{m, a, t, h\}$ if **(a)** no letter may be used more than once? **(b)** repetitions of letters are permitted?

2. Repeat Exercise 1 for four-letter words.

3. Bob has four sport shirts and three pairs of slacks. Assuming that he can wear any combination of these, how many different outfits can he assemble?

4. Jane has five dresses, three hats, and four pairs of shoes. Assuming that she can wear any combination of these, how many different outfits can she assemble?

5. A baseball team has six pitchers and four catchers. How many different batteries consisting of a pitcher and a catcher can they form?

6. Show, by means of a tree diagram, the number of different routes from New York to Los Angeles, via Chicago, if you can go from New York to Chicago by one train or one plane, and from Chicago to Los Angeles by one train, one plane, or one bus.

7. How many different two-digit numbers may be formed from the set of digits $D = \{1, 2, 3, 4, 5\}$ if repetitions of digits are allowed? How many numbers of two different digits can be formed from the set D?

8. Repeat Exercise 7 for $D = \{1, 2, 3, 4, 5, 6, 7\}$.

9. The Jonesboro Swim Club has 12 members. How many different sets of officers consisting of a president, a vice-president, and a secretary-treasurer can it form? No person can hold more than one of these three offices.

10. The Portland Swim Club has 250 members with 200 of these members eligible to hold any office. Repeat Exercise 9 for the Portland Swim Club.

11. How many five-digit numbers can be formed using the ten decimal digits without repetitions if zero is not to be used as the first digit?

12. How many two-digit even numbers can be formed without repetitions from the set $I = \{1, 2, 3, \ldots, 9\}$?

13. Find the number of "words" of three different letters that may be formed from the set $V = \{a, e, i, o, u\}$ if **(a)** the first letter must be i; **(b)** the first letter must be e and the last letter must be i.

14. How many license plates can be made using a letter from our alphabet followed by three decimal digits if the first digit must not be zero? How many are possible if the first digit must not be zero and no digit may be used more than once?

*15. Find the number of three-digit numbers that may be formed from the set $W = \{0, 1, 2, \ldots, 9\}$ if zero is not an acceptable first digit and

(a) the number must be even; (b) the number must be divisible by 5; (c) no digit may be used more than once; (d) the number must be odd and less than 500.

*16. In a certain so-called "combination" lock there are 50 different positions. To open the lock you move to a certain number in one direction, then to a different number in the opposite direction, and finally to a third number in the original direction. What is the total number of such "combinations":
(a) If the first turn must be clockwise?
(b) If the first turn may be either clockwise or counterclockwise?

*17. Find the number of different possible license plates if each one is to consist of two letters of the alphabet followed by three decimal digits, the first digit may not be zero, and no repetitions of letters or numbers are permitted.

*18. Repeat Exercise 17 for license plates consisting of three consonants followed by three decimal digits.

PEDAGOGICAL CONSIDERATIONS

Many interesting mathematical puzzles can be created to help students learn number facts and gain facility in computation in a novel and interesting way. Consider, for example, the following "dart" puzzles:

1. You are allowed to throw 4 darts, and we shall assume that there are no misses. In how many different ways can you obtain a score of 60?

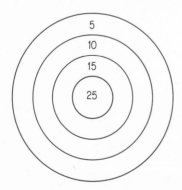

2. Again, use 4 darts and the figure at the top of the next page. Assume no misses and show two different ways of scoring 40.

3. Prepare several similar puzzles of your own.

14-2
definition
of
probability

When an ordinary coin is tossed, we know that there are two distinct and equally likely ways in which it may land, heads or tails. As in §13-5 we say that the probability of getting a head is one out of two, or simply $\frac{1}{2}$.

In rolling one of a pair of ordinary dice, there are six equally likely ways in which the dice may land. As in §13-5 we say that the probability of rolling a 5 on one toss of a die is one out of six, or $\frac{1}{6}$.

In each of these two examples, the events that may occur are said to be **mutually exclusive.** That is, one and only one of the events can occur at any given time. When a coin is tossed, there are two possible events (heads and tails); one and only one of these may occur. When a single die is rolled, there are six events (1, 2, 3, 4, 5, 6); one and only one of these may occur. Informally we define the probability of success as the ratio of the number of successes of an event to the number of possible outcomes of that event. More generally we define probability as follows:

> If an event can occur in any one of n mutually exclusive and equally likely ways, and if m of these ways are considered favorable, then the **probability** $P(A)$ that a favorable event A will occur is given by the formula
>
> $$P(A) = \frac{m}{n}.$$

The probability m/n satisfies the relation $0 \leq m/n \leq 1$, since m and n are integers and $m \leq n$. When success is inevitable, $m = n$ and the probability is 1; when an event cannot possibly succeed, $m = 0$ and the probability is 0. For example, the probability of getting either a head or a tail on a single toss of a coin is 1, assuming that the coin does not land on an edge. The probability of tossing a sum of 13 with a single toss of a pair of ordinary dice is 0. (Here, and in all future work, assume that normal dice are used unless otherwise instructed.)

The sum of the probability of an event's occurring and the probability of that same event's not occurring is 1.

$$\text{If} \quad P(A) = \frac{m}{n}, \qquad \text{then} \qquad P(\text{not } A) = 1 - \frac{m}{n}.$$

Example 1: A single card is selected from a deck of 52 bridge cards. What is the probability that it is a spade? What is the probability that it is not a spade? What is the probability that it is an ace or a spade?

Solution: Of the 52 cards, 13 are spades. Therefore, the probability of selecting a spade is $\frac{13}{52}$, that is, $\frac{1}{4}$. The probability that the card selected is not a spade is $1 - \frac{1}{4}$, that is, $\frac{3}{4}$. There are 4 aces and 12 spades besides the ace of spades. Therefore, the probability that the card selected is an ace or a spade is $(4 + 12)/52$; that is, $\frac{16}{52}$, which we express as $\frac{4}{13}$.

It is very important that only equally likely events be considered when the probability formula is applied; otherwise faulty reasoning can occur. Consider again the first question asked in Example 1. One might reason that any single card drawn from a deck of cards is either a spade or is not a spade. Thus there are two possible outcomes, and the probability of drawing a spade must therefore be $\frac{1}{2}$. Now it is correct to say that there are these two possible outcomes, but of course they are *not* equally likely since there are 13 spades in a deck of cards and 39 cards that are not spades.

Example 2: A committee of two is to be selected from the set

$$M = \{\text{Betty, Doris, Ellen, John, Tom}\}$$

by drawing names out of a hat. What is the probability that both members of the committee will be girls?

Solution: We solve this problem by first listing all of the possible committees of two that can be formed from the set M.

Betty – Doris	Doris – John
Betty – Ellen	Doris – Tom
Betty – John	Ellen – John
Betty – Tom	Ellen – Tom
Doris – Ellen	John – Tom

Of the ten possible committees, there are three (those boxed) that consist of two girls. Thus the probability that both members selected are girls is $\frac{3}{10}$. What is the probability of selecting a committee to consist of two boys?

EXERCISES

What is the probability of tossing on a single toss of one die:

1. An even number?
2. An odd number?
3. A number greater than 3?
4. A number less than 5?
5. A number different from 5?
6. A number different from 7?
7. The number 7?
8. A number less than 7?

In Exercises 9 through 12 what is the probability of drawing on a single draw from a deck of 52 bridge cards:

9. An ace?
10. A king?
11. A spade?
12. A red card?

In Exercises 13 through 16, we will consider naming a committee of two from the set N = {Alice, Bob, Carol, Doug}.

13. How many different committees of two can be formed?
14. What is the probability that a committee of two will consist of two boys?
15. What is the probability that a committee of two will consist of two girls?
16. What is the probability that a committee of two will consist of one boy and one girl?
17. What is the probability that the first-named author of this text was born in the month of December?
18. What is the probability that your instructor's telephone number has a 7 as its final digit?
19. The probability of obtaining all heads in a single toss of three coins is $\frac{1}{8}$. What is the probability that not all three coins are heads on such a toss?
20. What is the probability that the next person you meet was not born on a Sunday?

PEDAGOGICAL CONSIDERATIONS

1. Students enjoy cartoons that deal with mathematical ideas. Start a collection of such cartoons. A sample is shown on page 462.

NANCY®　　　　　　　　　　　　　　　　　　　　**By Ernie Bushmiller**

2. Prepare a bulletin board display showing the use of different concepts of probability that appear in newspaper and magazine articles.

3. Make a collection of interesting ways to initiate a discussion of probability with a class of junior high school students so as to motivate interest in the topic. Here is one example. Walk into a class tossing a coin and announce that you have just tossed the coin 19 times in a row and obtained 19 heads. You are about to toss the coin for the twentieth time and you ask the class to predict the outcome. Some will argue for tails on the basis that it's time for tails to turn up. Others will say that it makes no difference since the probability of heads for any given toss is $\frac{1}{2}$, and the probability for tails is also $\frac{1}{2}$. The most logical answer is that the coin must be biased, or the manner in which it is being tossed is biased. Thus the probability of obtaining 19 heads in a row is $(\frac{1}{2})^{19}$, or one chance out of 524,288 trials. This is so small a chance that we normally would conclude that it is virtually impossible to obtain 19 heads in a row by legal means. The motivating aspect of this problem is the heated discussion that normally arises as different students argue for their point of view.

4. You can perform some very simple experiments with early elementary school children that help develop concepts of probability. For example, prepare a bag that contains 10 red beads, 5 blue beads, and 3 yellow beads. Begin to select one bead from the bag at a time. Ask the youngsters to guess the color of the next bead to be drawn, and attempt to elicit some reason for their choice. Repeat the experiment using different assortments of beads.

5. If you are enrolled in a class of 30 students, what would you guess is the probability that there are two members of the group that celebrate their birthdays on the same date? Normally one would reason that the probability is low, inasmuch as there are 365 different days of the year on which an individual's birth date could occur, and 30 is small relative to 365.

　　Although the mathematical explanation is beyond our scope at this time, it can be shown that the actual probability of two birth dates occurring on the same day of the year in a random group of 30 students is 0.71. That is, you can safely predict that 71% of the time such an occurrence will take place. Or, alternatively, if you bet that two members of the group have the same birth date, then you can expect to win such a bet approximately 71 times out of 100.

The following graph shows the probability of two common birth dates for people in groups of different sizes. Note that the probability in a group of 23 is approximately 0.50. That is, for a group of 23 individuals, there is a 50% chance of having such an occurrence take place. For a group of 60 the probability is 0.99, or almost certainty!

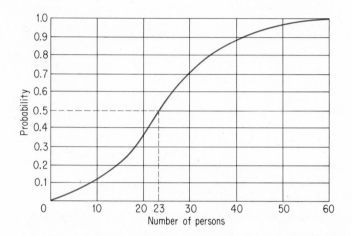

Number of persons

Use the members of your class to test the results shown above. You can also test these predicted results for groups of different sizes by deciding in advance to go to an encyclopedia and select the dates of birth of the first 30, or any other number, of individuals chosen at random.

14-3

sample spaces It is often convenient to solve problems of probability by making a list of all possible outcomes. Such a listing is called a **sample space.** Consider first the problem of tossing two coins. The sample space for this problem is given by the following set of all possible outcomes:

$$\{HH, HT, TH, TT\}$$

We may also summarize these data by means of a chart:

First coin	Second coin
H	H
H	T
T	H
T	T

463

Another convenient way to list all of the logical possibilities is by means of a tree diagram, where the branches represent the possible choices. The set of possible outcomes, $\{HH, HT, TH, TT\}$, may be read from the diagram. Note that there are four possible events: two heads occur in one event, one head and one tail occur in two events, no heads (that is, two tails) occur in one event. Thus we may list various probabilities regarding the tossing of two coins as follows:

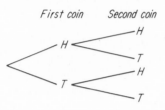

Event	Probabilty
2 heads	$\frac{1}{4}$
1 head	$\frac{2}{4}$
0 heads	$\frac{1}{4}$

Since all possibilities have been considered, the sum of the probabilities should be 1; $\frac{1}{4} + \frac{2}{4} + \frac{1}{4} = 1$. This provides a check on our computation. The list of probabilities is sometimes called a **probability distribution.**

For the case of three coins, the following tree diagram and array may be made.

First coin	Second coin	Third coin	First coin	Second coin	Third coin
		H	H	H	H
		T	H	H	T
		H	H	T	H
		T	H	T	T
		H	T	H	H
		T	T	H	T
		H	T	T	H
		T	T	T	T

From the tree diagram, we may list the probabilities of specific numbers of heads by means of the following distribution.

Event	Probabilty
0 heads	$\frac{1}{8}$
1 head	$\frac{3}{8}$
2 heads	$\frac{3}{8}$
3 heads	$\frac{1}{8}$

In each case, both for two coins and for three coins, the sum of the probabilities is 1; that is, all possible events have been listed and these events are mutually exclusive. Note also that for two coins there were four possible outcomes, for three coins there were eight possible outcomes, and, in general, for n coins there would be 2^n possible outcomes.

Example 1: A box contains two red and three white balls. Two balls are drawn in succession without replacement. List a sample space for this experiment.

Solution: To identify individual balls, we denote the red balls as R_1 and R_2; the white balls as W_1, W_2, W_3. Then the sample space is as follows:

$$
\begin{array}{ccccc}
R_1R_2 & R_2R_1 & W_1R_1 & W_2R_1 & W_3R_1 \\
R_1W_1 & R_2W_1 & W_1R_2 & W_2R_2 & W_3R_2 \\
R_1W_2 & R_2W_2 & W_1W_2 & W_2W_1 & W_3W_1 \\
R_1W_3 & R_2W_3 & W_1W_3 & W_2W_3 & W_3W_2
\end{array}
$$

Example 2: Use a tree diagram to show the possible selections for Example 1.

Solution:

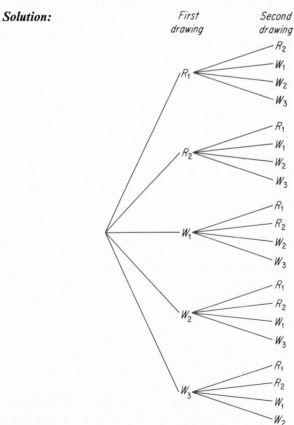

EXERCISES

Use the sample space of Example 1 to find the probability that:

1. Both balls are red.
2. Both balls are white.
3. The first ball is red.
4. The first ball is red and the second ball is white.
5. One ball is red and the other is white.

In Exercises 6 through 8 use the sample space for the outcomes when three coins are tossed to find the probability that:

6. All three coins are heads.
7. At least two coins are heads.
8. At most one coin is tails.
9. List a sample space for the outcomes when four coins are tossed.

In Exercises 10 through 12 use the sample space for the outcomes when four coins are tossed (Exercise 9) to find the probability that:

10. All four coins are heads.
11. At least three coins are heads.
12. At most two coins are tails.
13. List a sample space for the outcomes when a pair of dice is tossed. Represent each outcome by an ordered pair of numbers. For example, let $(1, 3)$ represent a 1 on the first die and a 3 on the second die.

In Exercises 14 through 22 use the sample space for the outcomes when a pair of dice is tossed (Exercise 13) to find the probability that:

14. The number on the first die is 2.
15. The number 2 is on both dice.
16. The same number is on both dice.
17. The sum of the numbers obtained is 11.
18. The sum of the numbers obtained is not 11.
19. The sum of the numbers obtained is 7.
20. The sum of the numbers obtained is not 7.
21. The number on one die is twice the number on the other die.
22. The number on one die is three more than the number on the other die.
23. A box contains two red balls R_1 and R_2 and two white balls W_1 and W_2. List a sample space for the outcomes when two balls are drawn in succession without replacement. Find the probability that both balls are red.

24. Repeat Exercise 23 for the case in which the first ball is replaced before the second ball is drawn.

25. Repeat Exercise 23 for a box that contains three red balls and two white balls.

PEDAGOGICAL CONSIDERATIONS

1. Many interesting tree diagrams can be drawn that reflect problems of interest to upper elementary grade students. The following problem illustrates this point of view. Complete the problem as instructed, and then attempt to find a similar one of current interest.

   Consider a World Series baseball contest between the Dodgers D and the Yankees Y. Assume that each team is equally likely to win each game. Remember that four victories are needed to win the series, each game must be won by one of the teams, and thus at most seven games will be needed. Make a tree diagram for the entire series under the assumption that the Dodgers win the first two games. Circle the entry and assign a probability to each position in the diagram that represents a completion (winning) of the series. Find the probability of the Dodgers' winning the series on the fourth game, on the fifth game, on the sixth game, and on the seventh game. Find the probability of the Yankees' winning the series on each of these games. Check your work by being sure that the sum of the probabilities of someone winning the series is 1.

2. There are many interesting probability questions whose answers are not intuitively obvious. Here are two popular ones that are best solved by means of a sample space.

   (a) Three cards are in a box. One is red on both sides, one is white on both sides, and one is red on one side and white on the other. A card is drawn at random and placed on a table. It has a red side showing. What is the probability that the side not showing is also red?

   As an aid to the solution of this problem, let us identify the three cards as in this diagram:

R_1	W_1	R_3
R_2	W_2	W_3

Now consider the set of possible outcomes when a card is drawn at random and placed on the table. In each of the following ordered pairs, the first card noted represents the one placed face up on the table whereas the second one indicates the face hidden from view.

$$R_1R_2 \qquad R_2R_1 \qquad R_3W_3 \qquad W_1W_2 \qquad W_2W_1 \qquad W_3R_3$$

Inasmuch as we are told that a red side is showing, we may narrow down the sample space to the first three pairs only. Of these three possibilities, if the first side is red can you now tell the probability that the second side is also red?

(b) One of the authors of this book has two children. One of these children is a boy. What is the probability that they are both boys? Believe it or not, the answer is *not* $\frac{1}{2}$!

14-4
computation of probabilities

If A and B represent two mutually exclusive events, then

$$P(A \text{ or } B) = P(A) + P(B).$$

That is, the probability that one event or the other will occur is the sum of the individual probabilities. Consider, for example, the probability of drawing an ace or a picture card (that is, a jack, queen, or king) from an ordinary deck of 52 bridge cards.

The probability of drawing an ace, $P(A)$, is $\frac{4}{52}$.
The probability of drawing a picture card, $P(B)$, is $\frac{12}{52}$.
Then $P(A \text{ or } B) = \frac{4}{52} + \frac{12}{52} = \frac{16}{52} = \frac{4}{13}$.

Example 1: A bag contains three red, two black, and five yellow balls. Find the probability that a ball drawn at random will be red or black.

Solution: The probability of drawing a red ball, $P(R)$, is $\frac{3}{10}$. The probability of drawing a black ball, $P(B)$, is $\frac{2}{10}$. Then

$$P(R \text{ or } B) = P(R) + P(B) = \frac{5}{10} = \frac{1}{2}.$$

This process can be extended to find the probability of any finite number of mutually exclusive events.

$$P(A \text{ or } B \text{ or } C \text{ or } \ldots) = P(A) + P(B) + P(C) + \cdots$$

Example 2: A single die is tossed. What is the probability that either an odd number or a number greater than 3 appears?

Solution: There are three odd numbers possible, $(1, 3, 5)$, so the probability of tossing an odd number is $\frac{3}{6}$. The probability of getting a number greater than 3 (that is, 4, 5, 6) is also $\frac{3}{6}$. Adding these probabilities gives $\frac{3}{6} + \frac{3}{6} = 1$. Something is obviously wrong, since a probability of 1 implies certainty and we can see that an outcome of 2 is neither odd nor greater than 3. The difficulty lies in the fact that the events are *not* mutually exclusive; a number may be both odd and also greater than 3 at the same time. In particular, 5 is both odd and greater than 3 at the same time. Thus $P(5)$ has been included twice. Since $P(5) = \frac{1}{6}$, our answer should be $\frac{3}{6} + \frac{3}{6} - \frac{1}{6}$, that is, $\frac{5}{6}$.

Example 2 can be illustrated by means of the following Venn diagram that lists all possible outcomes when a single die is tossed. Note that the outcome 5 is listed in the intersection of the two sets since it is both odd and greater than 3. On the other hand, the outcome 2 fits into neither of these descriptions and is thus placed outside the circles.

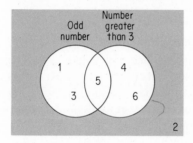

In situations like that of Example 2 we need to subtract the probability that both events occur at the same time. Thus, where A and B are not mutually exclusive events, we have

$$P(A \text{ or } B) = P(A) + P(B) - P(A \text{ and } B).$$

Note that in Example 2, we had the following probabilities:

$P(A)$, the probability of an odd number, is $\frac{3}{6}$.
$P(B)$, the probability of a number greater than 3, is $\frac{3}{6}$.
$P(A \text{ and } B)$, the probability that a number (in this case 5) is odd and greater than 3, is $\frac{1}{6}$.
$P(A \text{ or } B) = \frac{3}{6} + \frac{3}{6} - \frac{1}{6} = \frac{5}{6}$.

By an actual listing we can see that five of the six possible outcomes in Example 2 are either odd or greater than 3, namely 1, 3, 4, 5, and 6. The only "losing" number is 2. Thus the probability $P(A \text{ or } B)$ must be $\frac{5}{6}$.

In general, the two situations $P(A \text{ or } B)$ and $P(A \text{ and } B)$ that we have considered can be described by means of diagrams, where the points of the circular regions represent probabilities of events.

The first of the two diagrams at the bottom of the previous page shows mutually exclusive events:

$$P(A \text{ or } B) = P(A) + P(B)$$

In the second figure, the region that is shaded both horizontally and vertically represents $P(A \text{ and } B)$, $P(A \text{ and } B) \neq 0$, the events are **dependent events** (not mutually exclusive), and

$$P(A \text{ or } B) = P(A) + P(B) - P(A \text{ and } B).$$

We must subtract $P(A \text{ and } B)$, since we have counted it twice in the sum $P(A) + P(B)$.

Next we turn our attention to the probability that several events will occur, one after the other. Consider the probability of tossing a coin twice and obtaining heads on the first toss and tails on the second toss. From a sample space we see that the probability is $\frac{1}{4}$.

$$\{HH, \text{ } \textcircled{HT}, \text{ } TH, \text{ } TT\}$$

Furthermore, we see that the probability $P(A)$ that the first coin is heads is $\frac{1}{2}$. The probability $P(B)$ that the second coin is tails is $\frac{1}{2}$. Then $P(A \text{ and } B) = \frac{1}{2} \times \frac{1}{2} = \frac{1}{4}$. Note that these events are **independent**; that is, the outcome of the first toss does not affect the second toss.

In general, let the probability that an event A occurs be $P(A)$. Let the probability that an independent second event occurs, after A has occurred, be $P(B)$. Then

$$P(A \text{ and } B) = P(A) \times P(B).$$

This can be extended to any finite number of independent events.

$$P(A \text{ and } B \text{ and } C \text{ and } \ldots) = P(A) \times P(B) \times P(C) \times \cdots.$$

Example 3: Urn A contains three white and five red balls. Urn B contains four white and three red balls. One ball is drawn from each urn. What is the probability that they are both red?

Solution: Let $P(A)$ be the probability of drawing a red ball from urn A and $P(B)$ be the probability of drawing a red ball from urn B. Then

$$P(A) = \tfrac{5}{8}, \qquad P(B) = \tfrac{3}{7}, \qquad P(A \text{ and } B) = \tfrac{5}{8} \cdot \tfrac{3}{7} = \tfrac{15}{56}.$$

Example 4: Two cards are selected in succession, without replacement, from an ordinary bridge deck of 52 cards. What is the probability that they are both aces?

Solution: The probability that the first is an ace is $\frac{4}{52}$. Now the probability of success depends on the drawing of the first card. If it is an ace, then the probability that the second card is an ace is $\frac{3}{51}$. The probability that both cards are aces is $\frac{4}{52} \times \frac{3}{51}$, that is, $\frac{1}{221}$.

EXERCISES

What is the probability of tossing on a single toss of one die:

1. An even number or a number greater than 3?
2. An odd number or a number less than 5?
3. An odd number or a number greater than 4?
4. An even number or a number greater than 6?
5. An odd number or a number greater than 10?
6. An odd number or a number less than 6?
7. An even number or a number less than 10?

A single card is drawn from a deck of 52 bridge cards. In Exercises 8 through 14 find the probability that the card selected is:

8. An ace or a king.
9. A spade or a heart.
10. A spade or a king.
11. A spade and a king.
12. A spade and a queen.
13. A heart or a king or a queen.
14. A club or an ace or a king.

15. Two cards are drawn in succession from a deck of 52 bridge cards without the first card's being replaced. Find the probability that (a) both cards are spades; (b) both cards are aces of spades; (c) the first card is a spade and the second card is a heart; (d) the first card is an ace and the second card is the king of hearts; (e) both cards are of the same suit.

16. Repeat Exercise 15 if the first card is replaced in the deck after the first drawing.

17. A coin is tossed five times. What is the probability that all five tosses are heads?

18. A coin is tossed five times. What is the probability that at least one head is obtained? (*Hint:* First find the probability of getting no heads.)

19. A coin is tossed and then a die is rolled. Find the probability of obtaining (a) a head and a 3; (b) a head and an even number; (c) a head or a 3; (d) a head or an even number.

20. A bag contains three red balls and seven white balls. (a) If one ball is drawn at random, what is the probability that it is white? (b) If two balls are drawn at random, what is the probability that they are both white?

21. Five cards are drawn at random from an ordinary bridge deck of 52 cards. Find the probability that all five cards drawn are spades.

22. A box contains three red, four white, and five green balls. Three balls are drawn in succession, without replacement. Find the probability that (a) all three are red; (b) the first is red, the second is white, and the third is green; (c) none are green; *(d) all three are of the same color.

23. Repeat Exercise 22 if each ball is replaced after it is drawn.

24. A die is tossed three times. Find the probability that **(a)** a 6 is tossed on the first toss; **(b)** a 6 is tossed on the first two tosses; **(c)** a 6 is tossed on all three tosses; **(d)** a 6 is tossed on the first toss and not tossed on the second or third toss.

25. A die is tossed three times. Find the probability that **(a)** an even number is tossed on all three tosses; **(b)** an even number is tossed on the first two tosses and an odd number on the third toss; **(c)** an odd number is tossed on the first toss and an even number on the second and third tosses; **(d)** exactly one even number is tossed.

*26. A die is tossed three times. Find the probability that **(a)** at least one 6 is tossed; **(b)** exactly one 6 is tossed.

PEDAGOGICAL CONSIDERATIONS

Many interesting class experiments can be performed that illustrate basic concepts of probability. Several such experiments are suggested here that you should attempt to complete with a class of students. Also make a collection of any other such experiments that you can find.

1. Consider the network of streets shown in the following figure.

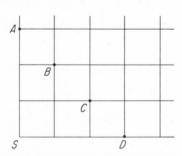

You are to start at the point S and move three "blocks." Each move is determined by tossing a coin. If the coin lands "tails," move one block to the right; for "heads," move one block up. Your terminal point will be at A, B, C, or D. Try to predict the number of times you will land at each point if the experiment is to be repeated 16 times. Then complete 16 trials, keeping a tally of the number of times you land at each point. Compare your actual results with your predictions as well as with the results obtained by your classmates. Finally, on the basis of your experimentation, revise your predictions if necessary.

2. Repeat Exercise 1 for the following network. This time you are to make four moves and will land at point *A*, *B*, *C*, *D*, or *E*.

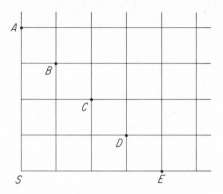

3. If a pair of normal dice is tossed repeatedly, one can theoretically expect a sum of 7 to occur in one out of six tosses. Thus if a pair of dice is tossed 36 times, on the average six of the tosses will give a sum of 7. Toss a pair of dice 36 times and count the frequency with which a sum of 7 appears. Compare your actual results with the theoretical probability of $\frac{1}{6}$. Collect this data for each member of your class and compare actual results with expected results.

4. Repeat Exercise 3 for 72 tosses of a pair of dice. Record each toss by placing an x over the numeral representing the sum, using graph paper. The final result should show a bar graph that is fairly "normal" in shape. The figure below shows the results obtained after one experiment of 18 tosses.

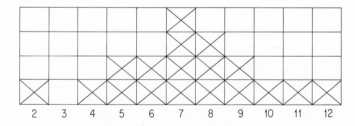

5. By "probability" we always mean "probability in the long run." That is, the fact that the probability of obtaining a head on one toss of a coin is $\frac{1}{2}$ does not imply that you will obtain exactly 4 heads in 8 tosses. Rather, it means that the longer you continue to toss a coin, the closer you can expect to come to having 50% of the tosses produce a head. This can be demonstrated by actually tossing a coin and recording the results. For example, the results obtained from 10 tosses of a single coin are summarized in the chart at the top of the next page. Note that the first two tosses were heads, the third was a tail, the fourth was a head, and so forth.

473

Tosses	1	2	3	4	5	6	7	8	9	10
Head	x	x		x			x		x	x
Tail			x		x	x		x		
Ratio of heads	$\frac{1}{1}$	$\frac{2}{2}$	$\frac{2}{3}$	$\frac{3}{4}$	$\frac{3}{5}$	$\frac{3}{6}$	$\frac{4}{7}$	$\frac{4}{8}$	$\frac{5}{9}$	$\frac{6}{10}$
Ratio of tails	$\frac{0}{1}$	$\frac{0}{2}$	$\frac{1}{3}$	$\frac{1}{4}$	$\frac{2}{5}$	$\frac{3}{6}$	$\frac{3}{7}$	$\frac{4}{8}$	$\frac{4}{9}$	$\frac{4}{10}$
Per cent of heads	100	100	67	75	60	50	57	50	56	60
Per cent of tails	0	0	33	25	40	50	43	50	44	40

The graphical presentation of the results obtained from these 10 tosses gives a visual interpretation of the meaning of probability. Despite minor fluctuations, in the long run the graphs of the per cent of heads and tails will come closer and closer to the 50% line. That is, in the long run we expect to come close to having one-half of our tosses produce heads and one-half produce tails.

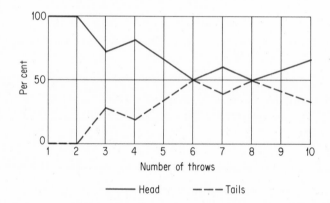

Repeat the experiment described here for 20 tosses of a coin.

6. Have each member of your class write down in advance the predicted outcome for five tosses of a coin. For example, one guess might be *H, T, T, H, T.* Then actually complete five tosses of a coin and count how many members of your class guessed correctly. Can you predict in advance how many members of the class can be expected to have perfect scores?

14-5

odds and mathematical expectation

One often reads statements in the sports section of the daily newspaper concerning "odds" in favor of, or against, a particular team or individual's winning or losing some encounter. For example, we may read that the odds in favor of the Cardinals' winning the pennant are "4 to 1." In this section we shall attempt to discover just what such statements really mean.

475

sec. 14-5
odds
and
mathematical
expectation

Consider the problem of finding the odds against obtaining a 3 in one toss of a die. Since the probability of obtaining a 3 is known to be $\frac{1}{6}$, most people would say that the odds are therefore 6 to 1 against rolling a 3. This is not correct—because out of every six tosses of the die, in the long run, one expects to toss one 3. The other five tosses are not expected to be 3's. Therefore, the correct odds against rolling a 3 in one toss of a die are 5 to 1. The odds in favor of tossing a 3 are 1 to 5. Formally we define odds as follows:

> The **odds in favor** of an event are defined as the ratio of the probability that an event will occur to the probability that the event will not occur. The reciprocal of this ratio gives the **odds against** the occurrence of the event.

Thus the odds in favor of an event that may occur in several equally likely ways is the ratio of the number of favorable ways to the number of unfavorable ways.

Notice that odds and probabilities are very closely related. Indeed, if either the odds for or the probability of an event is known, then the other can be found. For example, if the odds for an event are 1 to 2, then we have the ratio

$$\frac{\text{number of favorable ways}}{\text{number of unfavorable ways}} = \frac{1}{2}$$

and the probability is the ratio

$$\frac{\text{number of favorable ways}}{\text{total number of all ways (favorable and unfavorable)}} = \frac{1}{3}.$$

Similarly, if the probability is $\frac{2}{5}$, then the odds are 2 to 3.

Example 1: Find the odds in favor of drawing a spade from an ordinary deck of 52 bridge cards.

Solution: Since there are 13 spades in a deck of cards, the probability of drawing a spade is $\frac{13}{52} = \frac{1}{4}$. The probability of failing to draw a spade is $\frac{3}{4}$. The odds in favor of obtaining a spade are $\frac{1}{4} \div \frac{3}{4}$; that is, $\frac{1}{3}$.

The odds in favor of drawing a spade are stated as $\frac{1}{3}$. They may also be stated as "1 to 3" or as "1 : 3." Similarly, the odds against drawing a spade are $\frac{3}{1}$, which may be written as "3 to 1" or "3 : 1."

Mathematical expectation is closely related to odds and is defined as the product of the probability that an event will occur and the amount to be received upon such occurrence. Suppose that you are to receive $2.00 each time you obtain two heads on a single toss of two coins. You do not receive anything for any other outcome. Then your mathematical expectation will be one-fourth of $2.00; that is, $0.50. This means that you should be willing to pay $0.50 each time you toss the coins if the game is to be a fair one. In the long run, both you and the person who is running the game would break even. For example, if you played the game four times it would cost you 4 × $0.50; that is, $2.00. You expect to win, in the long run, once out of every four games played. Assuming that you do so, you will win $2.00 once every four times and thus be even.

If an event has several possible outcomes that occur with probabilities p_1, p_2, p_3, and so forth, and for each of these outcomes one may expect the amounts m_1, m_2, m_3, and so on, then the mathematical expectation E may be defined as

$$E = m_1p_1 + m_2p_2 + m_3p_3 + \cdots.$$

Whenever this formula is used, it is worthwhile to check that all possible outcomes have been considered by checking that the sum of the probabilities is 1.

Example 2: Suppose that you play a game wherein you are to toss a coin twice and are to receive 10 cents if two heads are obtained, 5 cents if one head is obtained, and nothing if both tosses produce tails. What is your expected value in this game?

Solution: The probabilities of obtaining two, one, and no heads, respectively, are $\frac{1}{4}, \frac{1}{2}$, and $\frac{1}{4}$. Therefore, the expected value E, in cents, is found to be

$$E = (10)(\tfrac{1}{4}) + (5)(\tfrac{1}{2}) + (0)(\tfrac{1}{4}) = 5.$$

This solution may be interpreted in several ways. For one thing, it is the price you should be willing to pay for the privilege of playing this game. It may also be interpreted as the average amount of winnings per game that one may expect when one is playing a large number of games.

EXERCISES

What are the odds in favor of obtaining:

1. Two heads in a single toss of two coins?

2. At least two heads in a single toss of three coins?

3. Two heads when a single coin is tossed twice?

4. At least two heads when a single coin is tossed three times?

In Exercises 5 through 8 what are the odds against obtaining:

5. Two heads in a single toss of two coins?

477

sec. 14-5
odds
and
mathematical
expectation

6. An ace in a single draw from a deck of 52 bridge cards?

7. An ace or a king in a single draw from a deck of 52 bridge cards?

8. A 7 or an 11 in a single toss of a pair of dice?

9. What are the odds in favor of rolling a 7 or an 11 in a single toss of a pair of dice?

10. One hundred tickets are sold for a lottery. The grand prize is $1000. What is a fair price to pay for a ticket?

11. Repeat Exercise 10 for the case in which 250 tickets are sold.

12. What is your mathematical expectation when you buy one of 300 tickets for a single prize worth $750?

13. What is your mathematical expectation in a game in which you will receive $10 if you toss a "double" (the same number on both dice) on a single toss of a pair of dice?

14. A box contains three dimes and two quarters. You are to reach in and select one coin, which you may then keep. Assuming that you are not able to determine which coin is which by its size, what would be a fair price for the privilege of playing this game?

15. There are three identical boxes on a table. One contains a five-dollar bill, one contains a one-dollar bill, and the third is empty. A man is permitted to select one of these boxes and to keep its contents. What is his expectation?

16. Three coins are tossed. What is the expected number of heads?

17. Two bills are to be drawn from a purse that contains three five-dollar bills and two ten-dollar bills. What is the mathematical expectation for this drawing?

18. If there are two pennies, a nickel, a dime, a quarter, and a half-dollar in a hat, what is the mathematical expectation of the value of a random selection of a coin from the hat?

PEDAGOGICAL CONSIDERATIONS

1. Make a worksheet of 20 problems to introduce odds to elementary school students.

2. Make a worksheet of 10 problems to introduce mathematical expectation to elementary school students.

14-6

permutations Suppose that three people, Ruth, Joan, and Debbie, are waiting to play singles at a tennis court. Two of the three can be selected in six different ways if the order in which they are named is significant, for example, if the first person named is to serve first. We may identify these six ways from a tree diagram.

We say that there are 3 × 2 (that is, 6) *permutations* of the set of three people selected two at a time. In each case the *order* in which the two people are named is significant. A **permutation** of a set of elements is an arrangement of certain of these elements in a specified order. In the problem just discussed, the number of permutations of three things taken two at a time is 6. In symbols we write

$$_3P_2 = 6$$

In general, we wish to find a formula for $_nP_r$, that is, the permutations of n things taken r at a time. To do this we note that we can fill the first of the r positions in any one of n different ways. Then the second position can be filled in $n - 1$ different ways, and so on.

Position: 1 2 3 4 ... r
$\downarrow$ $\downarrow$ $\downarrow$ $\downarrow$ $\downarrow$
Number of choices: n $n-1$ $n-2$ $n-3$ ... $n-(r-1)$
(i.e., $n - r + 1$)

The product of these r factors gives the number of different ways of arranging r elements selected from a set of n elements, that is, the permutation of n things taken r at a time.

$$_nP_r = (n)(n-1)(n-2)\ldots(n-r+1)$$

where n and r are integers and $n \geq r$.

Example 1: Find $_8P_4$.

Solution: Here $n = 8$, $r = 4$, and $n - r + 1 = 5$. Thus

$$_8P_4 = 8 \times 7 \times 6 \times 5 = 1680.$$

Note that there are r, in this case 4, factors in the product.

Example 2: How many different three-letter "words" can be formed from the 26 letters of the alphabet if each letter may be used at most once?

Solution: We wish to find the number of permutations of 26 things taken three at a time.

$$_{26}P_3 = 26 \times 25 \times 24 = 15,600$$

A special case of the permutation formula occurs when we consider the permutations of n things taken n at a time. For example, let us see in how many different ways we may arrange the five members of set M in a row. Here we have the permutation of five things taken five at a time:

$$_5P_5 = 5 \times 4 \times 3 \times 2 \times 1$$

In general, for n things n at a time, $n = n$, $r = n$, and $n - r + 1 = 1$;

$$_nP_n = (n)(n - 1)(n - 2) \ldots (3)(2)(1).$$

We use a special symbol, $n!$, read "n factorial," for this product of integers from 1 through n. The following examples should illustrate the use of the new symbol:

$1! = 1$
$2! = 2 \times 1$
$3! = 3 \times 2 \times 1$
$4! = 4 \times 3 \times 2 \times 1$

$5! = 5 \times 4 \times 3 \times 2 \times 1$
$6! = 6 \times 5 \times 4 \times 3 \times 2 \times 1$
$7! = 7 \times 6 \times 5 \times 4 \times 3 \times 2 \times 1$
$8! = 8 \times 7 \times 6 \times 5 \times 4 \times 3 \times 2 \times 1$

Also, we *define* $0! = 1$ so that $(n - r)!$ may be used when $r = n$.

Using this **factorial notation**, we are now able to provide a different, but equivalent, formula for $_nP_r$:

$$_nP_r = n(n - 1)(n - 2) \ldots (n - r + 1)$$
$$\times \frac{(n - r)(n - r - 1)(n - r - 2) \ldots (3)(2)(1)}{(n - r)(n - r - 1)(n - r - 2) \ldots (3)(2)(1)} = \frac{n!}{(n - r)!}$$

Example 3: Evaluate $_7P_3$ in two different ways.

Solution:

(a) $_7P_3 = 7 \times 6 \times 5$

(b) $_7P_3 = \dfrac{7!}{4!} = \dfrac{7 \times 6 \times 5 \times 4 \times 3 \times 2 \times 1}{4 \times 3 \times 2 \times 1} = 7 \times 6 \times 5$

Example 4: A certain class consists of 10 boys and 12 girls. They wish to elect officers in such a way that the president and treasurer are boys and the vice-president and secretary are girls. In how many ways can this be done?

479

Solution: The number of different ways of selecting the president and treasurer is $_{10}P_2$. The number of ways of selecting the vice-president and secretary is $_{12}P_2$. The total number of ways of choosing officers is

$$(_{10}P_2) \times (_{12}P_2) = (10 \times 9) \times (12 \times 11) = 11{,}880.$$

EXERCISES

Evaluate.

1. $5!$

2. $6!$

3. $\dfrac{8!}{6!}$

4. $\dfrac{11!}{7!}$

5. $_7P_2$

6. $_7P_3$

7. $_{10}P_1$

8. $_{10}P_{10}$

9. $_{12}P_{12}$

10. $_{12}P_3$

11. $_6P_2$

12. $_6P_4$

13. $_{10}P_3$

14. $_{10}P_7$

In Exercises 15 through 18 solve for n.

15. $_nP_1 = 6$

16. $_nP_2 = 6$

17. $_nP_2 = 20$

18. $_nP_3 = 24$

19. Find the number of different arrangements of the set of letters $V = \{a, e, i, o, u\}$ if they are taken **(a)** two at a time; **(b)** five at a time.

20. Find the number of four-digit numbers that can be formed using the digits 1, 2, 3, 4, 5 if no digit may be used more than once in a number. How many of these numbers will be even? What is the probability that such a four-digit number will be even? odd?

21. Find the number of different signals that can be formed by running up three flags on a flagpole, one above the other, if seven different flags are available.

22. Find the number of different ways that a disc jockey can arrange a musical program of seven selections. How many of the arrangements will have a particular one of the seven selections first? What is the probability that a particular one of the seven selections will be first on the program?

23. Find the number of different ways that a manager of a nine-man baseball team can arrange his batting order.

Find for any positive integer n a formula for:

24. $_nP_r \times (n - r)!$

25. $_nP_0$

1. Make a worksheet for elementary school students to use for exploring the number of permutations of five things taken three at a time.

2. Make a worksheet for elementary school students to use to develop a recognition that whenever permutations of elements taken three at a time are formed there are sets of six permutations with the same elements taken in different orders.

3. Make a worksheet for elementary school students to use in developing an understanding of whether or not the order of the elements in a set is important.

14-7

combinations A fictitious club with five members was considered in §14-1. The set of members was {Betty, Doris, Ellen, John, Tom}. The number of ways in which a president and a vice-president could be selected was essentially the number $_5P_2$ of permutations of five things taken two at a time. Here order is important in that Betty as president and Doris as vice-president is a different set of officers than Doris as president and Betty as vice-president.

Now, suppose we wish to select a committee of two members from set M without attaching any meaning to the order in which the members are selected. Then the committee consisting of Betty and Doris is certainly the same as the one consisting of Doris and Betty. In this case, we see that order is not important, and we call such an arrangement a **combination.** One way to determine the number of possible committees of two to be formed from the set M is by enumeration. We find that there are 10 possible committees, as follows:

Betty–Doris	Doris–John
Betty–Ellen	Doris–Tom
Betty–John	Ellen–John
Betty–Tom	Ellen–Tom
Doris–Ellen	John–Tom

We summarize this discussion by saying that the number of combinations of five things, taken two at a time, is 10. In symbols we write

$$_5C_2 = 10.$$

In general, we wish to find a formula for $_nC_r$; that is, the combinations of n things taken r at a time. This is written in symbols in the form

$$_nC_r \quad \text{or} \quad \binom{n}{r}.$$

To find a formula for $_nC_r$ let us first consider a specific problem, that of selecting committees of three from the set M. There are 10 such possibilities, and we list them below, using the first initial of each name only:

$$
\begin{array}{ll}
B, D, E \qquad & B, J, T \qquad M = \{B, D, E, J, T\} \\
B, D, J & D, E, J \qquad\quad _5C_3 = 10 \\
B, D, T & D, E, T \\
B, E, J & D, J, T \\
B, E, T & E, J, T
\end{array}
$$

Note that selecting committees of three is equivalent to selecting groups of two to be omitted. That is, omitting J and T is the same as selecting B, D, and E. Therefore, we find that $_5C_3 = {}_5C_2 = 10$.

Inasmuch as we wanted only committees, and assigned no particular jobs to the members of each committee, we see that order is not important. However, suppose that each committee is now to elect a chairman, secretary, and historian. In how many ways can this be done within each committee? This is clearly a problem in which order is important, involving permutations. The number of such possible arrangements within each committee is $_3P_3$; that is, $3!$. For example, the committee consisting of B, D, and E can rearrange themselves as chairman, secretary, and historian, respectively, as follows:

$$B, D, E; \quad B, E, D; \quad D, E, B; \quad D, B, E; \quad E, B, D; \quad E, D, B$$

All six of these permutations constitute only one combination.

We know that $_5C_3 = 10$. If each of these combinations is multiplied by $3!$ we then will have the total number of permutations of five things taken three at a time:

$$_5C_3 \times 3! = {}_5P_3 \quad \text{and} \quad _5C_3 = \frac{_5P_3}{3!}$$

In general, consider $_nC_r$. Each of these combinations, consisting of r elements each, may be used to form $r!$ permutations. Thus the number of combinations of n things taken r at a time is given by the formula

$$_nC_r \times r! = {}_nP_r \quad \text{and} \quad _nC_r = \frac{_nP_r}{r!}.$$

Since

$$_nP_r = (n)(n-1)(n-2)\ldots(n-r+1) = \frac{n!}{(n-r)!}$$

we also have the formula

$$_nC_r = \frac{(n)(n-1)(n-2)\ldots(n-r+1)}{r!} = \frac{n!}{r!(n-r)!}.$$

Example 1: Evaluate $_7C_2$ in two ways.

Solution:

(a) $_7C_2 = \dfrac{_7P_2}{2!} = \dfrac{7 \times 6}{2 \times 1} = 21$

(b) $_7C_2 = \dfrac{7!}{2!5!} = \dfrac{7 \times 6 \times 5 \times 4 \times 3 \times 2 \times 1}{2 \times 1 \times 5 \times 4 \times 3 \times 2 \times 1} = 21$

Example 2: In how many different ways can a hand of five cards be dealt from a deck of 52 cards? What is the probability that a particular hand contains four aces and the king of hearts?

Solution: The order of the five cards is unimportant, so this is a problem involving combinations.

$$_{52}C_5 = \dfrac{52!}{5!47!} = \dfrac{52 \times 51 \times 50 \times 49 \times 48 \times (47!)}{5! \quad (47!)} = 2{,}598{,}960$$

The probability of obtaining any one particular hand, such as that containing the four aces and the king of hearts, is $\dfrac{1}{2{,}598{,}960}$.

Many problems in probability are most conveniently solved through the use of the concepts of combinations presented here. For example, let us consider again Example 4 of §14-4. There we were asked to find the probability that both cards would be aces if two cards are drawn from a deck of 52 cards. This problem can be solved by noting that $_{52}C_2$ is the total number of ways of selecting two cards from a deck of 52 cards. Also $_4C_2$ is the total number of ways of selecting two aces from the four aces in a deck. The required probability is then given as

$$\dfrac{_4C_2}{_{52}C_2} = \dfrac{\dfrac{4!}{2!2!}}{\dfrac{52!}{2!50!}} = \dfrac{4 \times 3}{52 \times 51} = \dfrac{1}{221}.$$

EXERCISES

In Exercises 1 through 5 evaluate:

1. (a) $\dfrac{6!}{4!2!}$ (b) $\dfrac{8!}{5!3!}$

2. (a) $\dfrac{10!}{4!6!}$ (b) $\dfrac{12!}{7!5!}$

3. (a) $_7C_2$ (b) $_7C_3$

4. (a) $_8C_3$ (b) $_9C_4$

5. (a) $_{11}C_2$ (b) $_{15}C_3$

6. List the $_3P_2$ permutations of the elements of the set $\{a, b, c\}$. Then identify the permutations that represent the same combinations and find $_3C_2$.

483

7. List the $_4P_3$ permutations of the elements of the set $\{p, q, r, s\}$. Then identify the permutations that represent the same combinations and find $_4C_3$.

8. List the elements of each combination of $_4C_3$ for the set $\{w, x, y, z\}$. Then match each of these combinations with a combination of $_4C_1$ and thereby illustrate the fact that $_4C_3 = {}_4C_1$.

9. Find a formula for $_nC_n$ for any positive integer n.

10. Find the value and give an interpretation of $_nC_0$ for any positive integer n.

11. Evaluate $_3C_0$, $_3C_1$, $_3C_2$, $_3C_3$ and check that the sum of these combinations is 2^3, the number of possible subsets that can be formed from a set of three elements.

12. Evaluate $_5C_0$, $_5C_1$, $_5C_2$, $_5C_3$, $_5C_4$, $_5C_5$ and check the sum as in Exercise 11.

13. Use the results obtained in Exercises 11 and 12 and conjecture a formula for any positive integer n for

$$_nC_0 + {}_nC_1 + {}_nC_2 + {}_nC_3 + \cdots + {}_nC_{n-1} + {}_nC_n.$$

14. How many sums of money (include the case of no money) can be selected from a set of coins consisting of a penny, a nickel, a dime, a quarter, and a half-dollar?

15. A man has a penny, a nickel, a dime, a quarter, and a half-dollar in his pocket. In how many different ways can he give a tip if he wishes to use exactly two coins?

16. A class consists of 10 boys and 12 girls. How many different committees of four can be selected from the class if each committee is to consist of two boys and two girls?

17. In how many ways can a hand of 13 cards be selected from a bridge deck of 52 cards?

18. In how many ways can one choose three books to read from a set of seven books?

19. Explain why a so-called "combination" lock should really be called a permutation lock.

20. Urn A contains five balls and urn B contains ten balls. In how many ways can ten balls be selected if three are to be drawn from urn A and seven from urn B?

21. An urn contains seven black and three white balls. In how many ways can four balls be selected from this urn? How many of these selections will include exactly three black balls?

22. Repeat Exercise 21 for an urn that contains ten black balls and five white balls.

23. A group of six girls live in the same house. Three of the girls have blue eyes. If two of the girls are chosen at random, what is the proba-

bility that both will have blue eyes?

24. Three arbitors are to be chosen by lot from a panel of ten. What is the probability that a certain individual on the panel will be one of those chosen?

25. Five cards are drawn at random from an ordinary bridge deck of 52 cards. Find the probability that all four aces are among the cards drawn.

SUPPLEMENTARY EXERCISES

State whether each question involves a permutation, a combination, or a permutation and a combination. Then answer each question.

1. The 20 members of the Rochester Tennis Club are to play on a certain Saturday evening. In how many ways can pairs be selected for playing "singles"?

2. In how many different ways can a disc jockey with ten records: (a) select a set of four records? (b) present a program in which four records are played?

3. Students taking a certain examination are required to answer four out of eight questions. In how many ways can a student select the four questions that he tries to answer?

4. In how many ways can seven people line up at a single theater ticket window?

5. How many lines are determined by ten points if no three points are collinear?

6. In how many ways can the 15 boys in a certain class be selected in a group of nine to play baseball?

7. How many different hands can be dealt from a deck of 52 bridge cards if each hand contains: (a) four cards? (b) seven cards?

8. In how many different ways may five guests be seated in the five seats of a six-passenger car after the driver is seated?

9. In how many ways can four pictures be arranged with one in each of four given places on the walls of a room?

10. A class is to be divided into two committees. In how many different ways can some or all of eight of the students be assigned to one of the committees?

11. In how many different ways can eight students be divided into two groups of four students each?

12. In how many different ways can eight students be divided into two groups of students?

Recall that the notation $_nC_r$ may also be written in the form $\binom{n}{r}$. Consider the following array for sets of n elements where $n = 1, 2, 3, \ldots$:

$n = 1$: $\qquad\qquad \binom{1}{0} \quad \binom{1}{1}$

$n = 2$: $\qquad\qquad \binom{2}{0} \quad \binom{2}{1} \quad \binom{2}{2}$

$n = 3$: $\qquad\qquad \binom{3}{0} \quad \binom{3}{1} \quad \binom{3}{2} \quad \binom{3}{3}$

$n = 4$: $\qquad \binom{4}{0} \quad \binom{4}{1} \quad \binom{4}{2} \quad \binom{4}{3} \quad \binom{4}{4}$

$n = 5$: $\binom{5}{0} \quad \binom{5}{1} \quad \binom{5}{2} \quad \binom{5}{3} \quad \binom{5}{4} \quad \binom{5}{5}$

. . . .

If we replace each symbol by its equivalent number, we may write the following array, known as **Pascal's triangle.** Generally ascribed to the French mathematician Blaise Pascal (1623–1662), this array of numbers is said to have been known to the Chinese in the early fourteenth century.

```
n = 1:              1      1
n = 2:           1      2      1
n = 3:        1      3      3      1
n = 4:     1      4      6      4      1
n = 5:  1     5     10     10     5      1
```

. . . .

We read each row of this array by noting that the first entry in the nth row is $\binom{n}{0}$, the second is $\binom{n}{1}$, the third is $\binom{n}{2}$, and so on until the last entry, which is $\binom{n}{n}$. Since $\binom{n}{0} = \binom{n}{n} = 1$, each row begins and ends with 1.

There is a simple way to continue the array with very little computation. In each row the first number is 1 and the last number is 1. Each of the other numbers may be obtained as the sum of the two numbers appearing in the preceding row to the right and left of the position to be filled. Thus, to obtain the sixth row, begin with 1. Then fill the next position by adding 1 and 5 from the fifth row. Then add 5 and 10 to obtain 15, add 10 and 10 to obtain 20, and so forth as in this diagram:

Pascal's triangle may be used mechanically to compute probabilities as follows:

The elements of the second row are the numerators for the probabilities when two coins are tossed; the elements of the third row are the numerators when three coins are tossed; and so on. The denominator in each case is found as the sum of the elements in the row used. For example, when three coins are tossed, we examine the third row $(1, 3, 3, 1)$. The sum is 8. The probabilities of 0, 1, 2, and 3 heads are then given as

$$\frac{1}{8}, \frac{3}{8}, \frac{3}{8}, \frac{1}{8},$$

as in §14-3.

Note that the sum of the entries in the second row is 4, the sum in the third row is 2^3 or 8, the sum in the fourth row is 2^4 or 16, and, in general, the sum in the nth row will be 2^n.

1. Use the fourth row of Pascal's triangle to find the probability of 0, 1, 2, 3, 4 heads in a single toss of four coins.

2. Construct Pascal's triangle for $n = 1, 2, 3, \ldots, 10$.

3. List the entries in the eleventh row of Pascal's triangle, using the notation

$$\binom{n}{r}.$$

4. Repeat Exercise 3 for the twelfth row.

5. See how many different patterns of numbers you can find in Pascal's triangle. For example, do you see the sequence 1, 2, 3, 4, 5, ... ?

6. A recognizable pattern may be found by considering the number of ways in which each of the numbers 2, 3, 4, ... can be expressed as sums of counting numbers one at a time, two at a time, three at a time, and so forth. For example, 3 can be expressed in one way (3) as a single number, in two ways ($1 + 2$ and $2 + 1$) as a sum of two counting numbers, and in one way ($1 + 1 + 1$) as a sum of three counting numbers. Consider also numbers 4, 5, 6, ... until you recognize the pattern. Then state the number of ways in which 75 can be expressed as a sum of two numbers, and the number of ways in which 79 can be expressed as a sum of 78 numbers.

CHAPTER TEST

1. How many four-letter "words" can be formed from the set $B = \{p, e, n, c, i, l\}$ if **(a)** no letter may be used more than once? **(b)** repetitions of letters are permitted?

2. A single card is selected from a deck of 52 bridge cards. What is the probability that it is **(a)** not a spade; **(b)** a spade or a heart.

3. A certain florist has five different kinds of roses. In how many ways can one each of three different kinds be selected?

4. A man has a nickel, a dime, a quarter, and a half-dollar. In how many ways can he leave a tip of exactly two coins?

5. Use a sample space to represent the possible outcomes described in Exercise 4 and find the probability that the tip is at least thirty cents when the selection of the two coins is at random.

6. What is the probability of tossing on a single toss of two dice **(a)** a 7? **(b)** a number greater than 7?

7. What are the odds in favor of obtaining an ace when one card is drawn from a deck of 52 bridge cards?

8. What is your mathematical expectation when you buy three tickets out of a total of 6000 tickets for a single prize worth $1000?

9. A bag contains five red balls and three green balls. Two balls are drawn in succession, without replacement. Find the probability that **(a)** both balls are red; **(b)** the first ball is red and the second one is green.

10. A disc jockey has 25 records. In how many ways can he **(a)** select a set of five records? **(b)** arrange a program of five records?

Answers
to
Odd-Numbered
Exercises

Introduction: FUN WITH MATHEMATICS

1 EXPLORATIONS WITH NUMBER PATTERNS

1. The text shows the diagrams for 3×9, 7×9, and 4×9 respectively. Here are

3. $6^2 = 1 + 2 + 3 + 4 + 5 + 6 + 5$
$+ 4 + 3 + 2 + 1$
$7^2 = 1 + 2 + 3 + 4 + 5 + 6 + 7$
$+ 6 + 5 + 4 + 3 + 2 + 1$
$8^2 = 1 + 2 + 3 + 4 + 5 + 6 + 7$
$+ 8 + 7 + 6 + 5 + 4 + 3$
$+ 2 + 1$
$9^2 = 1 + 2 + 3 + 4 + 5 + 6 + 7$
$+ 8 + 9 + 8 + 7 + 6 + 5$
$+ 4 + 3 + 2 + 1$

7. (a) $9 \times 47 = 423$

(b) $9 \times 39 = 351$

(c) $9 \times 18 = 162$

(d) $9 \times 27 = 243$

9. (a) 40×81, that is, 3240;
(b) 100×201, that is, 20,100;
(c) $\frac{25}{2} \times 50$, that is, 625;
(d) 50×200, that is, 10,000;
(e) 100×402, that is, 40,200.

11. (a) $\frac{3}{4}$;
(b) $\frac{7}{8}$;
(c) $\frac{15}{16}$. In general: $\dfrac{2^n - 1}{2^n}$.

2 EXPLORATIONS WITH GEOMETRIC PATTERNS

1. There will be 4 holes produced by 4 folds, and 2^{n-2} holes with n folds.

3. $V + R = A + 2$: the sum of the number of vertices and the number of regions is 2 more than the number of arcs.

5.

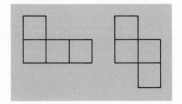

7. 35; *ABC, ABD, ABE, ABF, ABG, ABJ, BCD, BCE, BCF, BCG, BCH, ACD, ACE, ACI, CDE, CDG, CDH, CDI, ADE, ADG, BDE, BDJ, DEH, DEI, DEJ, AEF, AEI, AEJ, BEH, CEF, AFJ, BFG, CGH, DHI, EIJ.*

9. Dotted lines show segments removed.

11.

3 MATHEMATICAL RECREATIONS

1. (a) 12.
   (b) Only one, if it's long enough.
   (c) Only halfway; then you start walking out.
   (d) *One* of them is not a nickel, but the other one is.
   (e) There is no dirt in a hole.
   (f) Brother-sister.

3. There are eleven trips needed. First one Indian and one missionary go over; the missionary returns. Then two Indians go over and one returns. Next two missionaries go over; one missionary and one Indian return. Two missionaries go over next and one Indian returns. Then two Indians go over and one of them returns. Finally, the last two Indians go over.

5. After 27 days the cat still has 3 feet to go. It does this the next day and is at the top after 28 days.

7.

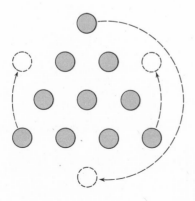

9. Use the matchsticks to form a triangular pyramid.

11. Both A's and B's will say that they are B's. Therefore, when the second man said that the first man said he was a B, the second man was telling the truth. Thus the first two men told the truth and the third man lied.

13. If the penny is in the left hand and the dime in the right hand, the computation will give $3 + 60 = 63$, an odd number. If the coins are reversed, we have $30 + 6 = 36$, an even number.

15. There really is no missing dollar. The computation may be done in one of two ways: $(30 - 3) - 2 = 25$, or $25 + 2 + 3 = 20$. In the problem the arithmetic was done in a manner that is not legitimate; that is, $(30 - 3) + 2$.

17. Eight moves are needed. The coins are identified in the following diagram as well as the squares which may be used. The moves are as follows, where the first numeral indicates the position of the coin and the second one tells you where to move it: D_1: 4–3; P_2: 2–4; P_1: 1–2; D_1: 3–1; D_2: 5–3; P_2: 4–5; P_1: 2–4; D_2: 3–2.

1 P_1	2 P_2	3	4 D_1	5 D_2

19. E; N. These are the first letters of the names (one, two, three, etc.) of the counting numbers.

Part I
NUMBER SYSTEMS

Chapter 1: SETS

1-1 SET NOTATION

1. $\in$

3. $\notin$

5. Well-defined.

7. Not well-defined.

9. Equal.

11. Equal.

13. {January, February, March, ..., December}.

15. $\{1, 2, 3, \ldots, 9\}$.

17. The set of counting numbers 1 through 6.

19. The set of counting numbers greater than 50.

21. The set of multiples of 10 from 10 through 150.

*23.** The set of numbers that can be expressed in the form $n(n-1)$, where n ranges from 1 through 10.

1-2 SUBSETS

1. Among the many correct answers are: the set of odd integers divisible by 2; the set of integers between 4 and 5; the set of even integers between 6 and 8; the human beings who traveled to Venus prior to 1968; the hens that lay glass eggs; the people born with ten left feet.

3. (a) $\{p, r, o, f, e, s\}$;
   (b) $\{f, s\}$;
   (c) $\{p, f\}$;
   (d) $\{f\}$.

5. (a) $\{2, 4, 6, 8\}$;
   (b) $\{1, 2, 4, 5, 7, 8\}$.

7.

Number of elements	0	1	2	3	4	5	6
Number of subsets	1	2	4	8	16	32	64

9. $N = 2^n$

1-3 EQUIVALENT SETS

1. $\{1, 2\}$ $\{1, 2\}$
$\{p, q\}$ $\{q, p\}$

$\{1, 2, 3, 4\}$ $\{1, 2, 3, 4\}$
$\{e, s, r, t\}$ $\{e, s, t, r\}$

3. $\{1, 2, 3, 4\}$ $\{1, 2, 3, 4\}$
$\{r, e, s, t\}$ $\{r, e, t, s\}$

$\{1, 2, 3, 4\}$ $\{1, 2, 3, 4\}$
$\{e, t, r, s\}$ $\{e, t, s, r\}$

$\{1, 2, 3, 4\}$ $\{1, 2, 3, 4\}$
$\{r, s, e, t\}$ $\{r, s, t, e\}$

$\{1, 2, 3, 4\}$ $\{1, 2, 3, 4\}$
$\{s, r, e, t\}$ $\{s, r, t, e\}$

$\{1, 2, 3, 4\}$ $\{1, 2, 3, 4\}$
$\{r, t, e, s\}$ $\{r, t, s, e\}$

$\{1, 2, 3, 4\}$ $\{1, 2, 3, 4\}$
$\{s, e, r, t\}$ $\{s, e, t, r\}$

$\{1, 2, 3, 4\}$ $\{1, 2, 3, 4\}$
$\{e, r, s, t\}$ $\{e, r, t, s\}$

$\{1, 2, 3, 4\}$ $\{1, 2, 3, 4\}$
$\{s, t, r, e\}$ $\{s, t, e, r\}$

$$\{1, 2, 3, 4\} \quad \{1, 2, 3, 4\}$$
$$\updownarrow \ \updownarrow \ \updownarrow \ \updownarrow \qquad \updownarrow \ \updownarrow \ \updownarrow \ \updownarrow$$
$$\{t, r, e, s\} \quad \{t, r, s, e\}$$

$$\{1, 2, 3, 4\} \quad \{1, 2, 3, 4\}$$
$$\updownarrow \ \updownarrow \ \updownarrow \ \updownarrow \qquad \updownarrow \ \updownarrow \ \updownarrow \ \updownarrow$$
$$\{t, e, r, s\} \quad \{t, e, s, r\}$$

$$\{1, 2, 3, 4\} \quad \{1, 2, 3, 4\}$$
$$\updownarrow \ \updownarrow \ \updownarrow \ \updownarrow \qquad \updownarrow \ \updownarrow \ \updownarrow \ \updownarrow$$
$$\{t, s, r, e\} \quad \{t, s, e, r\}$$

5. Among the many correct answers is: the set $\{n, u, m, b, e, r\}$ of letters in "number."

7. Among the many correct answers is: the empty set.

9. 3

11. 8

1-4 USES OF NUMBERS

1. Ordinal.

3. Ordinal.

5. Ordinal.

7. Identification.

9. Ordinal.

11. $\{1, 2, 3, 4, 5, \ldots, \quad n, \quad \ldots\}$
$$\updownarrow \ \updownarrow \ \updownarrow \ \updownarrow \ \updownarrow \qquad \updownarrow$$
$$\{1, 3, 5, 7, 9, \ldots, 2n - 1, \ldots\}$$

13. $\{2, 4, 6, 8, 10, \ldots, \quad 2n, \quad \ldots\}$
$$\updownarrow \ \updownarrow \ \updownarrow \ \updownarrow \ \updownarrow \qquad \updownarrow$$
$$\{1, 3, 5, 7, 9, \ldots, 2n - 1, \ldots\}$$

1-5 INTERSECTION AND UNION

1. (a) $\{1, 2, 3, 5, 7\}$;
(b) $\{1, 3\}$.

3. (a) $\{1, 3, 5, 7, 9\}$;
(b) $\{7\}$.

5. (a) $\{1, 3, 5, 7, 8, 10\}$;
(b) $\varnothing$

7. (a) $\{1, 2, 3, \ldots\}$;
(b) $\varnothing$

9. (a) $\{3, 4, 5\}$;
(b) $\{2, 4\}$;
(c) $\{2, 3, 4, 5\}$;
(d) $\{4\}$

11. (a) $\{2, 4, 6, \ldots\}$;
(b) $\{1, 3, 5, \ldots\}$;

13. 0

15. Among the many correct answers is: the set of counting numbers.

17. (a) Yes. Any two sets that consist of the same elements have the same number of elements.
(b) No. Two sets may have the same number of elements without having the same elements.

19. Among the many correct answers is:

$$\{a, \ e, \ i, \ o, \ u \}$$
$$\updownarrow \ \updownarrow \ \updownarrow \ \updownarrow \ \updownarrow$$
$$\{2, \ 4, \ 6, \ 8, \ 10\}$$

*21. $\{1, 3, 5, 7, 9, \ldots, 2n - 1, \ldots\}$
$$\updownarrow \ \updownarrow \ \updownarrow \ \updownarrow \ \updownarrow \qquad \updownarrow$$
$$\{5, 10, 15, 20, 25, \ldots, \quad 5n, \quad \ldots\}$$

15.
$$\{101, 103, 105, 107, 109, \ldots, \quad n, \quad \ldots\}$$
$$\updownarrow \ \updownarrow \ \updownarrow \ \updownarrow \ \updownarrow \qquad \updownarrow$$
$$\{102, 104, 106, 108, 110, \ldots, n + 1, \ldots\}$$

17. Infinite.

19. Finite.

21. Finite.

23. Infinite.

(c) $\{1, 2, 3, \ldots\}$;
(d) $\varnothing$.

13. (a) $\{2, 3\}$;
(b) $\{1, 2\}$;
(c) $\{1, 2, 3\}$;
(d) $\{2\}$.

15. $\{6, 8\}$

17. $\{2, 4, 5, 6, 8, 9, 10\}$

19. $\{4, 5, 9\}$

21. (a) A;
(b) B;
(c) $\varnothing$;
(d) U.

1. (a) 2;
 (b) 9;
 (c) 4;
 (d) 13.

3.

5.

7.

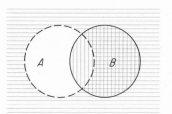

9. A' is shaded with horizontal lines; B is shaded with vertical lines. The union of these two sets is the subset of U that is shaded with lines in either or both directions.

11. A is shaded with vertical lines; B' is shaded with horizontal lines. The intersection of these two sets is the subset of U that is shaded with lines in both directions.

13.

$A \cup B = B \cup A$

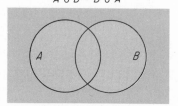

15.

$A \cap B \subseteq A$

17. In the following pair of diagrams, the final result is the same, showing the equivalence of the statements given.

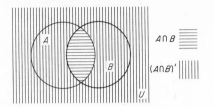

The set $(A \cap B)$ is shaded with horizontal lines. Its complement, $(A \cap B)'$, is the remaining portion of U shaded with vertical lines.

The set A' is shaded with vertical lines; the set B' is shaded with horizontal lines. Their union, $A' \cup B'$, is the portion of U shaded with lines in either or both directions.

19. (a) 1;
   (b) 5;
   (c) 4;
   (d) 12;
   (e) 21;
   (f) 25.

21. (a) 25;
   (b) 7;
   (c) 7;
   (d) 29;
   (e) 37;
   (f) 6.

23. There are 11 students who are not taking any of the three subjects, 10 taking only chemistry, and 2 taking physics and chemistry but not biology. The given data are shown in the following Venn diagram:

25. (a)

(b)

(c)

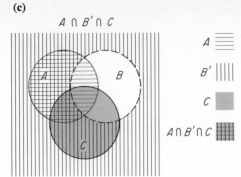

$A \cap B' \cap C$

(d)

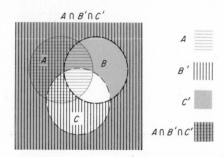

$A \cap B' \cap C'$

27.

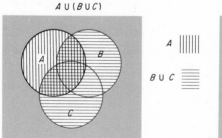

$A \cup (B \cup C)$ $(A \cup B) \cup C$

Set A is shaded with vertical lines; set $(B \cup C)$ is shaded with horizontal lines. Their union, $A \cup (B \cup C)$, is shaded with lines in either or both directions.

Set $(A \cup B)$ is shaded with vertical lines; set C is shaded with horizontal lines. Their union, $(A \cup B) \cup C$, is shaded with lines in either or both directions.

Chapter 2: CONCEPTS OF LOGIC

2-1 SETS OF STATEMENTS

1. (a) $(\sim p) \wedge (\sim q)$;
   (b) $(\sim p) \wedge q$;
   (c) $(\sim p) \wedge q$;
   (d) $\sim[p \wedge (\sim q)]$;
   (e) $p \vee (\sim q)$.

3. (a) $p \wedge (\sim q)$;
   (b) $p \vee (\sim q)$;
   (c) $(\sim p) \wedge q$;
   (d) $\sim[(\sim p) \wedge q]$;
   (e) $\sim[(p) \wedge q]$.

5. (a) I like this book and I like mathematics.
   (b) I do not like mathematics.
   (c) I do not like this book.
   (d) I do not like this book and I do not
   like mathematics.

7. 5(a), 6(b), and 6(d).

*9. (a) True;
   (b) true;
   (c) true;
   (d) true.

2-2 TRUTH VALUES OF STATEMENTS

1.

p	q	$(\sim p) \wedge q$
T	T	F
T	F	F
F	T	T
F	F	F

7.

p	q	$\sim$	$[p$	$\vee$	$(\sim q)]$
T	T	F	T	T	F
T	F	F	T	T	T
F	T	T	F	F	F
F	F	F	F	T	T

(d) (a) (c) (b)

3.

p	q	$(\sim p) \vee (\sim q)$
T	T	F
T	F	T
F	T	T
F	F	T

9.

p	q	$\sim$	$[(\sim p)$	$\wedge$	$(\sim q)]$
T	T	T	F	F	F
T	F	T	F	F	T
F	T	T	T	F	F
F	F	F	T	T	T

(d) (a) (c) (b)

5.

p	q	$\sim(p \wedge q)$
T	T	F
T	F	T
F	T	T
F	F	T

11.

p	q	$p \veebar q$
T	T	F
T	F	T
F	T	T
F	F	F

13. $\sim(p \wedge q)$

2-3 CONDITIONAL STATEMENTS

1. If you will study hard, then you will get an A.

3. If you do not study hard, then you will not get an A.

5. (a) If a triangle is equilateral, then the triangle is isosceles.
(b) If a triangle is isosceles, then the triangle is equilateral.
(c) If a triangle is not equilateral, then the triangle is not isosceles.

(d) If a triangle is not isosceles, then the triangle is not equilateral.

7. (a) T;
(b) T;
(c) T.

9.

p	q	p→q	(~p)	V	q
T	T	T	F	T	T
T	F	F	F	F	F
F	T	T	T	T	T
F	F	T	T	T	F
		(a)	(b)	(d)	(c)

Note that the truth values in columns (a) and (d) are the same.

11.

p	q	(p→q)	∧	(q→p)
T	T	T	T	T
T	F	F	F	T
F	T	T	F	F
F	F	T	T	T
		(a)	(c)	(b)

13.

p	q	[(~p)	∧	q]	→	(p V q)
T	T	F	F	T	T	T
T	F	F	F	F	T	T
F	T	T	T	T	T	T
F	F	T	F	F	T	F
		(a)	(c)	(b)	(e)	(d)

15. {7}

*17. The set of all real numbers.

*19. The set of all real numbers different from 2.

2-4 EQUIVALENT STATEMENTS

1.

p	q	[~ (p ∧ q)]	↔	[(~p)	V	(~q)]
T	T	F	T	T	F	F	F
T	F	T	F	T	F	T	T
F	T	T	F	T	T	T	F
F	F	T	F	T	T	T	T
		(b)	(a)	(f)	(c)	(e)	(d)

3.

p	q	(p→q)	↔	[q	V	(~p)]
T	T	T	T	T	T	F
T	F	F	T	F	F	F
F	T	T	T	T	T	T
F	F	T	T	F	T	T
		(a)	(e)	(b)	(d)	(c)

5. *Converse:* If we buy a new car, then we can afford it.

Inverse: If we cannot afford it, then we do not buy a new car.

Contrapositive: If we do not buy a new car, then we cannot afford it.

7. *Converse:* If the triangles are congruent, then two sides and the included angle of one are congruent to two sides and the included angle of the other.

Inverse: If two sides and the included angle of one triangle are not congruent to two sides and the included angle of another triangle, then the triangles are not congruent.

Contrapositive: If two triangles are not congruent, then two sides and the included angle of one are not congruent to two sides and the included angle of the other.

9. *Converse:* If $x = 1$, then $x(x - 1) = 0$.
Inverse: If $x(x - 1) \neq 0$, then $x \neq 1$.
Contrapositive: If $x \neq 1$, then
$$x(x - 1) \neq 0.$$

11. The statement is always true in Exercises 7 and 8.

13. The inverse is always true in Exercises 7 and 9.

1. If it is a duck, then it is a bird.

3. If two angles are complements of the same angle, then they are congruent.

5. If two lines are parallel, then they are coplanar.

7. If a geometric figure is a circle, then it is round.

9. If a person is a teacher, then he is boring.

11. If you like this book, then you like mathematics.

13. If you like mathematics, then you like this book.

15. If you like mathematics, then you will like this book.

17. $q \to p$

19. $q \to p$

21. $p \to (\sim q)$

23. $p \leftrightarrow q$

25. $p \to q$

27. $p \leftrightarrow q$

29. If $9 + 3 < 10$, then $11 - 3 > 8$; true.

31. If $5 + 3 = 8$, then $7 \times 4 = 25$; false.

33. If $7 \times 6 = 42$, then $8 \times 5 \neq 40$; false.

35. The assertion is equivalent to the statement: "If you give me $10,000, then I will marry your daughter." If he received the money, then he should have married the girl. Thus he should be sued for breach of promise.

2-6 THE NATURE OF PROOF

1. A tautology.

p	p	V	$(\sim p)$
T	T	T	F
F	F	T	T
	(a)	(c)	(b)

3. A tautology.

p	q	$[(p \to q)$	$\wedge$	$(\sim q)]$	$\to$	$(\sim p)$
T	T	T	F	F	T	F
T	F	F	F	T	T	F
F	T	T	F	F	T	T
F	F	T	T	T	T	T
		(a)	(c)	(b)	(e)	(d)

5. Not a tautology.

p	q	$[(p \to q)$	$\wedge$	$(\sim p)]$	$\to$	$(\sim q)$
T	T	T	F	F	T	F
T	F	F	F	F	T	T
F	T	T	T	T	F	F
F	F	T	T	T	T	T
		(a)	(c)	(b)	(e)	(d)

7. A tautology.

p	q	$[(p \vee q)$	$\wedge$	$(\sim p)]$	$\to$	q
T	T	T	F	F	T	T
T	F	T	F	F	T	F
F	T	T	T	T	T	T
F	F	F	F	T	T	F
		(a)	(c)	(b)	(e)	(d)

9. A tautology.

p	q	r	$[(p \to q)$	$\wedge$	$(q \to r)]$	$\to$	$(p \to r)$
T	T	T	T	T	T	T	T
T	T	F	T	F	F	T	F
T	F	T	F	F	T	T	T
T	F	F	F	F	T	T	F
F	T	T	T	T	T	T	T
F	T	F	T	F	F	T	T
F	F	T	T	T	T	T	T
F	F	F	T	T	T	T	T
			(a)	(c)	(b)	(e)	(d)

1. For

p: Elliot is a freshman.
q: Elliot takes mathematics.

the argument has the form

$$[(p \to q) \wedge p] \to q.$$

This statement is a tautology (§2-6, Example), and the argument is valid.

3. For

p: The Yanks win the game.
q: The Yanks win the pennant.

the argument has the form

$$[(p \to q) \wedge (\sim q)] \to (\sim p).$$

This statement is a tautology (§2-6, Exercise 3), and the argument is valid.

5. For

p: You work hard.
q: You are a success.

the argument has the form

$$[(p \to q) \wedge (\sim q)] \to (\sim p).$$

This statement is a tautology (§2-6, Exercise 3), and the argument is valid.

7. For

p: You are reading this book.
q: You like mathematics.

the argument has the form

$$[(p \to q) \wedge (\sim p)] \to (\sim q).$$

This statement is not a tautology (§2-6, Exercise 5) and the argument is not valid.

9. This argument is of the form $[(p \to q) \wedge (q \to r)] \to (r \to p)$ and is not valid.

11. You do not drink milk.

13. If you like to fish, then you are a mathematician.

15. If you like this book, then you will become a mathematician.

2-8 EULER DIAGRAMS

1. All juniors are intelligent.
All undergraduates are intelligent.
Some juniors are undergraduates.
Some intelligent people are juniors.
Some intelligent people are undergraduates.
Some undergraduates are juniors.

3. Valid.

5. Not valid.

7. (a) Valid;
(b) valid;
(c) valid;

(d) not valid.

9. (a) Valid;
(b) not valid;
(c) not valid;
(d) not valid;
(e) not valid;
(f) not valid;
(g) not valid.

11. (a) Valid;
(b) valid;
(c) valid;
(d) not valid.

2-9 QUANTIFIERS

1. 9

3. 6

5. (1) 7, or any value different from 9.
(2) 7, or any value different from 6.
(3) 5, or any value different from 6.
(4) 5, or any value different from 3 or −3.

7. In each case use $\forall_x$.

9. Not all numbers are whole numbers; that is, there exists at least one number that is not a whole number.

11. Not all fractions represent rational numbers; that is, there exists at least one fraction that does not represent a rational number.

13. All real numbers are rational numbers.

15. No rational numbers are integers; that is, all rational numbers are not integers.

3-1 AN ABSTRACT SYSTEM

1. #

3. Σ

5. Commutative property for multition.

7. Identity element (∗).

9. Associative property for multition.

11. Inverse element.

13. △

15. Yes.

17. Yes.

19. No; it must hold for all possible choices of three elements.

21. Yes, the inverse of △ is Q, the inverse of □ is □, the inverse of Q is △.

23. No; the entries within the table contain elements ∘ and ∼, which are not members of the original set of elements.

25. Yes; ·

SUPPLEMENTARY EXERCISES

1. One counterexample is sufficient; for example, $8 \div 3$ is not a counting number.

3. 1; $8 \times 1 = 8$, $13 \times 1 = 13$.

5. Yes. Yes.

*7. Regardless of the answer to $(b \# c)$, the result of $a \$ (b \# c)$ is always a, since $\$$ means to select the first number. Also $a \$ b = a$, $a \$ c = a$, and thus $a \# a = a$.

3-2 THE DISTRIBUTIVE PROPERTY

1. 8, 8, 3, 7.

3. 7, 7.

5. 3, 3.

7. No.

9. $7 \times 79 = 7 \times (80 - 1)$
$= 560 - 7 = 553.$

11. 3

13. 5

15. 7

17. 7

19. True for all replacements of n.

3-3 CLOCK ARITHMETIC

1. (a) Yes;
   (b) yes.

3. The identity with respect to addition is 12; the identity with respect to multiplication is 1.

5. 3

7. 9

9. 3

11. 5

13. 8

15. 6

17. 1, 5, or 9.

19. 4

*21. An impossible equation; that is, there is no value of t for which this equation is true.

*23. An identity; that is, this equation is true for all possible replacements of t.

25. 6

27. 9

29. 12

1.

X	0	1	2	3	4
0	0	0	0	0	0
1	0	1	2	3	4
2	0	2	4	1	3
3	0	3	1	4	2
4	0	4	3	2	1

3. Two specific cases are $(3 \cdot 2) \cdot 4 = 3 \cdot (2 \cdot 4) = 4$ and $(4 \cdot 3) \cdot 4 = 4 \cdot (3 \cdot 4) = 3$. There many others.

5. The inverse of 1 is 1, of 2 is 3, of 3 is 2,

and of 4 is 4. Note that 0 does not have an inverse with respect to multiplication.

7. 4

9. 4

11. 3

13. 3

15. 4

17. 4

19. 3

21. 4

23. An impossible equation; that is, there is no value of x for which this equation is true.

SUPPLEMENTARY EXERCISES

1.

+	0	1	2	3
0	0	1	2	3
1	1	2	3	0
2	2	3	0	1
3	3	0	1	2

X	0	1	2	3
0	0	0	0	0
1	0	1	2	3
2	0	2	0	2
3	0	3	2	1

This system is closed, commutative, and associative with respect to both addition and multiplication. It satisfies the distributive property for multiplication with respect to addition. There are identity elements for both addition (0) and multiplication (1). Each element has an inverse with respect to addition; not every element has an inverse with respect to multiplication.

3. 2

5. 5

7. 3

9. 3

11. 4

13. An impossible equation; that is, there is no value of x for which this equation is true.

15. 0

17. $2 \cdot 6 = 0; 3 \cdot 8 = 0; 4 \cdot 6 = 0; 4 \cdot 9 = 0; 6 \cdot 6 = 0; 6 \cdot 8 = 0; 6 \cdot 10 = 0; 8 \cdot 9 = 0$. Thus the zero divisors in arithmetic modulo 12 are 2 and 6, 3 and 4, 3 and 8, 4 and 9, 6 and 6, 6 and 8, 6 and 10, and 8 and 9.

Chapter 4: SYSTEMS OF NUMERATION

4-2 EGYPTIAN NUMERATION

1. ∩∩ | | | | |

3. ⌐⌐99999∩∩ | | | | | |

5. ⟋⌐⌐99999 | | | | | | |

7. 22

9. 1102

11. 1324

13.

$$
\begin{array}{r}
∩∩∩∩ || \\
+ \quad ∩∩ | \\
\hline
∩∩∩∩∩∩ |||
\end{array}
$$

15.

17.

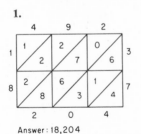

19. ①× 45 = ㊺
2 × 45 = 90
4 × 45 = 180
8 × 45 = 360
16 × 45 = ⑦⑳
17 = 1 + 16
17 × 45 = (1 + 16) × 45
= 45 + 720 = 765

21. ①× 31 = ㉛
②× 31 = ㊽
4 × 31 = 124
8 × 31 = 248
⑯× 31 = ㊍⑨⑥

27 = 1 + 2 + 8 + 16
27 × 31 = (1 + 2 + 8 + 16) × 31
= 31 + 62 + 248 + 496
= 837

23. 19 → �33
9 → ㊻
4   132
2   264
1 → ㊺㉘

19 × 33 = 33 + 66 + 528 = 627

25. 21 → ㊾
10   104
5 → ㉈⑧
2   416
1 → ㊳㉜

21 × 52 = 52 + 208 + 832 = 1092

27. We need a symbol for zero because the value of a numeral in our decimal system depends upon both the value and the position of the digits in the numeral. For example, 405 and 450 represent two entirely different numbers, yet the same digits are used in the two numerals. The ancient Egyptians were not dependent upon the position of a symbol in determining its value, so zero as a "place-holder" was not needed.

4-3 OTHER METHODS OF COMPUTATION

1.

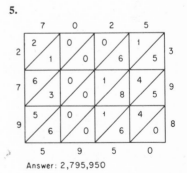

Answer: 18,204

3.

Answer: 119,232

5.

Answer: 2,795,950

7.

Answer: 2048

9.

	3	8	7
5	1/5	4/0	3/5

1 9 3 5

Answer: 1935

11.

	7	2	5	6
8	5/6	1/6	4/0	4/8

5 8 0 4 8

Answer: 58,048

4-4 DECIMAL NOTATION AND DIVISIBILITY

1. $(2 \times 10^2) + (3 \times 10) + (7 \times 10^0)$.

3. $(4 \times 10^3) + (2 \times 10^2) + (0 \times 10) + (5 \times 10^0)$.

5. $(1 \times 10^5) + (2 \times 10^4) + (5 \times 10^3) + (2 \times 10^2) + (0 \times 10) + (0 \times 10^0)$.

7. $(6 \times 10^5) + (0 \times 10^4) + (0 \times 10^3) + (1 \times 10^2) + (0 \times 10) + (0 \times 10^0)$.

9. 3253

11. 403,205

13. 9,000,007

15. 4,000,000,000

        (a) (b) (c) (d) (e) (f) (g) (h) (i)

17. Yes yes yes yes yes yes no no no

19. Yes yes yes no yes yes yes no no

21. Yes yes yes yes no yes no no no

23. Yes no yes yes no yes no yes yes

25. Yes no yes yes no yes no no no

27. A number is divisible by 6 if it is divisible by 2 and by 3.

29. A number is divisible by 9 if the sum of its digits is divisible by 9.

31. A number is divisible by 100 if its tens digit and its units digit are both 0.

4-5 OTHER SYSTEMS OF NUMERATION

1. 22_{five}

3. 30_{four}

5.

* * * * *
* * * * *
* * * *

7.

* * * * * * *
* * * * * * *
* * * * *

9.

* * *

11. 23

13. 26

15. 19

17. 32_{five}

19. 32_{six}

4-6 BASE FIVE NOTATION

1. 113

3. 124

5. 86

7. 141

9. 209

11. 3012_5

13. 11112_5

15. 10000_5

17. 1022_5

19. 4012_5

*21. 287

*23. 11

*25. 462

*27. 469

4-7 OTHER NUMBER BASES

1. 314

3. 1076

5. 50

7. 662

9. 2590

11. 3598_{12}

13. $87e0_{12}$

15. 11010_4

17. 1551_6

19. 11143_5

21. 655_9

23. 187_{12}

*25. $1e7_{12}$

*27. 101111_2

4-8 COMPUTATION IN BASE FIVE NOTATION

1. 111_5

3. 432_5

5. 11312_5

7. 112_5

9. 134_5

11. 413_5

13. 222_5

15. 3023_5

17. 3401_5

19. 1423_5

21. 21022_5

23. 22_5

25. $224_5 \, R \, 2$

27. 13_5

29. 41_5

31.

X	0	1	2	3	4
0	0	0	0	0	0
1	0	1	2	3	4
2	0	2	4	11_5	13_5
3	0	3	11_5	14_5	22_5
4	0	4	13_5	22_5	31_5

*33.

+	0	1	2	3	4	5	6	7
0	0	1	2	3	4	5	6	7
1	1	2	3	4	5	6	7	10_8
2	2	3	4	5	6	7	10_8	11_8
3	3	4	5	6	7	10_8	11_8	12_8
4	4	5	6	7	10_8	11_8	12_8	13_8
5	5	6	7	10_8	11_8	12_8	13_8	14_8
6	6	7	10_8	11_8	12_8	13_8	14_8	15_8
7	7	10_8	11_8	12_8	13_8	14_8	15_8	16_8

X	0	1	2	3	4	5	6	7
0	0	0	0	0	0	0	0	0
1	0	1	2	3	4	5	6	7
2	0	2	4	6	10_8	12_8	14_8	16_8
3	0	3	6	11_8	14_8	17_8	22_8	25_8
4	0	4	10_8	14_8	20_8	24_8	30_8	34_8
5	0	5	12_8	17_8	24_8	31_8	36_8	43_8
6	0	6	14_8	22_8	30_8	36_8	44_8	52_8
7	0	7	16_8	25_8	34_8	43_8	52_8	61_8

*35. 10322_6

*37. 226_8

*39. 14446_8

*41. $210_4 \ R \ 1$

4-9 BINARY NOTATION

1. 11100_2

3. 10011_2

5. 10011000_2

7. 15

9. 55

11. 38

13. 11000_2

15. 101_2

17. 100111_2

19. (a) 1000010_2;
(b) 1000100_2;
(c) 100111_2

21. $214 = 326_8 = 11010110_2$. When placed in groups of three, starting from the units digit, the binary representation can be translated into the octal system, and conversely. Thus

$$\underbrace{011}_{3} \ \underbrace{010}_{2} \ \underbrace{110_2}_{6} = 326_8.$$

23. 3531_8

25. $22,533_8$

27. $1,531,046_8$

29. $110,000,010,011_2$

31. $11,110,010,100_2$

33. $111,110,101,100,011_2$

Chapter 5: THE SET OF INTEGERS

5-1 COUNTING NUMBERS

1. $\{(1, 1), (1, 2), (1, 3), (1, 4), (2, 1), (2, 2), (2, 3), (2, 4)\}$.

3. Use any counter example such as $3 - 2 \neq 2 - 3$.

5. $8 - (3 - 2) = 8 - 1 = 7$,
$(8 - 3) - 2 = 3$; no, no.

7. Use any counter example such as
$$2 + (3 \times 5) \neq (2 + 3) \times (2 + 5);$$
$$17 \neq 5 \times 7.$$

9. Commutative, $+$.

11. Associative, $\times$.

13. Distributive, $\times$ with respect to $+$.

15. Commutative, $+$.

17. Distributive, $\times$ with respect to $+$.

19. (a) Associative, $+$;
(b) commutative, $+$;

(c) associative, $+$;
(d) associative, $+$;
(e) distributive, $\times$ with respect to $+$.

21. $92 + (50 + 8)$
$= 92 + (8 + 50)$ Commutative, $+$.
$= (92 + 8) + 50$ Associative, $+$.

23. $37 \times (1 + 100)$
$= 37 \times (100 + 1)$ Commutative, $+$.
$= 37 \times 100 + 37 \times 1$ Distributive,
$\times$ w.r.t. $+$.
$= 37 \times 100 + 37$ Identity, $\times$.
$= 3700 + 37$ Number system.

25. $(73 + 19) + (7 + 1)$
$= [(73 + 19) + 7] + 1$ Associative, $+$.
$= [73 + (19 + 7)] + 1$ Associative, $+$.
$= [73 + (7 + 19)] + 1$ Commutative, $+$.
$= [(73 + 7) + 19] + 1$ Associative, $+$.
$= (73 + 7) + (19 + 1)$ Associative, $+$.

27. No.

5-2 PRIME NUMBERS

1. Any number that is divisible by 15 is also divisible by 3.

3. The set of numbers divisible by 3 and 5 is the set of numbers divisible by 15.

5. Any number that is divisible by 12 is also divisible by 3.

7. $\{21, 22, 24, 25, 26, 27, 28, 30, 32, 33, 34, 35, 36, 38, 39\}$

507

sec. 5-4
equivalance
and
order
relations

9. No. No.

11. There are other possible answers in many cases. $4 = 2 + 2$, and

$$6 = 3 + 3; \qquad 8 = 3 + 5;$$
$$10 = 3 + 7; \qquad 12 = 5 + 7;$$
$$14 = 7 + 7; \qquad 16 = 3 + 13;$$
$$18 = 5 + 13; \qquad 20 = 7 + 13;$$
$$22 = 5 + 17; \qquad 24 = 7 + 17;$$
$$26 = 3 + 23; \qquad 28 = 5 + 23;$$
$$30 = 7 + 23; \qquad 32 = 3 + 29;$$
$$34 = 5 + 29; \qquad 36 = 7 + 39;$$
$$38 = 7 + 31; \qquad 40 = 3 + 37.$$

13. Any three consecutive odd numbers includes 3 or a multiple of 3. Each multiple of 3 is composite, and 1 is by definition not a prime. Therefore 3, 5, 7 is a set of three consecutive odd numbers that are all prime

numbers and, since any other such set contains a composite number that is a multiple of 3, this is the only prime triplet.

*15. (a) 13;
  (b) 19;
  (c) 31.

17. $1 \times 15; 3 \times 5$.

19. $1 \times 20; 2 \times 10; 4 \times 5$.

21. 1×29

23. $2^2 \times 17$

25. 3×41

27. $3 \times 5^2 \times 19$

29. 2×409

5-3 APPLICATIONS OF PRIME FACTORIZATIONS

1. $\{1, 2, 4, 5, 10, 20\}$

3. $\{1, 3, 9\}$

5. $\{1, 2, 4, 7, 14, 28\}$

7. $\{1, 2, 3, 4, 5, 6, 10, 12, 15, 20, 30, 60\}$

9. $68 = 2^2 \times 17; 76 = 2^2 \times 19$; G.C.F. $= 4$.

11. $76 = 2^2 \times 19; 1425 = 3 \times 5^2 \times 19$; G.C.F. $= 19$.

13. $215 = 5 \times 43; 1425 = 3 \times 5^2 \times 19$; G.C.F. $= 5$.

15. $12 = 2^2 \times 3; 15 = 3 \times 5$; $20 = 2^2 \times 5$; G.C.F. $= 1$.

17. $12 = 2^2 \times 3; 18 = 2 \times 3^2; 30 = 2 \times 3 \times 5$; G.C.F. $= 2 \times 3 = 6$.

19. $\{7, 14, 21, 28, 35\}$

21. $\{15, 30, 45, 60, 75\}$

23. See Exercise 9; L.C.M. $= 2^2 \times 17 \times 19 = 1292$.

25. See Exercise 11; L.C.M. $= 2^2 \times 3 \times 5^2 \times 19 = 5700$.

27. See Exercise 13; L.C.M. $= 3 \times 5^2 \times 19 \times 43 = 61,275$.

29. See Exercise 15; L.C.M. $= 2^2 \times 3 \times 5 = 60$.

31. See Exercise 17; L.C.M. $= 2^2 \times 3^2 \times 5 = 180$.

33. $\frac{123}{215}$

35. $\frac{19}{24}$

37. $\frac{13}{30}$

39. $\frac{5}{12}$

41. $\dfrac{1,259}{26,445}$

*43. 1

5-4 EQUIVALENCE AND ORDER RELATIONS

1. (a), (b), (c), (e), (f), (i).

3. (a), (d), (e), (g), (h), (i).

5. (a) Yes; (b) yes; (c) yes.

7. (a) Yes; (b) no; (c) yes.

9. (a) No; (b) no; (c) yes.

11. (a) Yes; (b) no; (c) yes.

13. (a) $>$; (b) $<$; (c) $=$; (d) $>$; (e) $<$.

15. (a) As in logic p or $\sim p$; exactly one must hold. Here $a = b$ or $a \neq b$, exactly one must hold.

  (b) By the trichotomy law exactly one

of the relations $a = b$, $a < b$, $a > b$ must hold. If the first of these relations does not hold

 $(a \neq b)$, then exactly one of the other two must hold.

5-5 WHOLE NUMBERS

1.

3.

5.

7.

9. The graph is the empty set; there are no points in the graph.

11. $W = \{0, 1, 2, 3, 4\}$
   $C = \{1, 2, 3, 4, 5\}$

13. If $a \times b = 1$, then both $a = 1$ and $b = 1$. If $a \times b = 2$, then either $a = 1$ and $b = 2$ or $a = 2$ and $b = 1$. If $a \times b = 3$,

then either $a = 1$ and $b = 3$ or $a = 3$ and $b = 1$. If $a \times b = 4$, then $a = 1$ and $b = 4$, or $a = 2$ and $b = 2$, or $a = 4$ and $b = 1$.

15. None.

17. $W = \{0, 1, 2, 3, 4, \ldots, \quad n, \ldots\}$
   $C = \{1, 2, 3, 4, 5, \ldots, n + 1, \ldots\}$

5-6 INTEGERS

1. $\{ \ 1, \ 2, \ 3, \ 4, \ 5, \ldots, \quad n, \ldots\}$
   $\{-1, -2, -3, -4, -5, \ldots, \ -n, \ldots\}$

3.

5.

7. True.

9. False.

11. True.

13. False; -0 is not a negative integer.

15. False; for example, $4 \div 3$ is not an integer.

17. No, zero is an integer but is neither positive nor negative and thus is not in the union of the set of positive integers and the set of negative integers.

19. $\{1, 2, \ 3, 4, \ 5, 6, \ 7, \ldots, 2n, 2n+1, \ldots\}$
   $\{0, 1, -1, 2, -2, 3, -3, \ldots, n, - \ n, \ldots\}$

21. Let $2k$ and $2m$ represent two even integers. Then

$$2k \times 2m = 2 \times 2 \times k \times m = 2(2km),$$

where $2km$ is an integer. Thus $2(2km)$ is an even integer.

23. Any integer is either even or odd. If an integer is even, then its square is even (as in Example 5). If an integer is odd, then its square is odd (as in Exercise 22). If the square of an integer is odd, the integer cannot be even and thus must be odd. If the square of an integer is even, the integer cannot be odd and thus must be even.

5-7 OPERATIONS WITH INTEGERS

1. $+11$

3. -5

5. $+3$

7. -14

9. $(+8) + (-3) = +5$

11. $(-8) + (-3) = -11$

13. $(-3) + (+8) = +5$

15. $(-8) + (-8) = -16$

17. $+20$

19. -20

21. 0

23. $+42$

25. $+4$

27. -4

29. 0

31. -7

33. -15

35. $+24$

37. -4

39. Division of integers is not associative;

$$24 \div (6 \div 2) = 24 \div 3 = 8,$$
$$(24 \div 6) \div 2 = 4 \div 2 = 2.$$

5-8 ABSOLUTE VALUE

1. $+7$

3. -6

5. $+8$

7. $+8$

9. $+8$

11. -8

13. -8

15. $+14$

17. $+4$

19. $+42$

21. -42

23. The largest of $+5$ and -5 is $+5$; $+5 = |5|$. The largest of -2 and $+2$ is $+2$; $+2 = |-2|$. The largest of 0 and 0 is 0; $0 = |0|$. (Many other examples may be given.)

25. $5, -9.$

27. $9, -5.$

Chapter 6: THE SET OF RATIONAL NUMBERS

6-1 SYMBOLS AND DEFINITIONS

1. $<$

3. $>$

5. $<$

7. (a) 2; (b) 1; (c) 2; (d) 0.

9. (a) $\frac{3}{12} + \frac{8}{12}$;   (b) $\frac{9}{12} + \frac{4}{12}$;   (c) $\frac{10}{12} + \frac{21}{12}$.

11. $\frac{2}{15}$

13. $\frac{14}{33}$

15. $\frac{6}{35}$

17. $\frac{10}{21}$

19. $\frac{13}{12}$

21. $\frac{11}{15}$

1. True.

3. True.

5. False.

7. **9.**

Set	−5	$\frac{2}{3}$
Counting numbers	x	x
Whole numbers	x	x
Integers	✓	x
Rational numbers	✓	✓

11. Every rational number except 0 has a multiplicative inverse; the set of rational numbers is dense.

13. $\frac{1}{200}$, $\frac{1}{400}$, $\frac{1}{800}$. There are many other correct answers.

15. No, the density property of the set of rational numbers states that between any two elements of the set there is always another element of the set.

	Sentence	Counting numbers	Whole numbers	Integers	Rational numbers
17.	$n + 3 = 3$	none	0	0	0
19.	$3n = 5$	none	none	none	$\frac{5}{3}$

6-3 RATIONAL NUMBERS AS FRACTIONS

1. $\frac{7}{6}$

3. $\frac{10}{21}$

5. $\frac{5}{6}$

7. $\frac{3}{4}$

9. $\frac{1}{2}$

11. $\frac{9}{8}$

13. $-\frac{3}{10}$

15. $-\frac{10}{27}$

17. $-\frac{21}{16}$

19. $-\frac{1}{28}$

21. $\frac{55}{24}$

23. $\frac{5}{8}$

25. $\frac{3}{10}$

27. $\frac{9}{32}$

29. $\frac{2}{3}$

31. 1

33. $\frac{14}{9}$

35. $\frac{13}{14}$

*37. $\dfrac{adf + bcf + bde}{bdf}$

*39. $\dfrac{acf + ade}{bdf}$

*41. $\dfrac{adf + bcf}{bde}$

6-4 RATIONAL NUMBERS AS DECIMALS

1. 0.001

3. 0.04

5. 0.007

*7. 0.0002

9. $(4 \times 10^0) + (2 \times 10^{-1}) + (3 \times 10^{-2})$

11. $(4 \times 10^2) + (2 \times 10) + (3 \times 10^0)$
$+ (8 \times 10^{-1}) + (3 \times 10^{-2})$

13. $(0 \times 10^0) + (0 \times 10^{-1})$
$+ (0 \times 10^{-2}) + (7 \times 10^{-3})$

15. $(0 \times 10^0) + (0 \times 10^{-1}) + (0 \times 10^{-2})$
$\qquad + (0 \times 10^{-3}) + (1 \times 10^{-4})$

17. $(0 \times 10^0) + (1 + 10^{-1})$
$\qquad + (0 \times 10^{-2}) + (1 \times 10^{-3})$

19. 0.235

21. 70.082

23. 5705.5

25. 37.29

27. 2.0019

29. Two hundred seventy-three and eighty-seven hundredths.

31. Nine and eighteen ten-thousandths.

33. 1.6

35. 0.11

37. 0.7

39. 0.07

41. (a)

43. (c)

45. (a)

47. (a)

6-5 RATIONAL NUMBERS AND REPEATING DECIMALS

1. 0.8

3. 0.875

5. 1.8

7. $0.\overline{428571}$

9. $0.\overline{692307}$

11. $\frac{5}{11}$

13. $\frac{9}{11}$

15. $\frac{58}{111}$

17. $\frac{19}{9}$

19. $\frac{212}{99}$

21. $\frac{41}{333}$

23. (a) 0.7272727272;
(b) 0.3571571571;
(c) 0.8000000000;
(d) 0.4237373737.

25. $2.08, $2.37, $2.57, $2.64, $2.98.

27. $1.\overline{4}$, 1.45, $1.\overline{45}$, $1.4\overline{5}$, $1.\overline{5}$.

Among the many correct answers for Exercises 29 and 31 are:

*29. 0.2342

*31. 0.231

*33. (a) 5; (b) 120.

6-6 RATIONAL NUMBERS AS PER CENTS

1. 0.37

3. 0.05

5. 0.0095

7. 1.50

9. 1.00

11. $\frac{1}{2}$

13. $\frac{17}{20}$

15. $\frac{99}{100}$

17. $\frac{3}{2}$

19. $\frac{1}{20}$

21. 45%

23. 65%

25. 70%

27. 235%

29. 0.1%

31. 3%

33. 92%

35. 150%

37. 36%

39. 130%

41. $\frac{40}{100} = \frac{n}{80}$; 32.

43. $\frac{35}{100} = \frac{n}{60}$; 21.

45. $\dfrac{40}{160} = \dfrac{n}{100}$; 25%.

47. $\dfrac{120}{160} = \dfrac{n}{100}$; 75%.

49. $\dfrac{15}{n} = \dfrac{20}{100}$; 75.

51. $\dfrac{60}{n} = \dfrac{125}{100}$; 48.

53. $\dfrac{40}{100} = \dfrac{n}{2.00}$; $0.80.

55. $\dfrac{1.50}{n} = \dfrac{30}{100}$; $5.00.

57. 20%

6-7 PROPERTIES OF THE SET OF RATIONAL NUMBERS

	Counting numbers	Whole numbers	Integers	Positive rationals	Rational numbers
1.	√	√	√	√	√
3.	x	√	√	x	√
5.	√	√	√	√	√
7.	√	√	√	√	√
9.	√	√	√	√	√
11.	√	√	√	√	√
13.	√	√	√	√	√

Chapter 7: THE SET OF REAL NUMBERS

7-1 INTRODUCING THE REAL NUMBERS

1. True.

3. True.

5. True.

7. 9.

Set	$-\dfrac{2}{3}$	0
Counting numbers	x	x
Whole numbers	x	√
Integers	x	√
Rational numbers	√	√
Real numbers	√	√

15.   17.

Property	Positive integers	Negative integers
Closure, +	√	√
Closure, x	√	x
Inverse, +	x	x
Inverse, x	x	x
Dense	x	x

	Sentence	Counting numbers	Whole numbers	Integers	Rational numbers	Real numbers
11.	$n + 3 = 1$	none	none	-2	-2	-2
13.	$n \times n = 2$	none	none	none	none	$\sqrt{2}, -\sqrt{2}$

513

sec. 7-2
graphs
of sets
of real
numbers

19. If $\sqrt{2} + 2$ were a rational number a/b, then $\sqrt{2}$ would be a rational number $\dfrac{a}{b} - 2$ (impossible). Therefore $\sqrt{2} + 2$ cannot be a rational number and must be an irrational number.

21.

7-2 GRAPHS OF SETS OF REAL NUMBERS

1.

3.

5.

7.

9.

11.

13.

15.

17.

19 and 21.

23. $(-4, 3)$

25. $(2, 5)$

27. $(-3, -4)$

7-3 DECIMAL REPRESENTATION OF REAL NUMBERS

1. Rational number.

3. Irrational number.

5. Irrational number.

7. A terminating decimal, 0.75.

9. A terminating decimal, 9.

11. A nonterminating, nonrepeating decimal.

13. 0.757575757575757

15. 0.353553555355553

17. 2

19. 7

25. 2.6

27. 3.3

29. 0.35; $0.\overline{35}$; 0.35355; 0.353553555...; $0.3\overline{5}$.

31. 0.07; $0.\overline{07}$; 0.078; 0.0787887888...; $0.07\overline{8}$.

33. (c) and (d).

35. 0.385385538555...; 0.383883888...; many other answers are possible.

37. 0.3565; p.$\overline{35655}$; many other answers are possible.

7-4 PROPERTIES OF THE REAL NUMBERS

1. False.

3. True.

5. True.

7. True.

9. True.

	Counting numbers	Whole numbers	Integers	Rational numbers	Real numbers
11.	√	√	√	√	√
13.	√	√	√	√	√
15.	√	√	√	√	√
17.	x	√	√	√	√
19.	x	x	√	√	√
21.	√	√	√	√	√
23.	x	x	x	√	√

25. True, $\frac{2}{3} \times \frac{7}{13} = \frac{14}{39}$.

27. False, $\sqrt{2} \cdot \sqrt{2} = 2$.

29. False, 2 does not have a multiplicative inverse.

1. $2 + 3i$

3. $3 + 0i$

5. $5 + \sqrt{3}i$

7. True.

9. True.

11. 13. 15.

Set	−7	0	$\sqrt{-9}$
Counting numbers	x	x	x
Whole numbers	x	✓	x
Integers	✓	✓	x
Rational numbers	✓	✓	x
Irrational numbers	x	x	x
Real numbers	✓	✓	x
Imaginary numbers	x	x	✓
Complex numbers	✓	✓	✓

	Sentence	Counting numbers	Integers	Rational numbers	Real numbers	Complex numbers
17.	$n + 5 = 2$	none	−3	−3	−3	−3
19.	$n^2 = 4$	2	2,−2	2,−2	2,−2	2,−2

7-6 TRANSFINITE CARDINAL NUMBERS

1. $\{1, 2, 3, 4, 5, \ldots, \quad n, \quad \ldots\}$
$\updownarrow \updownarrow \updownarrow \updownarrow \updownarrow \qquad \updownarrow$
$\{1, 3, 5, 7, 9, \ldots, 2n - 1, \ldots\}$

3. The set of whole numbers has transfinite cardinal number $\aleph_0 + 1$; the set of counting numbers has cardinal number $\aleph_0$; the two sets are equivalent, $\aleph_0 + 1 = \aleph_0$.

5. $\aleph_0 + 3$: $\{-2, -1, 0, 1, 2, 3,$
$\qquad\quad \updownarrow \quad \updownarrow \updownarrow \updownarrow \updownarrow \updownarrow$
$\aleph_0$: $\{ \quad 1, \quad 2, 3, 4, 5, 6,$

$\qquad\qquad \ldots, \quad n, \quad \ldots\}$
$\qquad\qquad\qquad \updownarrow$
$\qquad\qquad \ldots, n + 3, \ldots\}$

7. $\aleph_0 - 1$: $\{2, 3, 4, 5, 6, \ldots, \quad n, \quad \ldots\}$
$\qquad\quad \updownarrow \updownarrow \updownarrow \updownarrow \updownarrow \qquad \updownarrow$
$\aleph_0$: $\{1, 2, 3, 4, 5, \ldots, n - 1, \ldots\}$

9. $\aleph_0 + 10$: $\{-9, -8, -7, -6, -5,$
$\qquad\qquad\quad \updownarrow \quad \updownarrow \quad \updownarrow \quad \updownarrow \quad \updownarrow$
$\aleph_0$: $\{ \quad 1, \quad 2, \quad 3, \quad 4, \quad 5,$

$\qquad\quad -4, -3, -2, -1, 0, 1,$
$\qquad\quad \updownarrow \quad \updownarrow \quad \updownarrow \quad \updownarrow \updownarrow \updownarrow$
$\qquad\quad 6, \quad 7, \quad 8, \quad 9, 10, 11,$

$\qquad\qquad \ldots, \quad n, \quad \ldots\}$
$\qquad\qquad\qquad \updownarrow$
$\qquad\qquad \ldots, n + 10, \ldots\}$

Chapter 8: SENTENCES IN ONE VARIABLE

8-1 SENTENCES AND STATEMENTS

1. $x + 1 = x$; $x - 3 = x$; $x + 2 = x - 2$.
There are many other correct answers.

3. (a) iii;   (b) iii;   (c) iv;   (d) iii;
   (e) i;   (f) i;   (g) iv;   (h) iv.

5. Identity.

7. Empty set.

9. Identity.

11. Identity.

13. $\{5\}$

15. $\{0, 1\,2, 3, 4\}$

17. $\{6, 7, 8, \ldots\}$

19. $\{\ldots, -2, -1, 0, 1, 2, 3, 4\}$

21. $\{4, 5, 6, \ldots\}$

23. All real numbers greater than 3.

25. All real numbers greater than 5.

27. All real numbers except 0.

8-2 GRAPHS OF SENTENCES

1. Point.

3. Ray.

5. Line segment.

7. Line.

9. Half-line.

11.

13.

15.

17.

19.

21.

23.

25.

27.

29.

*31.

*33.

*35.

516

1. {1, 2, 3, 4}   3. {2}

5.

7.   9.

11.

13.   15.

17.

19.

21.

23.

*25.

*27.

8-4   SENTENCES INVOLVING ABSOLUTE VALUE

1. {2, −2}   11. 5

3. {−4, −3, −2, −1, 0, 1, 2, 3, 4}   13. 25

5. {−2, 4}   15. Line segment.

7. {−3, −1}   17. Union of two rays.

9. 7   19.

21.

23.

25.

27.

29.

*31.

*33.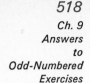

*35.

8-5 SENTENCES OF THE FIRST DEGREE

1. (a) Addition, $=$;
   (b) associative, $+$;
   (c) addition;
   (d) zero, $+$.

3. (a) Addition, $=$;
   (b) associative, $+$;
   (c) addition;
   (d) zero, $+$;
   (e) multiplication, $=$:
   (f) associative, $\times$:
   (g) multiplication;
   (h) one, $\times$.

5. $x = 3$

7. $x = -2$

9. $x = 14$

11. $x < 2$

13. $x > 2$

15. $x > -2$

17. $x > -6$

19. $x < -4$

8-6 PRODUCTS OF EXPRESSIONS OF THE FIRST DEGREE

1. $\{3, 5\}$

3. $\{-1, -5\}$

5. $\{-2, 1, -3\}$

7. $\{0, -5, 2\}$

9. $\{\frac{1}{2}, 2, -5\}$

11. $\{5, -5\}$

13. $\{6, -6\}$

15. $\{3, -3\}$

17. $x^2 + 3x + 2$

19. $x^2 - 6x + 9$

21. Multiplication, $=$.

23. Addition, $=$.

25. Addition.

27. Addition, $=$.

29. Equivalent fractions, $\dfrac{-c}{a} = \dfrac{-4ac}{4a^2}$.

31. Addition, $=$.

33. Definition of subtraction.

35. $\{5, -3\}$

37. $\left\{ \dfrac{-1 + \sqrt{157}}{6}, \dfrac{-1 - \sqrt{157}}{6} \right\}$

Chapter 9: SENTENCES IN TWO VARIABLES

9-1 SOLUTION SETS

1. $\{(1, 2), (2, 1)\}$

3. $\{(1, 1), (1, 2), (1, 3), (2, 1), (2, 2), (3, 1)\}$

5. $\varnothing$

7. $\{(1, 4), (2, 3), (3, 2), (4, 1)\}$

9. $\{(1, 1), (1, 2), (2, 1)\}$

11. $\{(-4, 4), (-3, 3), (-2, 2), (-1, 1), (0, 0), (1, 1), (2, 2), (3, 3), (4, 4)\}$

13. $\{(-4, -4), (-3, -3), (-2, -2), (-1, -1), (0, 0), (1, -1), (2, -2), (3, -3), (4, -4)\}$

15. $\{(-2, 4), (-1, 1), (0, 0), (1, 1), (2, 4)\}$

1. {(1, 1)}

3.

9.

5.

11.

7.

13. {(1, 4), (2, 3), (3, 2), (4, 1)}

15. {(1, 3), (1, 4), (2, 4)}

*17. {(1, 1, 1), (1, 1, 2), (1, 2, 1), (2, 1, 1), (1, 2, 2), (2, 1, 2), (2, 2, 1), (2, 2, 2)}

9-3 GRAPHS ON A PLANE

1. (a) 5; (b) 5.

3. (a) 3; (b) −2.

5. (a) −1$\frac{1}{2}$; (b) 3.

7. (a) −2; (b) 6.

9. (a) 1$\frac{4}{5}$; (b) 2$\frac{1}{4}$.

11. $y = -2x + 7$

13. $y \geq -3x + 6$

15. $y \geq 2x - 4$

17.

19.

21.

23.

25.

27.

29.

1.

7.

3.

9.

5.

11.

13.

***17.**

15.

***19.**

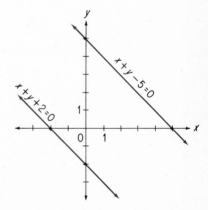

9-5 LINEAR SYSTEMS

1.

3.

5.

The graph consists of the union of the two lines.

7.

The lines are parallel; the solution set is the empty set.

9.

11.

13.

15.

*17.

*19.

1.

3.

5.

7.

9.

11.

13.

15.

17.

19.

21. $y = (x + 3)^2 - 4$; vertex: $(-3, -4)$;
axis: $x = -3$.

23. $y = (x - 1)^2 + 2$;   vertex: $(1, 2)$;
axis: $x = 1$.

25. $y = (x - 4)^2 - 6$;   vertex: $(4, -6)$;
axis: $x = 4$.

27. $y = (x - \frac{5}{2})^2 - \frac{1}{4}$;   vertex: $(\frac{5}{2}, -\frac{1}{4})$;
axis: $x = \frac{5}{2}$.

***29. (a)**

(b)

(c)

(d)

1. (a) −1;
   (b) 0;
   (c) 15.

3. (a) 1;
   (b) −8;
   (c) −15.

5.

A function.
Domain: {1, 2, 3}
Range: {1, 2, 3}

7.

Not a function.

9.

A function. Both the domain and the range are the set of real numbers.

11.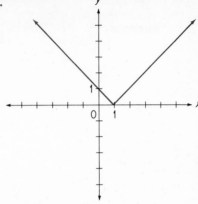

A function. The domain is the set of real numbers; the range is the set of non-negative real numbers.

13.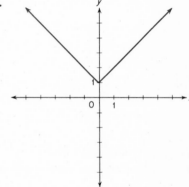

A function. The domain is the set of real numbers; the range is the set of real numbers greater than or equal to 1.

15.

Not a function.

17.

27.

A function. The domain is the set of real numbers; the range is the set of real numbers less than or equal to 0.

A function. (Note that the inverse is the same as the given function.)

19. $\{(1, 1), (2, 2), (3, 3)\}$: a function.

21. $\{(3, 1), (4, 2), (4, 3)\}$: not a function.

23. $x = y + 1$: a function.

25.

29.

A function.

Not a function.

9-8 LINEAR PROGRAMMING

1.

3. (a) Maximum value of 8 at $(2, 3)$;
   (b) minimum value of 2 at $(2, 0)$.

5. (a) Maximum value of 5 at $(3, 2)$;
   (b) minimum value of 0 at $(0, 0)$.

7. The conditions are $0 \le x$, $0 \le y$, $x + y \le 180$, and $5x + 3y \le 750$. The maximum of $3x + 2y$ occurs at $(105, 75)$; thus under the assumptions of this exercise, 105 minutes of regular teaching and 75 minutes of television instruction would be best for the students.

9. The maximum of $x + 2y$ occurs at $(30, 150)$; thus under the assumptions of this exercise, 30 minutes of regular teaching and 150 minutes of television instruction would be best for the students.

Chapter 10: NONMETRIC GEOMETRY

10-1 POINTS, LINES, AND PLANES

1.

3. *ABC, ABD, ACD, BCD.*

5. **(a)** 1;
  (b) 3;
  (c) 6;

(d) 10;
(e) 15;
*(f)** 45;
*(g)** $n(n - 1)/2$.

7. True.

9. False.

11. False.

13. False.

15. False.

17. True.

10-2 RAYS, LINE SEGMENTS, AND ANGLES

1. $\overline{AD}$

3. $\overline{BC}$

5. $\overline{BC}$

7. $\overline{BC}$

9. $\overrightarrow{BC}$ (or $\overrightarrow{BD}$).

11. B

13. Empty set.

15. Interior of $\angle ABD$.

17. $\overset{\circ}{\overrightarrow{BA}}$

19. $\overset{\circ}{\overrightarrow{PA}}$

21. $\angle BPC$

23. $\overleftrightarrow{AC} \cup \overset{\circ}{\overrightarrow{PB}}$

SUPPLEMENTARY EXERCISES

1. $\overline{BD}$

3. $\overline{BC}$

5. $\overline{DA}$

7. $\overline{BD}$

9. $\overleftrightarrow{AB}$

11. $\overline{DB}$

13. $\overrightarrow{DB}$

15. Empty set.

17. $\overline{BD}$

19. (Interior $\angle ADC$) $\cup$ D.

21. Exterior $\angle ADB$.

10-3 PLANE FIGURES

There are many correct answers for Exercises 1 through 7.

1.

3.

5.

7.

9. Empty set.

11. M

13. P

15. $\overset{\circ—\circ}{MN}$

17. $\triangle RMN$

19. $\overrightarrow{MP} \cup \overset{\circ}{NV}\rightarrow$

21. $\overline{MN}$

23. Interior $\triangle MRN$.

25. Interior of quadrilateral $STNM$.

27. Interior $\triangle RST$.

SUPPLEMENTARY EXERCISES

There are many correct answers for Exercises 1 and 3.

1.

3.

5. Empty set.

7. $\{B, D\}$

9. $\overline{BF}$

11. Interior $\triangle CBD$.

13. $\overset{\circ—\circ}{FG}\rightarrow$

15. $\overset{\circ—\circ}{BD}$

10-4 SPACE FIGURES

1. (a) M, N, O, P;
   (b) $\overline{MN}, \overline{MO}, \overline{MP}, \overline{ON}, \overline{NP}, \overline{PO}$;

(c) the triangular regions MNO, PMN, PNO, PMO.

3. (a) *A, B, C, D, E, F, G, H;*
 (b) $\overline{AB}, \overline{BC}, \overline{CD}, \overline{AD}, \overline{AE}, \overline{BF}, \overline{CG},$ $\overline{DH}, \overline{EF}, \overline{FG}, \overline{GH}, \overline{HE};$
 (c) the square regions *ABCD, ABFE, BCGF, CDHG, DAEH, EFGH.*

5. (a) *J, K, L, M, N, O;*
 (b) $\overline{JK}, \overline{KL}, \overline{JL}, \overline{JM}, \overline{KN}, \overline{LO}, \overline{MN},$ $\overline{NO}, \overline{OM};$
 (c) the triangular regions *JKL, MNO* and the rectangular regions *JKNM, KLON, LJMO.*

7. Pentagonal pyramid.

9. Hexagonal prism.

11. (Given figure.)

13. (Given figure.)

15. (b) and (c).

17. (a)

V	E	F
6	10	6
12	18	8
5	9	6
9	16	9

19.

21.

23.

25.

27.

29.

Number of painted faces	6	5	4	3	2	1	0
31. Number of cubes	0	0	0	8	12	6	1
*33. Number of cubes	0	0	0	8	36	54	27

10-5 PLANE CURVES

Exercises 1, 3, 5, and 7 may be done in many ways.

1.

3.

5.

7.

	V	A	R
9. (a)	4	4	2
(b)	4	5	3
11. (a)	2	2	2
(b)	2	3	3
13. (a)	5	4	1
(b)	16	24	10

15. $A = V + R - 2$

10-6 NETWORKS

1. (a) 4;
   (b) 0;
   (c) traversable, A, B, C, D.

3. (a) 2;
   (b) 2;
   (c) traversable, K, M.

5. (a) 4;
   (b) 0;
   (c) traversable, U, V, W, X.

7. (a) 0;
   (b) 4;
   (c) not traversable.

9. The inspector can use a map for the highways involved as a network, determine the number of odd vertices, and know that each section can be traversed exactly once in a single trip if there are at most two odd vertices.

***13.** Add one more bridge joining any two of the points A, B, C, D.

11.

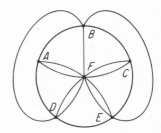

Think of the six regions as labeled and note the line segments that are needed as represented by arcs in the network. Since the network has four odd vertices ($B, D, E,$ and F), the network is not traversable and the suggested broken line cannot be drawn.

Chapter 11: METRIC GEOMETRY

11-1 LINEAR MEASURES

1. 4

3. 3

5. 6

7. $m(\overline{P_2P_5}) + m(\overline{P_5P_7}) = m(\overline{P_2P_7})$

9. $m(\overline{P_1P_7}) = m(\overline{P_1P_4})$
   $+ m(\overline{P_2P_7}) - m(\overline{P_1P_4} \cap \overline{P_2P_7})$

11. $m(\overline{P_{-3}P_4}) = m(\overline{P_1P_{-3}})$
   $+ m(\overline{P_{-2}P_4}) - m(\overline{P_1P_{-3}} \cap \overline{P_{-2}P_4})$

13. $m(\overline{P_4P_{-1}}) = m(\overline{P_{-1}P_2}) + m(\overline{P_2P_4})$

15. $m(\overline{P_{-3}P_5}) = m(\overline{P_2P_{-3}})$
   $+ m(\overline{P_0P_5}) - m(\overline{P_2P_{-3}} \cap \overline{P_0P_5})$

17. 14 feet.

19. 73 feet.

21. $3\frac{1}{2}$

23. (20) $\frac{1}{2}$ unit; (21) $\frac{1}{4}$ unit; (22) $\frac{1}{8}$ unit.

25. Approximate.

27. Exact.

1. Among the correct answers is:

3.

5. Among the correct answers is:

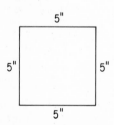

7. Impossible, any two straight angles each have degree-measure 180 and thus are congruent.

9. ∠FEG and ∠DCE, ∠FEC and ∠DCB, ∠GEH and ∠ECA, ∠CEH and ∠BCA.

11. ∠GEF and ∠ACB, ∠GEH and ∠DCB.

13.

15.

11-3 CLASSIFICATIONS OF POLYGONS

1. 20 inches.

3. 18 feet.

5. 4e

7. 2(b + h)

9. The linear measure of the square is doubled.

11. The linear measure of the triangle is multiplied by k.

13. $\overline{PQ}$ and $\overline{SR}$, $\overline{QR}$ and $\overline{SP}$.

15. ∠P and ∠Q, ∠Q and ∠R, ∠R and ∠S, ∠S and ∠P.

17.

19.

21. A rectangle *ABCD* is a parallelogram with at least one right angle. Suppose that $m \angle A = 90°$. Then $\angle B$ and $\angle D$ are each supplementary to $\angle A$ (consecutive angles of a parallelogram are supplementary) and thus are right angles; $\angle C$ is supplementary to $\angle B$ (also $\angle D$) and therefore a right angle. Thus all four angles are right angles.

SUPPLEMENTARY EXERCISES

Note: All of the possible sketches may be done in several ways.

1.

3.

5.

7.

9.

11.

13.

15.

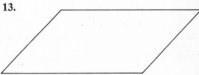

17. Impossible, a rhombus is defined as a special type of parallelogram.

19.

11-4 TRIANGLES

1. $\triangle ABC$ and $\triangle ADC$, $\triangle ABC \cong \triangle CDA$.

3. $\triangle AOD$ and $\triangle BOC$, $\triangle AOD \cong \triangle BOC$ (also $\triangle AOD \cong \triangle COB$).

5. $\triangle ABC$ and $\triangle CDE$, $\triangle ABC \cong \triangle EDC$.

7. (a) From the result in Exercise 6,

$$m \angle Q + 95 + 60$$
$$= m \angle T + 95 + 60;$$

$m \angle Q = m \angle T$, that is, $\angle Q \cong \angle T$.

(b) The corresponding angles are congruent.

9. $\triangle CDE \sim \triangle CAB$

11. $\overline{BC} \cong \overline{AC}$ as corresponding parts of the congruent triangles; thus $\triangle ABC$ is an isosceles triangle.

13. As in Exercise 4 any parallelogram $EFGH$ and a diagonal $\overline{FH}$ determine congruent triangles; $\triangle EFH \cong \triangle GHF$.

15. For any triangle ABC with $\overline{AC} \cong \overline{BC}$ we have $\triangle ACB \cong \triangle BCA$ by s.a.s. $\cong$ s.a.s.; then $\angle A \cong \angle B$ as corresponding parts of the congruent triangles.

17. If the triangles are congruent, the corresponding angles are congruent and the ratio of the measures of the corresponding sides is 1 (since the measures are equal).

19. A rhombus is a parallelogram with at least one pair of consecutive sides congruent; suppose $\overline{AB} \cong \overline{BC}$. For any parallelogram and thus any rhombus $ABCD$, $\triangle ABC \cong \triangle CDA$ (a.s.a. $\cong$ a.s.a.). Then $\overline{AB} \cong \overline{CD}$, $\overline{BC} \cong \overline{DA}$ by corresponding parts of congruent triangles and all four sides are congruent.

21. **(a)** Quadrilateral $AEFD$ has two pairs of opposite sides parallel;

(b) $CDEF$ and $BFDE$.

23. If two parallel lines are cut by a transversal, the corresponding angles are congruent; therefore

$$\angle BAC \cong \angle EAD \cong \angle BEF \cong \angle FDC,$$
$$\angle ABC \cong \angle AED \cong \angle EBF \cong \angle DFC,$$

and

$$\angle BCA \cong \angle EDA \cong \angle BFE \cong \angle FCD.$$

Also alternate interior angles are congruent;

$$\angle FDC \cong \angle DFE, \quad \angle DFC \cong \angle FDE,$$
$$\angle FCD \cong \angle DEF.$$

The triangles are similar since there is a correspondence of vertices such that the corresponding angles are congruent.

11-5 AREA MEASURES

1. 96 square inches.

3. 36 square inches.

5. 56 square inches.

7. 69 square inches.

9. $S = 6e^2$

11. The area measure of the square is multiplied by 4.

13. The area measure of the rectangle is multiplied by **(a)** 4; **(b)** 9; **(c)** $\frac{1}{4}$; **(d)** k^2.

15. The area measure of the polygon is multiplied by **(a)** 4; **(b)** k^2.

SUPPLEMENTARY EXERCISES

1. For the parallel lines $\overleftrightarrow{AD}$ and $\overleftrightarrow{BC}$ with transversal $\overleftrightarrow{AB}$, corresponding angles are congruent.

3. The opposite sides of a rectangle are congruent.

5. From Exercises 4 and 3, $m(\overline{AB}) = m(\overline{DC}) = m(\overline{EF})$.

7. s.a.s. $\cong$ s.a.s.

9. Definition of the area of a rectangle and Exercises 1 through 8.

11. Formula for area of a triangle.

11-6 VOLUME MEASURES

1. **(a)** 24 square feet; **(b)** 8 cubic feet.

3. **(a)** 96 square centimeters; **(b)** 64 cubic centimeters.

5. **(a)** 132 square feet; **(b)** 80 cubic feet.

7. 180 cubic feet.

9. The volume measure of the cube is multiplied by 8.

11. The volume measure is multiplied by (a) 8; (b) 27; (c) $\frac{1}{8}$; (d) k^3.

*13. $B_1 = 0$, $M = \pi r^2$, $B_2 = 0$, $h = 2r$;
$$V = \frac{2r}{6}(0 + 4\pi r^2 + 0) = \frac{4}{3}\pi r^3.$$

*15. $B_1 = \pi r^2$, $M = \frac{1}{4}\pi r^2$, $B_2 = 0$;
$$V = \frac{h}{6}[\pi r^2 + 4(\tfrac{1}{4}\pi r^2) + 0] = \tfrac{1}{3}\pi r^2 h.$$

Chapter 12: COORDINATE GEOMETRY

12-1 COORDINATES ON A PLANE

1. (a) 4, (2, 3); (b) 3, $(3\frac{1}{2}, 1)$;
(c) 4, (2, 3); (d) 9, $(5, 2\frac{1}{2})$;
(e) 5, $(-3, 4\frac{1}{2})$; (f) 4, $(-5, 2)$.

3. (a) $(4, -2)$; (b) $(2, 4)$; (c) $(-3, -6)$.

5. $(-5, 8)$

7. $C: (a, a)$

9. $S: (a - b, c)$

11.

13.

*15.

Consider $A: (0, 0)$, $B: (a, 0)$, $C: (b, c)$, and $D: (d, e)$. The midpoints of the sides are $E: \left(\frac{a}{2}, 0\right)$, $F: \left(\frac{a+b}{2}, \frac{c}{2}\right)$, $G: \left(\frac{b+d}{2}, \frac{c+e}{2}\right)$, and $H: \left(\frac{a}{2}, \frac{b}{2}\right)$. The line segments $\overline{EG}$ and $\overline{HF}$ both have $\left(\frac{a+b+d}{2}, \frac{c+e}{2}\right)$ as midpoint and thus bisect each other.

12-2 LINES ON A PLANE

1. $\frac{6}{3}$; that is, 2.

3. $\frac{0}{3}$; that is, 0.

5. (a) $\dfrac{y-1}{x-1} = \dfrac{7-1}{4-1}$;

(b) $y - 1 = 2(x - 1)$;

(c) $y = 2x - 1$.

7. (a) $\dfrac{y-5}{x-3} = \dfrac{5-5}{6-3}$;

(b) $y - 5 = 0(x - 3)$;

(c) $y = 0x + 5$.

9. $\overleftrightarrow{AB}$ has slope $\dfrac{0 - (-2)}{6 - 1}$, that is, $\dfrac{2}{5}$; $\overleftrightarrow{BC}$ is parallel to the y-axis; $\overleftrightarrow{CD}$ has slope $\dfrac{2 - 4}{1 - 6}$, that is, $\dfrac{2}{5}$; $\overleftrightarrow{DA}$ is parallel to the y-axis. Since $\overleftrightarrow{AB} \parallel \overleftrightarrow{CD}$ and $\overleftrightarrow{BC} \parallel \overleftrightarrow{DA}$, $ABCD$ is a parallelogram.

11. $\overleftrightarrow{AB}$ has slope $\dfrac{-3 - 1}{7 - 3}$, that is, -1; $\overleftrightarrow{BC}$ has slope $\dfrac{2 - (-3)}{5 - 7}$, that is, $-\dfrac{5}{2}$; $\overleftrightarrow{CD}$ has slope $\dfrac{6 - 2}{1 - 5}$, that is, -1; $\overleftrightarrow{DA}$ has slope $\dfrac{6 - 1}{1 - 3}$, that is, $-\dfrac{5}{2}$. Since $\overleftrightarrow{AB} \parallel \overleftrightarrow{CD}$ and $\overleftrightarrow{BC} \parallel \overleftrightarrow{DA}$, $ABCD$ is a parallelogram.

*13. Coordinate axes may be placed upon the plane of any quadrilateral $ABCD$ such that we have the figure for the answer of Exercise 15 of §12-1. The midpoints are as stated in that exercise. Then $\overleftrightarrow{EF}$ and $\overleftrightarrow{HG}$ each have slope c/b, and are parallel; $\overleftrightarrow{FG}$ and $\overleftrightarrow{EH}$ each have slope $e/(d - a)$ and are parallel. Thus $EFGH$ is a parallelogram. In other words, the midpoints of the sides of any quadrilateral are the vertices of a parallelogram.

12-3 DISTANCES ON A PLANE

1. 5

3. 25

5. $\sqrt{18}$; that is, $3\sqrt{2}$.

7. The points of the circle with center (0, 0) and radius 3.

9. The exterior points of the circle with center (0, 0) and radius 8.

11. The points of the circular region with center (1, 2) and radius 2.

13. The points of the circular region with center $(-1, -2)$ and radius 5.

15. $(x - 2)^2 + (y - 5)^2 = 9$

17. $(x + 3)^2 + (y + 1)^2 \leq 16$

19. $(x + 4)^2 + (y - 3)^2 > 16$

*21. The points of the solution set are the points of the circular region that are on or below the line.

12-4 PARABOLAS, ELLIPSES, HYPERBOLAS

1. Circle.

3. Hyperbola.

5. Parabola.

7. $x^2 + (y - 4)^2 = (y + 2)^2$, that is, $x^2 = 12y - 12$.

9. $(x - 3)^2 + y^2 = (x + 3)^2$, that is, $y^2 = 12x$.

11. $\sqrt{(x - 3)^2 + y^2} + \sqrt{(x + 3)^2 + y^2} = 10$

12-5 COORDINATES IN SPACE

1. A: (4, 0, 0), B: (4, 2, 0), C: (4, 2, 3), D: (4, 0, 3), E: (0, 2, 0), F: (0, 0, 0), G: (0, 2, 3), H: (0, 0, 3).

3. A: (4, -2, 0), B: (4, 0, 0), C: (4, 0, 3), D: (4, -2, 3), E: (0, 0, 0), F: (0, -2, 0), G: (0, 0, 3), H: (0, -2, 3).

5. A: (7, 0, 0), B: (7, 5, 0), C: (7, 5, 2), D: (7, 0, 2), E: (0, 5, 0), F: (0, 0, 0), G: (0, 5, 2), H: (0, 0, 2).

7. A: (7, -5, 0), B: (7, 0, 0), C: (7, 0, 2), D: (7, -5, 2), E: (0, 0, 0), F: (0, -5, 0), G: (0, 0, 2), H: (0, -5, 2).

537

*sec. 13-1
uses
and
misuses
of
statistics*

9.

11.

13.

15.

17. The yz-plane.

19. The xy-plane.

21. The plane that is parallel to the xz-plane and intersects the y-axis at b.

23. The z-axis.

25. The y-axis.

27. $z = 5$

29. $z \geq 2$

*31. $|x| \leq 2$

12-6 DISTANCES IN SPACE

1. $\sqrt{(1 - 1)^2 + (4 - 0)^2 + (5 - 2)^2}$; that is, 5.

3. $\sqrt{(3 - 2)^2 + [1 - (-1)]^2 + (2 - 0)^2}$; that is, 3.

5. $(x - 1)^2 + (y + 2)^2 + (z - 5)^2 = 4$

7. $(x + 1)^2 + (y + 2)^2 + (z - 3)^2 = 16$

9. A sphere with center $(0, 0, 0)$ and radius $\sqrt{5}$.

11. A sphere with center $(1, 2, 3)$ and radius 2.

13. A sphere with center $(-3, 1, -1)$ and radius 1.

15. $(x - 1)^2 + (y + 2)^2 + (z - 3)^2 = 4$

17. $(x - 2)^2 + (y - 3)^2 + (z - 4)^2 \geq 4$

*19. $4 \leq (x - 1)^2 + (y - 6)^2 + (z - 4)^2 \leq 25$

Chapter 13: AN INTRODUCTION TO STATISTICS

13-1 USES AND MISUSES OF STATISTICS

1. The fact that most accidents occur near home probably means that most of the miles driven are near the home of the driver. It does not necessarily mean that long trips are safer than short trips.

3. The conclusion that mathematics is very popular is not justified by the previous statement even though the conclusion may be true. The fact that more students are studying mathematics reflects the fact that there are more students in college and that colleges are requiring more mathematics.

5. The 100% sale probably indicates that the book was the required text in a course that all students had to elect, but the given evidence does not justify the given conclusion.

7. The survey may not have been a representative sample of the voting population. Details of the sampling procedure would be needed to feel confident of the conclusion.

9. One wonders how such a count could possibly have been made. Details of the sampling procedure would be needed to feel confident of the conclusion.

11. Many people who have asthma go to Arizona because of the climate. Probably a larger part of the people in Arizona have asthma than for any other state. Thus the conclusion is not justified.

13. People do not necessarily vote for the candidate of the party in which they are registered.

15. It should be stated what average means in this statement. If average denotes the median of the sample itself and the sample had an odd number of elements, there would always be 50% of the sample below average.

17. How short? How tall? What do you mean by aggressive? How do you measure aggressiveness?

19. Effective for what conditions? How does brand A compare to other brands? What do you mean by effectiveness? How do you determine effectiveness?

21. Fewer than whom? What age groups are under consideration? How much swimming is considered?

23. What do you mean by success? What is implied by "student"?

25. Are those who marry during their last year of college included? What is the source of this information?

27. The small changes in temperature appear very large due to the fact that the temperature scale does not start at zero.

29. No; a scale is needed before any such conclusions can be made.

13-2 COLLECTING AND PRESENTING DATA

7.

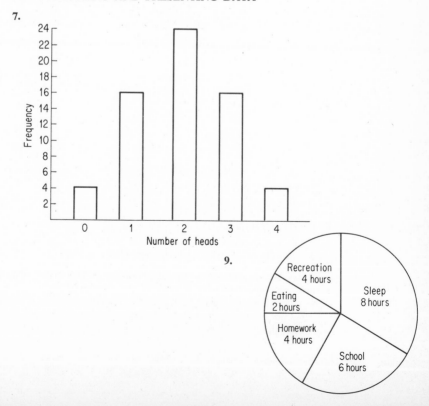

9.

11. 47%

13. (a) Yes, 8 cents; (b) no, per cent but not al amounts are shown.

15. 169°

17. $80,000

19. 59 and 88.

21. The number of deaths each year decreases after age 75 because there are fewer people still alive then. However, on a percentage basis, the number of deaths per 1,000 increases. Compare the last two columns in the mortality table after age 75.

23. Wyoming, South Dakota, North Dakota, West Virginia, and Mississippi.

25. California.

27. Total population is completely different from per cent change in population. For example, a state with 400,000 population could have a 25% increase with 100,000 people whereas for a state such as New York or California an increase of 1,000,000 people would still only be less than a 6% increase.

29. No region showed a decrease.

13-3 MEASURES OF CENTRAL TENDENCY

1. (a) 83; (b) 82; (c) 85.

3. (a) 17; (b) 17; (c) none.

5. (a) 79; (b) 75; (c) 70.

7. Mean.

9. Mode is often used as related to style; that which is worn or done most often by most people.

11. 2460

13. 770

15. (a) 3400; (b) 77.

17. 125; she cannot obtain a 90 average if the tests are on a 100 point basis.

13-4 MEASURES OF DISPERSION

1. 43

3. {50, 60, 80, 85, 90, 92, 103}. There are many correct answers.

5. {57, 75, 80, 85, 90, 95, 100}. There are many correct answers.

7. Normal.

9. (a) 16%; (b) 2.5%; (c) 250; (d) 1600; (e) 8150.

11. Scores:

Scores:	d	d^2
7	-3	9
9	-1	1
10	0	0
11	1	1
13	3	9
5)50	—	20
$\overline{X} = 10$		

$\sigma = \sqrt{\frac{20}{5}} = \sqrt{4} = 2$

13-5 BINOMIAL DISTRIBUTIONS

1. 120

3. 150

5. 100 to 140.

7. The mean is 32; $\sigma = 4$; $32 \pm 2\sigma$ provides a range of 24 to 40.

9. 45

11. 20 to 40.

Chapter 14: AN INTRODUCTION TO PROBABILITY

14-1 COUNTING PROBLEMS

1. (a) 24; (b) 64.

3. 12

5. 24

7. 25; 20.

9. 1320

11. 27,216

13. (a) 12;

(b) 3.

*15. (a) 450;
(b) 180;
(c) 648;
(d) 200.

*17. $26 \times 25 \times 9 \times 9 \times 8$; that is, 421,200.

14-2 DEFINITION OF PROBABILITY

1. $\frac{1}{2}$

3. $\frac{1}{2}$

5. $\frac{5}{6}$

7. 0

9. $\frac{1}{13}$

11. $\frac{1}{4}$

13. 6

15. $\frac{1}{6}$

17. $\frac{1}{12}$

19. $\frac{7}{8}$

14-3 SAMPLE SPACES

1. $\frac{1}{10}$

3. $\frac{2}{5}$

5. $\frac{3}{5}$

7. $\frac{1}{2}$

9.

First coin	Second coin	Third coin	Fourth coin
H	H	H	H
H	H	H	T
H	H	T	H
H	H	T	T
H	T	H	H
H	T	H	T
H	T	T	H
H	T	T	T
T	H	H	H
T	H	H	T
T	H	T	H
T	H	T	T
T	T	H	H
T	T	H	T
T	T	T	H
T	T	T	T

11. $\frac{5}{16}$

13.
(1, 1), (1, 2), (1, 3), (1, 4), (1, 5), (1, 6),
(2, 1), (2, 2), (2, 3), (2, 4), (2, 5), (2, 6),
(3, 1), (3, 2), (3, 3), (3, 4), (3, 5), (3, 6),
(4, 1), (4, 2), (4, 3), (4, 4), (4, 5), (4, 6),
(5, 1), (5, 2), (5, 3), (5, 4), (5, 5), (5, 6),
(6, 1), (6, 2), (6, 3), (6, 4), (6, 5), (6, 6).

15. $\frac{1}{36}$

17. $\frac{1}{18}$

19. $\frac{1}{6}$

21. $\frac{1}{6}$

23.
$R_1R_2, \ R_2R_1, \ W_1R_1, \ W_2R_1,$
$R_1W_1, \ R_2W_1, \ W_1R_2, \ W_2R_2,$
$R_1W_2, \ R_2W_2, \ W_1W_2, \ W_2W_1; \frac{1}{6}.$

25.
$R_1R_2, \ R_2R_1, \ R_3R_1, \ W_1R_1, \ W_2R_1,$
$R_1R_3, \ R_2R_3, \ R_3R_2, \ W_1R_2, \ W_2R_2,$
$R_1W_1, \ R_2W_1, \ R_3W_1, \ W_1R_3, \ W_2R_3,$
$R_1W_2, \ R_2W_2, \ R_3W_2, \ W_1W_2, \ W_2W_1; \frac{3}{10}.$

1. $\frac{2}{3}$

3. $\frac{2}{3}$

5. $\frac{1}{2}$

7. 1

9. $\frac{1}{2}$

11. $\frac{1}{52}$

13. $\frac{19}{52}$

15. (a) $\frac{1}{17}$; (b) 0; (c) $\frac{13}{204}$; (d) $\frac{1}{663}$; (e) $\frac{4}{17}$.

17. $\frac{1}{32}$

19. (a) $\frac{1}{12}$; (b) $\frac{1}{4}$; (c) $\frac{7}{12}$; (d) $\frac{3}{4}$.

Note: The following sample space is useful:

$H1, H2, H3, H4, H5, H6,$
$T1, T2, T3, T4, T5, T6.$

21. (a) $\frac{_{13}C_5}{_{52}C_5}$, that is, $\frac{33}{66,640}$;

(b) $\frac{_4C_4 \times _{48}C_1}{_{52}C_5}$, that is, $\frac{1}{54,145}$.

23. (a) $\frac{1}{64}$; (b) $\frac{5}{144}$; (c) $\frac{343}{1728}$; *(d) $\frac{1}{8}$.

25. (a) $\frac{1}{8}$; (b) $\frac{1}{8}$; (c) $\frac{1}{8}$; (d) $\frac{3}{8}$.

14-5 ODDS AND MATHEMATICAL EXPECTATION

1. 1 to 3.

3. 1 to 3.

5. 3 to 1.

7. 11 to 2.

9. 2 to 7.

11. $4

13. $1.67

15. $2.00

17. The probability that both of the bills drawn will be tens is $\frac{2}{5} \times \frac{1}{4} = \frac{1}{10}$. The probability that both will be fives is $\frac{3}{5} \times \frac{2}{4} = \frac{1}{10}$. The probability that one will be a five and one a ten is found as $(\frac{3}{5} \times \frac{2}{4}) + (\frac{2}{5} \times \frac{3}{4}) = \frac{3}{5}$. The mathematical expectation is then found to be

$$(\$20)(\tfrac{1}{10}) + (\$10)(\tfrac{3}{10}) + (\$15)(\tfrac{3}{5}) = \$14.$$

14-6 PERMUTATIONS

1. 120

3. 56

5. 42

7. 10

9. 12!, that is, 479,001,600.

11. 30

13. 720

15. 6

17. 5

19. (a) 20;
(b) 120.

21. 210

23. 9!; that is, 362,880.

25. $_nP_0 = 1$

14-7 COMBINATIONS

1. (a) 15;
(b) 56.

3. (a) 21;
(b) 35.

5. (a) 55;
(b) 455.

7. *pqr, prq, qpr, qrp, rpq, rqp;*
pqs, psq, qps, qsp, spq, sqp;
prs, psr, rps, rsp, spr, srp;
qrs, qsr, rqs, rsq, sqr, srq.
Thus $_4C_3 = 4$.

9. $_nC_n = 1$

11. $_3C_0 + {_3}C_1 + {_3}C_2 + {_3}C_3 = 1 + 3 + 3 + 1 = 8 = 2^3$

13. $_nC_0 + {_n}C_1 + \cdots + {_n}C_{n-1} + {_n}C_n = 2^n$

15. $_5C_2$; that is, 10.

17. $_{52}C_{13}$; that is, 635,013,559,600.

19. Order is important.

21. $_{10}C_4$; that is, 210; $_7C_3 \times {_3}C_1$; that is, 105.

23. $\frac{1}{5}$

25. $\dfrac{_4C_4 \times {_{48}}C_1}{_{52}C_5}$, that is, $\dfrac{1}{54,145}$.

SUPPLEMENTARY EXERCISES

1. $_{20}C_2$; that is, 190.

3. $_8C_4$; that is, 70.

5. $_{10}C_2$; that is, 45.

7. (a) $_{52}C_4$; that is, 270,725.
   (b) $_{52}C_7$; that is, 133,784,560.

9. $_4P_4$; that is, 24.

11. $\frac{1}{2} \times {_8}C_4$; that is, 35.

Index

544

Venn diagram, 47
Vertex:
of a cone, 411
of a curve, 354
even, 358
of a network, 358
odd, 358
of a parabola, 311
of a polygon, 343
of a polyhedron, 348
of a pyramid, 347
of a tetrahedron, 347

Vertical angles, 371
Volume measures, 392

W

Well-defined sets, 26
Whole numbers, 174
  properties of, 175

X

x-intercept, 296

Y

y-form of a sentence, 299
y-intercept, 296

Z

Zeno's paradoxes, 200
Zero, 174
  addition property of, 174
  as identity element, 174
  multiplication property of, 175